International Series in Operations Research & Management Science

Volume 252

More information about this series at http://www.springer.com/series/6161

Tsan-Ming Choi • Jianjun Gao
James H. Lambert • Chi-Kong Ng
Jun Wang
Editors

Optimization and Control for Systems in the Big-Data Era

Theory and Applications

Editors
Tsan-Ming Choi
Institute of Textiles and Clothing
The Hong Kong Polytechnic University
Hung Hom, Kowloon, Hong Kong

James H. Lambert
Department of Systems and Information Engineering
University of Virginia
Charlottesville, VA, USA

Jun Wang
Department of Management Science and Engineering, Business School
Qingdao University
Shandong, People's Republic of China

Jianjun Gao
School of Information Management and Engineering
Shanghai University of Finance and Economics
Shanghai, People's Republic of China

Chi-Kong Ng
Department of Systems Engineering and Engineering Management
The Chinese University of Hong Kong
Shatin, N.T., Hong Kong

ISSN 0884-8289 ISSN 2214-7934 (electronic)
International Series in Operations Research & Management Science
ISBN 978-3-319-85171-6 ISBN 978-3-319-53518-0 (eBook)
DOI 10.1007/978-3-319-53518-0

Softcover reprint of the hardcover 1st edition 2017

Printed on acid-free paper

This Springer imprint is published by Springer Nature
The registered company is Springer International Publishing AG
The registered company address is: Gewerbestrasse 11, 6330 Cham, Switzerland

Preface

Nowadays, for both business operations and engineering applications, there are huge amounts of data that can overwhelm computing resources of large-scale systems. These "big data" provide new opportunities to improve decision making and address risk for individuals as well as organizations. For example, the presence of market and sales data will yield better inventory planning for retail companies; massive and timely financial data will help improve portfolio management; the security holes of the Internet and the availability of data affect cryptography and privacy. Undoubtedly, utilizing big data smartly can enhance decision making. However, how to use and incorporate data into the decision making framework to yield a scientifically sound optimal decision is a challenging topic.

Motivated by the importance of big data and the respective challenges in optimization and control, we have compiled and developed this edited volume on scientific innovations and reviews in optimization, control, and resilience management in the big data era.

This book includes several important parts, namely, (1) Reviews on Optimization and Control Theories, (2) Reviews on Optimization and Control Applications, (3) Financial Optimization Analysis, (4) Operations Analysis, and (5) Concluding Remarks. All the featured papers are peer-refereed, and the specific topics covered include the following:

- Optimization and control for systems in the big data era: an introduction
- Dual control in big data era
- Time inconsistency and self-control optimization
- Quadratic convex reformulations for integer and mixed-integer quadratic programs
- Measurements of financial contagion
- Asset-liability management in continuous time
- Modern cryptography from the World War II era to the big data era
- Supply risk in the new business era
- A parameterized method for optimal multi-period portfolio selection

- Sparse and multiple risk measures approach for data-driven portfolio optimization
- Multistage optioned portfolio selection
- Multi-period portfolio selection with stochastic investment horizon
- A new model and method for order selection problems in flow-shop production
- Quick response fashion supply chains in the big data era
- Optimization and control for systems in the big data era: concluding remarks and future research.

We would like to take this opportunity to express our hearty thanks to Matthew Amboy and John Wolfe of Springer for their kindest support. We are indebted to all the reviewers who have provided timely review reports on the manuscripts. We are grateful for all the authors who have contributed their important and interesting research to this book.

This book is dedicated to our mentor Professor Duan Li, the Patrick Huen Wing Ming Professor of Systems Engineering and Engineering Management at The Chinese University of Hong Kong, to honor his great achievements in both systems control and optimization and celebrate his 65th birthday in July 2017. In the bottom of our hearts, he is always a distinguished scholar, a kind gentleman, an excellent professor, and an outstanding teacher. We are very proud of being his students. As a remark, the royalty received by the editorial team from this book project is 100% fully donated to Department of Systems Engineering and Engineering Management, The Chinese University of Hong Kong.

Hung Hom, Kowloon, Hong Kong — Tsan-Ming Choi, PhD
Shanghai, People's Republic of China — Jianjun Gao, PhD
Charlottesville, VA, USA — James H. Lambert, PhD
Shatin, N.T., Hong Kong — Chi-Kong Ng, PhD
Shandong, People's Republic of China — Jun Wang, PhD
November 2016

Contents

Contributors

Mei Choi Chiu Department of Mathematics and Information Technology, The Education University of Hong Kong, Tai Po, Hong Kong

Tsan-Ming Choi Institute of Textiles and Clothing, The Hong Kong Polytechnic University, Hung Hom, Kowloon, Hong Kong

Xiangyu Cui School of Statistics and Management, Shanghai University of Finance and Economics, Shanghai, People's Republic of China

Peilin Fu Department of Applied Engineering, School of Engineering and Computing, National University, San Diego, CA, USA

Jianjun Gao School of Information Management and Engineering, Shanghai University of Finance and Economics, Shanghai, People's Republic of China

Rujun Jiang Department of Systems Engineering and Engineering Management, The Chinese University of Hong Kong, Shatin, N.T., Hong Kong

James H. Lambert Department of Systems and Information Engineering, University of Virginia, Charlottesville, VA, USA

Xiang Li College of Economic and Social Development, Nankai University, Tianjin, People's Republic of China

Xun Li Department of Applied Mathematics, The Hong Kong Polytechnic University, Hong Kong, China

Yongjian Li Business School, Nankai University, Tianjin, People's Republic of China

Zhongfei Li Department of Finance and Investment, Sun Yat-Sen Business School, Sun Yat-Sen University, Guangzhou, China

Jianfeng Liang Department of Finance, Lingnan (University) College, Sun Yat-sen University, Guangzhou, People's Republic of China

Bojun Lu Portfolio Management Department, Foresea Life Insurance Co., Ltd., Shenzhen, China

Chi-Kong Ng Department of Systems Engineering and Engineering Management, The Chinese University of Hong Kong, Shatin, N.T., Hong Kong

Xi Pei Department of Finance and Investment, Sun Yat-Sen Business School, Sun Yat-Sen University, Guangzhou, China

Yun Shi School of Management, Shanghai University, Shanghai, People's Republic of China

Jun Wang Department of Management Science and Engineering, Business School, Qingdao University, Shandong, People's Republic of China

Baiyi Wu School of Finance, Guangdong University of Foreign Studies, Guangzhou, People's Republic of China

Weiping Wu Department of Automation, Shanghai Jiao Tong University, Shanghai, People's Republic of China

Xianping Wu School of Mathematical Sciences, South China Normal University, Guangzhou, China

Haixiang Yao School of Finance, Guangdong University of Foreign Studies, Guangzhou, China

Lan Yi Management School, Jinan University, Guangzhou, People's Republic of China

Linghua Zhao College of Economic and Social Development, Nankai University, Tianjin, People's Republic of China

Shushang Zhu Department of Finance and Investment, Sun Yat-Sen Business School, Sun Yat-Sen University, Guangzhou, China

Xiaoxia Zhuang Department of Management Science and Engineering, Business School, Qingdao University, Shandong, People's Republic of China

Chapter 1
Optimization and Control for Systems in the Big Data Era: An Introduction

Tsan-Ming Choi, Jianjun Gao, James H. Lambert, Chi-Kong Ng, and Jun Wang

Abstract The big data era is characterized by the presence of many Vs in terms of data and data usage. In this introductory chapter, we first discuss some challenges in optimization and control for systems in the presence of massive amount of data. We then introduce the papers featured in this book.

Keywords Introduction • Big data • Optimization • Control

T.-M. Choi (✉)
Institute of Textiles and Clothing, The Hong Kong Polytechnic University, Hung Hom, Kowloon, Hong Kong
e-mail: jason.choi@polyu.edu.hk

J. Gao
School of Information Management and Engineering, Shanghai University of Finance and Economics, Shanghai, People's Republic of China
e-mail: gao.jianjun@shufe.edu.cn

J.H. Lambert
Department of Systems and Information Engineering, University of Virginia, Charlottesville, VA, USA
e-mail: lambert@virginia.edu

C.-K. Ng
Department of Systems Engineering and Engineering Management, The Chinese University of Hong Kong, Shatin, N.T., Hong Kong
e-mail: ckng@se.cuhk.edu.hk

J. Wang
Department of Management Science and Engineering, Business School, Qingdao University, Shandong, People's Republic of China
e-mail: jwang@qdu.edu.cn

T.-M. Choi et al. (eds.), *Optimization and Control for Systems in the Big-Data Era*, International Series in Operations Research & Management Science 252,
DOI 10.1007/978-3-319-53518-0_1

1.1 Optimization and Control in Big Data Era

We are now in the big data era. The popularity of social media, mobile devices, cloud storage and application services, etc., all means that the amount of data which can be collected and used by organizations and companies is increasing everyday (Chan et al. 2016; Wang et al. 2016).

Traditionally, the term "big data" is associated with many Vs (Choi et al. 2016), such as volume (the amount of data), velocity (speed of data collection and processing), variety (the structured and unstructured data; complex data), veracity (the data accuracy and uncertainty) and value (the value associated with data). Undoubtedly, the presence of a massive amount of data means that we have to rethink about the strengths and weaknesses of the existing optimization and control methods.

In the recent literature, big data related optimization and control problems have been examined. For instance, Facchinei and Scutari (2015) propose a decomposition framework for achieving parallel optimization for a class of nonconvex problems with a massive amount of data. The authors demonstrate that their proposed method outperforms other existing methods. Daneshmand et al. (2015) develop a novel hybrid and parallel decomposition scheme for solving convex and nonconvex big data optimization problems. For very big problems, the authors show that their proposed decomposition scheme works well compared to other random or deterministic schemes. Bhattacharya et al. (2016) explore how an evolutionary optimization based algorithm can handle optimization problems with a high volume dataset. The authors claim that they have successfully applied the proposed algorithm in real world financial portfolio management. Richtarik and Takac (2016) explore how parallel randomized block coordinate descent methods can be used for developing a big data optimization algorithm. The authors show that their proposed algorithm can solve a class of large scale problems efficiently. Most recently, Boone et al. (2016) develop a framework for exploring service parts performance optimization problems with big data. They propose how, where and why big data applications can be applied in their proposed framework. For more recent developments of big data optimization, refer to the books by Emrouznejad (2016) and Japkowicz and Stefanowski (2016).

In this introductory chapter, we briefly review the papers featured in this book. According to the sectioning of the book, we present the papers in four sections: Sect. 1.2 reports the papers in the "Reviews on Theories" section, Sect. 1.3 examines the papers in the "Reviews on Applications" section, Sect. 1.4 introduces the "Financial Optimization Analysis" papers and Sect. 1.5 describes the papers in "Operations Analysis".

1.2 Reviews on Theories

For the review on optimization and control theories, this book features three papers. First, Fu examines the dual control theory in the big data era. The author first presents an overview of dual control and its probable applications. She explores different dual and non-dual controllers and highlights their complexity and limitations. In particular, she focuses on a class of discrete-time LQG problems with unknown parameters. She shows the optimal dual control, active open-loop feedback control via variance minimization (Li et al. 2002) and optimal nominal dual control for this class of problems. Finally, the author also discusses the probable usage of dual control in economic systems, information retrieval as well as mechanical engineering.

Shi and Cui review the time inconsistency and self control optimization problems, which are commonly found in financial optimization and conflict decision making (Cui et al. 2012). The authors examine different approaches which can effectively deal with the time inconsistency challenge in decision making. Moreover, they report the recent progress in the area and mention the challenges of time inconsistency optimization in the big data era.

Wu and Jiang examine the recent developments in the quadratic convex reformulation method. In fact, the quadratic convex reformulation method is commonly used to derive efficient equivalent reformulations for mixed-integer quadratically constrained quadratic optimization problems. The authors comment that even though the proposed problems can be solved by a standard mixed-integer quadratic solver using the branch-and-bound method, the solver's performance is far from satisfactory. They thus argue that the quadratic convex reformulation approach provides a systematic way to solve the above-mentioned challenging optimization problems. The authors also review some recent extensions of the quadratic convex reformulation method for problems such as the challenging semi-continuous quadratic optimization problems.

1.3 Reviews on Applications

For the review on applications of optimization and control methods in the big data era, there are four related papers in this book.

Pei and Zhu explore measurements of financial contagion. The authors propose that the financial contagion is a timely issue which is closely related to financial systemic risk. In their paper, they first clarify and summarize various critical concepts and measurements of financial contagion and then highlight their common features and differences. Since the structural break is known to be especially crucial, the authors review and discuss the respective financial contagion measurements. The authors conclude that the big data technology may be helpful for advancing risk management relevant to financial contagion in, e.g., information acquisition as well as model specification.

Dynamic portfolio optimization is a critical application area of advanced optimization models and methods. Chiu reports the summary of recent advances in the optimal asset-liability management system, which employs the dynamic portfolio optimization technique, in a continuous-time domain. From the stochastic optimal control perspective, the author derives a new asset-liability management solution for the case with insurers possessing a constant absolute risk averse utility function and Poisson-type insurance liabilities.

Cryptography is a very interesting area for secret communication. Traditionally, cryptography focuses on confidentiality, integrity and authentication of information. Lu reports in her paper a review of modern cryptography. The author chooses two important systems in cryptography, namely the Merkle–Hellman knapsack cryptography system and the subset-sum problem based cryptography system in the review. She examines the encryption and decryption processes of the above two systems.

In supply risk analysis, Li et al. conduct a systematic review on supply risk modeling. From the operations management perspective, the authors present various recent developments in supply risk modeling which include vertical supply chain interaction, horizontal supply chain competition and network competition problems. The authors present analytical models for each scenario. They also discuss future research directions in the big data era.

1.4 Financial Optimization Analysis

In addition to reviews on optimization and control theories and applications, this book also features many technical papers with novel insights and new findings. In the area of financial optimization, there are four related papers.

First, Li et al. develop a parameterized method for achieving optimal multi-period mean-variance portfolio selection with liability considerations. The authors note that the financial market generates a massive amount of data which are related to portfolio management. To effectively select the optimal portfolio requires very careful planning and the use of an efficient method. The authors hence propose a new method to help derive the analytical optimal portfolio strategies and efficient frontiers accurately. The authors demonstrate the applicability of their work by presenting a numerical example.

Gao and Wu explore the data driven mean-CVaR (DDMC) portfolio optimization problem. The authors consider the case when the out of sample performance of the DDMC portfolio optimization model is unstable, which occurs in practice owing to the availability of historical data. To deal with this challenge, the authors propose a novel method by adding a penalty on the sparsity of the portfolio weight and combine the variance term in the DDMC model. The authors run a numerical analysis and confirm that the out-of-sample performance fragility is being well mitigated using the new method.

Options are critical in the financial market. Liang conducts an analysis on the multi-stage optioned portfolio selection problem. The author introduces the mean-variance models first and then establishes the target tracking model for the optioned portfolio selection problem in both dynamic and static settings. She proposes two different solution schemes, namely the stochastic programming approach with optimality condition scheme, and the stochastic control with dynamic programming scheme. She also reveals the closed form relationship between the mean-variance model and the target tracking model.

Yi studies the multi-period portfolio selection problem with a stochastic investment horizon. She argues that the problem is commonly seen in the real world as an investor may suddenly terminate an investment (which leads to the stochastic investment horizon) owing to various factors. The author finds that the formulated problem is non-separable in the sense of dynamic programming. She thus employs an embedding technique to derive the optimal policy.

1.5 Operations Analysis

Finally, in this book, two papers on operations analysis are featured. First, Wang et al. present a new model and a novel method to solve the order selection problems in flow-shop operations. The authors notice that traditional order selection models separate the production scheduling process and the order selection problem, and the performance of order selection solely depends on production scheduling. The authors hence propose a new model which simultaneously considers both production scheduling and order selection. By computational experiments, the authors demonstrate that the proposed new model is much better than the traditional ones.

Quick response is a well-established strategy in supply chain management (Choi et al. 2003, 2004, 2006). Under quick response, companies can better utilize market information to improve their forecasts. In the big data era, it is easier for companies to collect and use a large amount of data. Motivated by the presence of massive amount of data and the importance of quick response in supply chain operations, Choi explores quick response in fashion supply chains. He focuses his attention on how the number of observations affects the expected values of quick response for the centralized supply chain system as well as the individual supply chain agents in the decentralized setting. He proves that quick response is a beneficial practice for the centralized supply chain system as well as the fashion retailer, and the benefit is increasing in the number of observations. If the number of observations goes to infinity, resembling the presence of "big data", the expected values of quick response will go to the steady states and the analytical expressions are found. However, there exist cases in which the fashion supplier suffers from a loss after adopting quick response. As such, the author derives a wholesale pricing markdown contract to achieve win–win coordination, which means both the supplier and the fashion retailer are better off (i.e. "win–win") and the supply chain is simultaneously globally optimized (i.e. "coordinated").

References

M. Bhattacharya, R. Islam, J. Abawajy, Evolutionary optimization: a big data perspective. J. Network Comput. Appl. **59**, 416–426 (2016)

C.A. Boone, B.T. Hazen, J.B. Skipper, R.E. Overstreet. A framework for investigating optimization of service parts performance with big data. Ann. Oper. Res. (2016), in press

H.K. Chan, T.M. Choi, X. Yue, Big data analytics: risk and operations management for industrial applications. IEEE Trans. Ind. Inf. **12**, 1214–1218 (2016)

T.M. Choi, D. Li, H. Yan, Optimal two-stage ordering policy with Bayesian information updating. J. Oper. Res. Soc. **54**(8), 846–859 (2003)

T.M. Choi, D. Li, H. Yan, Optimal single ordering policy with multiple delivery modes and Bayesian information updates. Comput. Oper. Res. **31**, 1965–1984 (2004)

T.M. Choi, D. Li, H. Yan, Quick response policy with Bayesian information updates. Eur. J. Oper. Res. **170**, 788–808 (2006)

T.M. Choi, H.K. Chan, X. Yue, Recent development in big data analytics for business operations and risk management. IEEE Trans. Cybern. **47**, 81–92 (2016). doi:10.1109/TCYB.2015.2507599

X.Y. Cui, D. Li, S.Y. Wang, S.S. Zhu, Better than dynamic mean-variance: Time inconsistency and free cash flow stream. Math. Financ. **22**(2), 346–378 (2012)

A. Daneshmand, F. Facchinei, V. Kungurtsev, G. Scutari, Hybrid random/deterministic parallel algorithms for convex and nonconvex big data optimization. IEEE Trans. Signal Process. **63**(15), 3914–3929 (2015)

A. Emrouznejad (ed.), *Big Data Optimization: Recent Developments and Challenges*, Series on Studies in Big Data (Springer, Switzerland, 2016)

F. Facchinei, G. Scutari, Parallel selective algorithms for nonconvex big data optimization. IEEE Trans. Signal Process. **63**(7), 1874–1889 (2015)

N. Japkowicz, J. Stefanowski, *Big Data Analysis: New Algorithms for a New Society*, Series on Studies in Big Data (Springer, Switzerland, 2016)

D. Li, F.C. Qian, P.L. Fu, Variance minimization approach for a class of dual control problems. IEEE Trans. Autom. Control **47**(12), 2010–2020 (2002)

P. Richtarik, M. Takac, Parallel coordinate descent methods for big data optimization. Math. Program. Ser. A **156**, 433–484 (2016)

X. Wang, Y. He, Learning from uncertainty for big data. IEEE Syst. Man Cybernet. Mag. **2**(2), 26–32 (2016)

Part I
Reviews on Optimization and Control Theories

Chapter 2
Dual Control in Big Data Era: An Overview

Peilin Fu

Abstract This paper provides an overview of dual control and its applications in the big data era. Different non-dual and dual controllers as well as their attributes, complexity, and limitations are analyzed. As an example, dual control of a class of discrete-time LQG problems with unknown parameters in both the state and observation equations is discussed in depth. Optimal dual control, open-loop feedback control, active open-loop feedback control via variance minimization approach, and optimal nominal dual control are demonstrated for this type of problems. The optimal nominal dual control, taking into account the effect of future learning, is the best possible (partial) closed-loop feedback control that can be achieved. Applications of dual control in economic systems, manufacturing processes, information retrieval, robotics, etc. are also introduced.

Keywords Dual control • Stochastic control • Dynamic programming • LQG control problem

2.1 Introduction

Most real-world processes are very complex and are not well understood. As such, the control of systems whose dynamics are not completely known is a problem of major theoretical and practical importance. Feldbaum, in his seminal work in 1960s (Feldbaum 1965), pointed out that, when implementing the optimal control strategy for stochastic systems with parameter uncertainty, the controller usually pursues two often conflicting objectives: to drive the system toward a desired state, and to perform active learning to reduce the system's uncertainty. Such a control scheme, which affects not only the states of the system but also the quality of estimation, is known as dual control. In 2000, IEEE Control Systems Society listed the dual control as one of the 25 most prominent subjects in the last century which had significantly impacted the development of control theory.

P. Fu (✉)
Department of Applied Engineering, School of Engineering and Computing,
National University, San Diego, CA, USA

T.-M. Choi et al. (eds.), *Optimization and Control for Systems in the Big-Data Era*,
International Series in Operations Research & Management Science 252,
DOI 10.1007/978-3-319-53518-0_2

Except for a few ideal situations, the optimal dual control cannot be achieved both analytically and numerically. Feldbaum showed that the optimal dual control is the solution to a functional equation known as Bellman equation based on dynamic programming. Solving this functional equation is intractable due to the "curse of dimensionality" inherent in dynamic programming. The two subproblems of stochastic control, estimation and control in most situations are intercoupled. The future uncertainties of the parameters are functions of the control signals applied to the system. The loss function, which has to be minimized with respect to the control signal, thus contains some information of the future observations through the statistics of the observations given the present information (Bar-Shalom and Tse 1974). The efforts in dual control have thus mainly been devoted to developing certain suboptimal solution schemes, such as the certainty equivalence scheme and open-loop feedback control, by bypassing this essential feature of coupling between estimation and control.

The control policies were categorized into the following classes in Bar-Shalom and Tse (1974) according to their information patterns—the availability of past observation and the possible usage of information about the future observation:

(1) The Open-loop Policy. In this case no measurement knowledge is available for the controller.
(2) The Feedback Policy. At every time the current information set is available for the computation of the control but no knowledge about the future measurements is available. The open-loop optimal feedback (OLOF) control belongs to the feedback class. It assumes that no observations will be made in the future, the control law is obtained by using the observations already acquired.
(3) The Closed-Loop Policy. This policy incorporates with the remaining observation program, i.e. the knowledge that the loop will stay closed through the end of the process is fully utilized.

There are two aspects in which the closed-loop policy differs from the feedback policy (Bar-Shalom and Tse 1974).

(1) Caution: In a stochastic control problem, due to the inherent uncertainties, the controller has to be "cautious" not to increase the effect of the existing uncertainties on the cost. However, the closed-loop controller, since it "knows" that future observations will be available and corrective actions based upon them will be taken, will exercise less "caution."
(2) Probing or Active Learning: When the dual effect is present, the control can "help" in learning (estimation) by decreasing the uncertainty about the state. Therefore, the closed-loop control, which takes into account the future observation program and statistics, has the capability of active learning when the dual effect exists. A feedback controller, even though it "learns" by using the measurements, does not actively "help" the learning. This learning can be called, therefore, passive, or accidental, and the corresponding control policy is passively adaptive, as opposed to the closed-loop control which is actively adaptive.

Most resulting suboptimal control laws are of a nature of passive learning, since the function of future active probing of the control is purposely deprived in order to achieve analytical attainability in the solution process. A central problem in dual control, and indeed a key barrier to its development, is to power a control law with the property of active learning.

Prominent features and fundamental properties of dual control have been extensively studied in the literature (Bar-Shalom 1981; Bar-Shalom and Tse 1974; Tse et al. 1973). An analysis of various approximations in dual control was given by Lindoff et al. (1999). Filatov and Unbehauen (2000) developed a bi-criteria approach to cope with the two conflicting goals in dual control. Surveys on dual control can be found in Wittenmark (1975c) and Filatov and Unbehauen (2000).

Li, Qian, and Fu in a series of papers studied the dual control of discrete-time LQG problems with unknown parameters. A variance minimization approach was proposed for discrete-time LQG problems with parameters uncertainty in the observation equation (Li et al. 2002). Minimizing a covariance term at the final stage introduced a feature of active learning for the derived control law. The optimal degree of active learning was determined for achieving an optimality. Fu et al. (2002) further applied the variance minimization approach to discrete-time LQG problems with parameters uncertainty in both the state and observation equations, an optimal open-loop feedback control law with active learning property was developed. The same problem was revisited in Li et al. (2008), in which the optimal nominal dual control was proposed. By exploring the future nominal posterior probabilities, the control law takes into account the function of future learning, thus is the best possible closed-loop feedback control that can be achieved. Some of these results are summarized in Sect. 2.3 as an example of dual control problems.

In Sect. 2.2, classification of controllers is introduced. Different non-dual and dual controllers as well as their attributes, complexity, and limitations are analyzed. As an example, dual control of a class of discrete-time LQG problems with unknown parameters in both the state and observation equations is discussed in depth in Sect. 2.3. Optimal dual control, open-loop feedback control, active open-loop feedback control, and optimal nominal dual control are demonstrated. Section 2.4 provides successful applications of dual control in economic systems, manufacturing processes, information retrieval etc. in the big data era. The paper concludes in Sect. 2.5.

2.2 Classification of Controllers

2.2.1 Non-dual Controller

If the performance index only takes into account the previous measurements and does not assume that future information will be available, then the resulting controller will be called non-dual in Feldbaum's terminology. In this situation the control law does not facilitate the identification. The non-dual controllers can be divided into three classes: certainty equivalence controller, one-step cautious controller, and open-loop optimal feedback controller.

2.2.1.1 Certainty Equivalence Controller

One widely used non-dual approach is developed using the concept of certainty equivalence. The certainty equivalence holds if it is possible to first solve the deterministic problem with known parameters and then obtain the optimal controller for unknown parameters by substituting the true parameter values with the estimated values (Wittenmark 1975c). One well-known class of problems for which the certainty equivalence principle holds is the linear-quadratic-gaussian control problems. In adaptive control there are very few cases where the certainty equivalence principle is applicable. The controller obtained by enforcing the certainty equivalence principle does not take into consideration the fact that the estimated parameters are not equal to the true ones and are inaccurate. Although the simplicity of the control law, it ignores the confidence level of the parameter estimates in deriving the adaptive control scheme. Such a control scheme would result in a control system that is extremely sensitive to stochastic variations.

A method based on process parameter estimation was first described by Kalman (1958) using least squares to determine the unknown parameters in the model. This type of method works well for constant or slow time-varying parameters. Different approximation methods (Hasting-James and Sage 1969; Panuska 1968; Young 1968) have been suggested for the models of maximum likelihood type. Methods using state space models were given in Jenkins and Roy (1966) and Luxat and Lees (1973).

2.2.1.2 One-Step Cautious Controller

Minimizing over a single time period leads to the one-step cautious controller. This controller takes the parameter uncertainties into account, in contrast to the certainty equivalence controller. However, the controller of this type may generate the turn-off phenomenon. If the estimates are very poor, the magnitude of the control signal will become very small. The control is thus unintentionally turned off for some period of time until the noise excites the system in such a way that better estimates are achieved. This makes the one-step cautious controller unsuitable for control of systems with quickly varying parameters.

A one-step minimization where the unknown parameters are modeled by a stochastic process was discussed in Aoki (1967) and Astrom and Wittenmark (1971). The unknown parameters can be first estimated using a Kalman filter, which then give the one-step ahead estimates and covariance matrix based on the current information set. Using a fundamental lemma in stochastic optimal control (Astrom 1970), it is possible to find the control law to solve the one-step minimization problem. The control law clearly shows the influence of the uncertainties of the estimates. Examples with turn-off were given, for instance, by Astrom and Wittenmark (1971).

2.2.1.3 The Open-Loop Feedback Optimal Controller

The open-loop optimal feedback (OLOF) control is derived at distinct time instants under the assumption that no future measurements will be available. Thus an open-loop control sequence is determined. The first step in the control sequence is then used and the performance of the system is measured. Based on the new information (feedback), a new minimization is performed again. In this open-loop feedback approach, the fact that the estimated parameter may not be exact is therefore taken into consideration, but the knowledge of the future observation program is completely ignored. According to the theory of dual control, introduced by Feldbaum (1965), the open-loop feedback control is, from the estimation point of view, passive, since it does not take into account that learning is possible in the future.

Many suboptimal controllers achieved in the current literature are open-loop optimal feedback controllers (Florentine 1962; Tse and Athans 1972; Aoki 1967). Lainiotis, Deshpande, and Upadhyay wrote a series of papers (Deshpande et al. 1973; Lainiotis et al. 1972) on an open-loop feedback optimal approach to the stochastic control of linear systems with unknown parameters. The controller is designed to minimize the average performance-to-go conditioned on the present measurements and past control actions and without any active anticipation of new measurements. The result is a feedback control law similar to the optimal LQG one, but averaged over the space of the unknown parameters. The algorithm is straightforward and easy to implement. It may be generated by computing the average of the specific controllers for some value of the parameters weighted by the a posteriori probability densities which are Gaussian (Deshpande et al. 1973). Casiello and Loparo proved in Casiello and Loparo (1989) that these types of passive control laws are optimal for certain quadratic functionals.

The OLOF controller might be overly cautious because of the assumption that no further measurements will be available to correct for erroneous control actions. The properties of the OLOF controller were further discussed by Bar-Shalom and Sivan (1969) and Tse and Athans (1972).

2.2.2 *Dual Controller*

If besides the previous measurements the performance index is also considered to be dependent on the future observations, a dual controller will be constructed. In this case, the future uncertainties of the parameters are functions of the control applied to the system. The control law must compromise between the two conflicting tasks: control and identification. The dual controllers can be classified as optimal dual controller and suboptimal dual controller.

2.2.2.1 Optimal Dual Controller

There are very few cases where it is possible to obtain an analytical representation of the optimal dual control law. The imposed assumptions are usually unrealistic. In Gorman and Zaborszky (1968) and Grammaticos and Horowitz (1970), the problem of controller synthesis was considered under the assumption that the entire state is measurable. Moreover, it was assumed that the poles of the system are known while the zeros are unknown. In both cases, it is possible to find the optimal dual control by solving a set of differential or difference equations corresponding to the Riccati equation in the standard linear quadratic case. Sternby (1978) discussed a Markov chain with four states. The transition probabilities are functions of the control. In that particular example it is possible to find the analytical expression of the optimal dual controller.

Some results can be seen in the literature to achieve the optimal dual controller numerically. Florentine (1962) considered a first-order system where the gain is fixed but unknown with a given a priori distribution. The problem was solved by discretizing the state and control. Another numerically solved problem was given by Jacobs and Langdon (1970). The absolute value of the state can be measured through the observation while the sign is unknown. Introducing the probability for the state to be positive, it is possible to derive the corresponding functional equation. A zero-order system with an unknown gain was considered in Astrom and Wittenmark (1971) where the gain was assumed to be described by a known stochastic process. A more general treatment of the problem was given by Griffiths and Loparo (1985).

A variance minimization approach for dual control of discrete-time LQG problems with parameter uncertainty in the observation equation was proposed by Li et al. (2002). Minimizing a covariance term at the final stage introduced a feature of active learning for the derived control low. The optimal degree of active learning was derived for achieving the optimality.

2.2.2.2 Suboptimal Dual Controller

Since it is difficult to determine the optimal dual controllers, much effort has been devoted to finding suboptimal solutions with dual properties. The approaches can be classified as follows (Wittenmark 1975c; Astrom and Wittenmark 1989):

(1) Perturbation signals

Employing a cautious controller can give rise to "turn-off" of the control if the unknown parameters are strongly time-varying. Several ways have been suggested to avoid the turn-off phenomenon. The turn-off is due to a lack of excitement. The perturbation signal, which can be a square-wave or pseudo random signal, etc., can be used to excite the system in order to get good estimation (Wieslander and Wittenmark 1971). The addition of the extra signal will naturally increase the probing loss, but may make it possible to improve the total performance.

(2) Constrained one-step-ahead minimization

Another way to avoid turn-off is to minimize the loss function one-step ahead under certain constraints. The constraints such as limiting the minimum value of the control signal or limiting the variance of the parameter estimates can prevent the control signal from being too small and impose extra probing (Hughes and Jacobs 1974; Alster and Belanger 1974). These controllers have the advantage that the control signal can be easily computed, but the algorithm will contain application-dependent parameters that have to be chosen by the user.

(3) Approximations of the loss function

Suboptimal dual controls can also be obtained by extending the loss function in order to prevent the shortsightedness of the cautious controller (Astrom and Wittenmark 1989). For state space models, one approach is to make a serial expansion of the loss function in the Bellman equation (Gorman and Zaborszky 1968). Such an expansion can be done around the certainty equivalence or the cautious controllers. But due to its computational complexity, this approach has been limited to situations where the control horizon is rather short, usually less than 10.

Another way is to try to solve the two-step minimization problem. The derived suboptimal control has correction terms which depend on the sensitivity functions of the expected future cost, which can avoid the turn-off. But in most cases, it is not possible to get an analytical solution.

(4) Modifications of the loss function

Adding terms that are reflecting quality of the parameter estimate in the loss function can prevent the cautious controller from turning off. Solution proposals (Alster and Langer 1974; Wittenmark 1975b; Milito et al. 1982) have been seen in the literature to incorporate certain variance terms of the state or the innovation process into the objective function in order to force the control to perform active learning. These solution schemes, however, truncate the time horizon into shorter time periods of one stage, prompting a concern of possible myopic behaviors.

2.2.2.3 Optimal Nominal Dual Controller

Although the optimal nominal dual controller is also suboptimal, the author would like to list it as a separate category to distinguish it from other suboptimal dual controllers. The reason is that it is the best possible closed-loop dual control if the optimal dual control cannot be achieved. The optimal nominal dual control was first proposed by Li, Qian and Fu in Li et al. (2008). They pointed out that a major difficulty in solving dual control for discrete-time LQG problems with unknown parameters is that the optimal control cannot be determined when the future posterior probabilities are unknown, while at the same time the future posterior probabilities depend on the control applied at the early stages. In order to break this loop, a possible solution scheme is to derive the relationship between the posterior probability and the control. A control which satisfies a deterministic version of this relationship is defined as the nominal control. The expected posterior probabilities when applying the nominal control are called nominal future posterior probabilities. Applying the nominal future posterior probabilities generated by the nominal control in the Bellman equation, the effect of future learning can be taken into account. Since in this situation, all the achievable future information is used in terms of its expected value, the control law obtained can be considered to be the best possible closed-loop control law in this sense.

2.3 An Example: LQG Problems with Unknown Parameters

Consider the following class of linear-quadratic stochastic optimal control problems where there exist parameter uncertainties in both the state and the observation equations,

$$(P)\ \min E\left\{x'(N)Q(N)x(N) + \sum_{k=0}^{N-1}[x'(k)Q(k)x(k) + u'(k)R(k)u(k)] \mid I^0\right\}$$

$$\text{s.t. } x(k+1) = A(k,\theta)x(k) + B(k,\theta)u(k) + w(k),\ \ k = 0,1,\cdots,N-1$$

$$y(k) = C(k,\theta)x(k) + v(k),\ k = 1,2,\cdots,N,$$

where $x(k) \in R^n$ is the state, $u(k) \in R^p$ is the control, $y(k) \in R^m$ is the measured output, and I^0 is the initial information set that includes information about the probability distribution of the initial state $x(0)$, the statistics of the random sequences $\{w(k)\}$ and $\{v(k)\}$, and the initial probability distribution of the unknown parameter θ. $\{w(k)\} \in R^n$ and $\{v(k)\} \in R^m$ are the two independent Gaussian white noise sequences with zero mean, and variances σ_w^2 and σ_v^2, respectively. The random initial state $x(0)$ is assumed to be of Gaussian distribution $N(\hat{x}(0), P(0))$ and is assumed to be independent of the process and observation noises: The quantities $A(k,\theta), B(k,\theta)$, and $C(k,\theta)$ are matrices of appropriate dimensions whose values depend on an unknown parameter θ. It is assumed that θ belongs to a finite set $\Theta = \{\theta_1, \theta_2, \ldots, \theta_s\}$ and is a constant over the entire time horizon. The a priori probabilities of the parameter θ are

$$q_i(0, I^0) = P(\theta = \theta_i \mid I^0),\ i = 1, 2, \ldots, s.$$

Furthermore, $\{Q(k)\}$ and $\{R(k)\}$ are sequences of positive semidefinite and positive definite symmetric matrices of appropriate dimensions, respectively. Define the information set at stage k, $k = 0, 1, \ldots, N$, to be I^k,

$$I^k = \{u(0), \ldots, u(k-1), y(1), \ldots, y(k), I^0\}.$$

The dual control problem for (P) is to find a closed-loop control law,

$$u(k) = f_k\left(I^k\right),\ k = 0, 1, \ldots, N-1,$$

such that the expected performance index in (P) is minimized.

Notice that two kinds of uncertainty are involved in (p): irreducible uncertainty caused by Gaussian white noise sequences $\{w(k)\}$ and $\{v(k)\}$, and reducible uncertainty caused by an unknown parameter θ. If there is no parameter uncertainty about θ, the above problem reduces to the conventional linear-quadratic Gaussian stochastic control problem which is not a dual control problem since the control does not have an effect on the system's uncertainty. The certainty equivalence principle then can be applied to determine the optimal control. Note that the certainty equivalence principle may not hold even for some stochastic control problems with only irreducible uncertainty, for example, linear Gaussian systems with an exponential performance criterion (Jacobson 1973).

2.3.1 *Optimal Dual Control*

Define $\hat{x}_i(k|k)$ to be the state estimate at stage k when assuming $\theta = \theta_i$:

$$\hat{x}_i(k|k) = E\left\{x(k)|\theta = \theta_i, I^k\right\}.$$

$\hat{x}_i(k|k)$ can be obtained using the Kalman filters as stated in Casiello and Loparo (1989):

$$\hat{x}_i(k|k) = \hat{x}_i(k|k-1) + F_i(k)\left[y(k) - C(k,\theta_i)\hat{x}_i(k|k-1)\right] \quad (2.1)$$

$$\hat{x}_i(k|k-1) = A(k-1,\theta_i)\hat{x}_i(k-1|k-1) + B(k-1,\theta_i)u(k-1) \quad (2.2)$$

$$F_i(k) = P_i(k|k-1)C'(k,\theta_i)[C(k,\theta_i)P_i(k|k-1)C'(k,\theta_i) + \sigma_v^2(k)]^{-1} \quad (2.3)$$

$$P_i(k|k-1) = A(k-1,\theta_i)P_i(k-1|k-1)A'(k-1,\theta_i) + \sigma_w^2 \quad (2.4)$$

$$P_i(k|k) = \left[I - F_i(k)C(k,\theta_i)\right]P_i(k|k-1), \quad (2.5)$$

with the initial condition of $\hat{x}_i(0|0) = \hat{x}(0)$ and $P_i(0|0) = P(0)$.

Define $q_i(k, I^k)$ to be the posterior probability of model i at stage k,

$$q_i(k, I^k) = P(\theta = \theta_i \mid I^k),\ k = 0, 1, \ldots, N-1.$$

The posterior probabilities, $q_i(k, I^k)$, $i = 1, 2, \ldots, s$, can be calculated recursively based on the observation (Casiello and Loparo 1989) as follows:

$$q_i(k, I^k) = \frac{L_i(k)}{\sum_{j=1}^{s} q_j(k-1, I^{k-1})L_j(k)} q_i(k-1, I^{k-1}),\ k = 1, 2, \ldots, N, \quad (2.6)$$

with the initial condition $q_i(0, I^0)$, where

$$\begin{aligned} L_i(k) &= |P_y(k|k-1,\theta_i)|^{-\frac{1}{2}} \exp[-\frac{1}{2}\tilde{y}(k|k-1,\theta_i)' P_y(k|k-1,\theta_i)^{-1} \\ &\quad \times \tilde{y}(k|k-1,\theta_i)] \qquad (2.7) \\ \tilde{y}(k|k-1,\theta_i) &= y(k) - C(k,\theta_i)\hat{x}_i(k|k-1) \qquad (2.8) \\ P_y(k|k-1,\theta_i) &= C(k,\theta_i)P_i(k|k-1)C'(k,\theta_i) + \sigma_v^2(k). \qquad (2.9) \end{aligned}$$

Define for $i = 1, 2, \ldots, s$,

$$\begin{aligned} J_i(k, I^k) &= E\{x'(k)Q(k)x(k) + u'(k)R(k)u(k) \mid \theta_i, I^k\}, \quad k = 0, \ldots, N-1 \\ J_i(N, I^N) &= E\{x'(N)Q(N)x(N) \mid \theta_i, I^N\}. \end{aligned}$$

Then the following is obvious,

$$\begin{aligned} J(k, I^k) &= E\{x'(k)Q(k)x(k) + u'(k)R(k)u(k) \mid I^k\} \\ &= \sum_{i=1}^{s} q_i(k, I^k) J_i(k, I^k) \quad k = 0, 1, \ldots, N-1 \\ J(N, I^N) &= E\{x'(N)Q(N)x(N) \mid I^N\} \\ &= \sum_{i=1}^{s} q_i(N, I^N) J_i(N, I^N). \end{aligned}$$

By the principle of stochastic dynamic programming, the closed-loop control that minimizes the performance index in problem (P) can be obtained by solving the following recursive relation,

$$\begin{aligned} &\min_{u(0)} E\Bigg\{ \sum_{i=1}^{s} q_i(0, I^0) J_i(0, I^0) \\ &\quad + \min_{u(1)} E\Bigg\{ \sum_{i=1}^{s} q_i(1, I^1) J_i(1, I^1) + \ldots \\ &\quad + \min_{u(k)} E\Bigg\{ \sum_{i=1}^{s} q_i(k, I^k) J_i(k, I^k) + \ldots \\ &\quad + \min_{u(N-1)} E\Bigg[\sum_{i=1}^{s} q_i(N-1, I^{N-1}) J_i(N-1, I^{N-1}) \\ &\quad + \sum_{i=1}^{s} q_i(N, I^N) J_i(N, I^N) | I^{N-1} \Bigg] \ldots | I^k \Bigg\} \ldots | I^1 \Bigg\} | I^0 \Bigg\}. \qquad (2.10) \end{aligned}$$

In principle, the optimal dual control problem (P) can be solved via (2.10). However, the difficulty and complexity in solving (P) hide deeply behind these seemingly straightforward equations. In fact, in dual control problems, all of the posterior probabilities at later stages are affected by previous controls. The curse of uncertainty of the posterior probabilities in later stages is further compounded by the required expectation operations. Therefore, to derive the cost-to-go functions in stochastic dynamic programming from (2.10) is a formidable task, as long as the posterior probabilities at later stages are previously control-dependent.

2.3.2 Open-Loop Feedback Control

Suppose that future learning will not be performed, the open-loop feedback control can be obtained by fixing all the posterior probabilities in the later stages at $q_i(k, I^k), i = 1, 2, \ldots, s$. As a result, the following optimal open-loop feedback control problem is considered at stage k,

$$\begin{aligned}\min_{u(k)} E\Bigg\{ & \sum_{i=1}^{s} q_i(k, I^k) J_i(k, I^k) + \ldots \\ & + \min_{u(N-2)} E\Bigg\{ \sum_{i=1}^{s} q_i(k, I^k) J_i(N-2, I^{N-2}) \\ & + \min_{u(N-1)} E\Bigg[\sum_{i=1}^{s} q_i(k, I^k) J_i(N-1, I^{N-1}) \\ & + \sum_{i=1}^{s} q_i(k, I^k) J_i(N, I^N) | I^{N-1} \Bigg] | I^{N-2} \Bigg\} \ldots | I^k \Bigg\}. \end{aligned} \tag{2.11}$$

A controller that uses observations to update online the estimation of the uncertain parameter is said to have a learning feature. The learning policies can be further classified into two types—active learning and passive learning. We can always expect an improving knowledge about the system's uncertainty when future observations are utilized. A controller that takes the future uncertainty reduction is said to have a property of active learning. In return, a controller with an active learning property affects the degree of future uncertainty reduction. To power a control law with a property of active learning is, in general, needed to achieve an optimality in dual control (Griffiths and Loparo 1985). The open-loop feedback control law is a passive scheme that does not possess an active learning feature (as it does not take into account any impact from the future learning) and thus can never be optimal.

2.3.3 Active Open-Loop Feedback Control: Variance Minimization Approach

A degree of success of active learning can be measured by the variance of the final state. Therefore minimizing a variance term of the final state will add a feature of active learning to the derived control law. In this section, we consider a modified problem $(M_a(\mu))$ in which a variance term at the final stage is attached to the performance index of (P),

$$
\begin{aligned}
(M_a(\mu)) \min\ & E\Big\{x'(N)Q(N)x(N) + \sum_{k=0}^{N-1}\big[x'(k)Q(k)x(k) + u'(k)R(k)u(k)\big] \mid I^0\Big\} \\
& +\mu \mathrm{Tr}[Cov(x(N) \mid I^0)] \\
\text{s.t.}\ & x(k+1) = A(k,\theta)x(k) + B(k,\theta)u(k) + w(k) \quad k = 0,1,\cdots,N-1 \\
& y(k) = C(k,\theta)x(k) + v(k),\ k = 1,2,\cdots,N
\end{aligned}
$$

Parameter $\mu \in [0,\infty)$ is a weighting coefficient of active learning. A larger μ implies that more importance has been placed on active learning.

Problem $(M_a(\mu))$ is difficult to be solved directly, since the recursive equations of dynamic programming involve certain nonlinear terms of the state estimates that introduces a nonseparability in the sense of dynamic programming. In order to overcome this difficulty, problem $(M_a(\mu))$ is embedded into a tractable auxiliary problem in which the optimal open-loop feedback control can be found. Solving the auxiliary problem and investigating the relationship between the solution sets of problem $(M_a(\mu))$ and the auxiliary problem, the optimal control of problem $(M_a(\mu))$ can be identified.

Define $S(N) = Q(N) + \mu I$, the performance index of $(M_a(\mu))$ can be written as

$$
\begin{aligned}
J = & E\Big\{x'(N)S(N)x(N) + \sum_{k=0}^{N-1}[x'(k)Q(k)x(k) + u'(k)R(k)u(k)] \mid I^0\Big\} \\
& -\mu E(x(N) \mid I^0)'E(x(N) \mid I^0). \qquad (2.12)
\end{aligned}
$$

Let

$$
\begin{aligned}
J^I &= E\Big\{x'(N)S(N)x(N) + \sum_{k=0}^{N-1}[x'(k)Q(k)x(k) + u'(k)R(k)u(k)] \mid I^0\Big\} \\
J^{II} &= E(x(N) \mid I^0).
\end{aligned}
$$

It is easy to see that the performance index in $(M_a(\mu))$, J, is a concave function of J^I and J^{II},

$$
J(J^I, J^{II}) = J^I - \mu\left(J^{II}\right)' J^{II}. \qquad (2.13)
$$

The following auxiliary parametric problem is now constructed for the problem $(M_a(\mu))$ with a fixed multiplier vector $r \in R^n$,

$$(A(r,\mu)) \ \min E\Big\{x'(N)S(N)x(N) + \sum_{k=0}^{N-1}[x'(k)Q(k)x(k) + u'(k)R(k)u(k)] - 2r'x(N) \mid I^0\Big\}$$

$$\text{s.t. } x(k+1) = A(k,\theta)x(k) + B(k,\theta)u(k) + w(k) \ \ k = 0, 1, \cdots, N-1$$

$$y(k) = C(k,\theta)x(k) + v(k), \ k = 1, 2, \cdots, N.$$

Theorem 1 *Suppose that $\{u^*(k)\}$ is an optimal open-loop feedback control of problem $(M_a(\mu))$, then $\{u^*(k)\}$ is also an optimal open-loop feedback control of the auxiliary parametric problem $(A(r^*,\mu))$ where r^* satisfies*

$$r^* = \mu E(x(N) \mid I^0) \mid_{\{u^*(k)\}} . \tag{2.14}$$

The implication of Theorem 1 is that any optimal open-loop feedback solution to problem $(M_a(\mu))$ is in the set of optimal open-loop feedback solutions to auxiliary problem $(A(r,\mu))$. Note that the auxiliary problem is strictly convex with respect to $\{u(k)\}$. Thus the optimal open-loop feedback solution to problem $(A(r,\mu))$ is unique for a given r. As a result, if r satisfies the optimality condition in (2.14), then the optimal open-loop feedback control to $(A(r^*,\mu))$ becomes a possible candidate for the optimal open-loop feedback control to $(M_a(\mu))$.

Define for $i = 1, 2, \ldots, s$,

$$J_i(k, I^k) = E\{x'(k)Q(k)x(k) + u'(k)R(k)u(k) \mid \theta_i, I^k\}, \quad k = 0, \ldots, N-1, \tag{2.15}$$

$$J_i(N, I^N) = E\{x'(N)S(N)x(N) - 2r'x(N) \mid \theta_i, I^N\}. \tag{2.16}$$

Then the following is obvious,

$$J(k, I^k) = E\{x'(k)Q(k)x(k) + u'(k)R(k)u(k) \mid I^k\} = \sum_{i=1}^{s} q_i(k, I^k)J_i(k, I^k) \qquad k = 0, 1, \ldots, N-1 \tag{2.17}$$

$$J(N, I^N) = E\{x'(N)S(N)x(N) - 2r'x(N) \mid I^N\} = \sum_{i=1}^{s} q_i(N, I^N)J_i(N, I^N). \tag{2.18}$$

Since at stage k, all the posterior probabilities at later stages are unknown, a closed-loop optimal control cannot be computed analytically. Suppose that future learning is suspended, then the open-loop feedback control can be obtained by fixing all the posterior probabilities at later stages at $q_i(k, I^k), i = 1, 2, \ldots, s$. As a result, the following optimal open-loop feedback control problem is considered at stage k,

$$
\begin{aligned}
&\min_{u(k)} \sum_{i=1}^{s} q_i(k, I^k)\Big\{E\Big\{J_i(k, I^k) + \ldots \\
&\quad + \min_{u(N-2)} E\{J_i(N-2, I^{N-2}) \\
&\quad + \min_{u(N-1)} E\left[J_i(N-1, I^{N-1}) + J_i(N, I^N)|I^{N-1}\right]|I^{N-2}\} \ldots |I^k\Big\}\Big\}. \quad (2.19)
\end{aligned}
$$

Define $\lambda = [\lambda_1, \ldots, \lambda_s]' = [q_1(0, I^0), \ldots, q_s(0, I^0)]'$. Thus the open-loop feedback control problem for $(A(r, \mu))$ at stage 0 is as follows:

$$
\begin{aligned}
(OFC(\lambda)) \min E &\left\{\sum_{k=0}^{N}(\sum_{i=1}^{s} \lambda_i J_i(k))\right\} \\
\text{s.t. } x_i(k+1) &= A_i(k)x_i(k) + B_i(k)u(k) + w(k), \\
k &= 0, 1, \cdots, N-1,\ i = 1, 2, \cdots, s \\
y_i(k) &= C_i(k)x_i(k) + v(k), \\
k &= 1, 2, \cdots, N,\ i = 1, 2, \cdots, s,
\end{aligned}
$$

where $A_i(k) = A(k, \theta_i)$, $B_i(k) = B(k, \theta_i)$, $C_i(k) = C(k, \theta_i)$, and $x_i(k)$ and $y_i(k)$ are the state and observation of the ith fictitious system, respectively, when assuming $\theta = \theta_i$. Note that as all the posterior probabilities in the later stages are fixed at $q_i(0, I^0)$, the optimal control to $(OFC(\lambda))$ is the optimal open-loop feedback control to problem $(A(r, \mu))$ at stage 0.

Problem $(OFC(\lambda))$ is a multiple-model formulation with $\lambda_i(= q_i(0, I^0)), i = 1, 2, \ldots, s$ serving as the weighting coefficients. Let

$$
\begin{aligned}
X(k) &= [x_1'(k), x_2'(k), \ldots, x_s'(k)]' \\
Y(k) &= [y_1'(k), y_2'(k), \ldots, y_s'(k)]' \\
\bar{A}(k) &= diag(A_1(k), A_2(k), \ldots, A_s(k)) \\
\bar{B}(k) &= [B_1'(k), B_2'(k), \ldots, B_s'(k)]' \\
\bar{C}(k) &= diag(C_1(k), C_2(k), \ldots, C_s(k)) \\
\bar{Q}(k) &= diag(\lambda_1 Q(k), \lambda_2 Q(k), \ldots, \lambda_s Q(k)) \\
\bar{S}(N) &= diag(\lambda_1 S(N), \lambda_2 S(N), \ldots, \lambda_s S(N)) \\
\bar{r} &= [\lambda_1 r', \lambda_2 r', \ldots, \lambda_s r']'
\end{aligned}
$$

$$D_1 = [I_n, I_n, \ldots, I_n]'$$
$$D_2 = [I_p, I_p, \ldots, I_p]',$$

where *diag* denotes a block diagonal matrix. We thus obtain a compact form for the multi-model formulation,

$$\min E\Big\{X'(N)\bar{S}(N)X(N) + \sum_{k=0}^{N-1}[X'(k)\bar{Q}(k)X(k) + u'(k)R(k)u(k)]$$
$$-2\bar{r}'X(N) \mid I^0\Big\} \quad (2.20)$$
$$\text{s.t. } X(k+1) = \bar{A}(k)X(k) + \bar{B}(k)u(k) + D_1 w(k) \quad (2.21)$$
$$k = 0, 1, \cdots, N-1$$
$$Y(k) = \bar{C}(k)X(k) + D_2 v(k), k = 1, 2, \cdots, N. \quad (2.22)$$

Define

$$\hat{X}(k) = [\hat{x}_1'(k|k), \hat{x}_2'(k|k), \ldots, \hat{x}_s'(k|k)]'.$$

The solution to $(OFC(\lambda))$ can be obtained by using dynamic programming. We give the results in the following theorem.

Theorem 2 *For a given r, the optimal control of the auxiliary problem* $(OFC(\lambda))$ *is*

$$u^*(k) = -\Gamma_1(k)\hat{X}(k) + \Gamma_2(k)\bar{r} \quad (2.23)$$

where for $k = N-1, N-2, \ldots, 1, 0$,

$$\bar{S}(k) = \bar{A}'(k)[\bar{S}(k+1) - T(k+1)]\bar{A}(k) + \bar{Q}(k) \quad (2.24)$$
$$T(k) = \Gamma_1'(k)\bar{B}'(k)[\bar{S}(k+1) - T(k+1)]\bar{A}(k) \quad (2.25)$$
$$G(k) = \bar{B}'(k)[\bar{S}(k+1) - T(k+1)]\bar{B}(k) + R(k) \quad (2.26)$$
$$\Gamma_1(k) = G(k)^{-1}\bar{B}'(k)[\bar{S}(k+1) - T(k+1)]\bar{A}(k) \quad (2.27)$$
$$\Gamma_2(k) = G(k)^{-1}\bar{B}'(k)L'(k+1) \quad (2.28)$$
$$L(k) = L(k+1)\left[\bar{A}(k) - \bar{B}(k)\Gamma_1(k)\right] \quad (2.29)$$

with the boundary conditions $T(N) = 0$ *and* $L(N) = I$.

Recall from Theorem 1 that the optimal open-loop feedback control to problem $(A(r, \mu))$ may also be the optimal open-loop feedback control to problem $(M_a(\mu))$

only when condition (2.14) is satisfied. The following theorem is given to show how to determine parameter r^* at stage 0. Define

$$\Phi(k) = I - \mu H(k)\left\{ \sum_{s=k+1}^{N-1} \prod_{i=s}^{N-1} [\bar{A}(i) - \bar{B}(i)\Gamma_1(i)]\bar{B}(s-1)\Gamma_2(s-1) + \bar{B}(N-1)\Gamma_2(N-1)\right\} H^T(k), \quad (2.30)$$

$$\Psi(k) = \mu H(k) \prod_{i=k}^{N-1} [\bar{A}(i) - \bar{B}(i)\Gamma_1(i)]. \quad (2.31)$$

where $H(k) = \left[q_1(k, I^k)I_n, q_2(k, I^k)I_n, \ldots, q_s(k, I^k)I_n\right]$.

Theorem 3 *Assume that Φ is invertible. Then the optimal r^* with which the optimal open-loop feedback solution to $(A(r^*, \mu))$ also solves $(M_a(\mu))$ is equal to*

$$r^* = \Phi^{-1}(0)\Psi(0)\hat{X}(0). \quad (2.32)$$

Substitute (2.32) into the control law in (2.23), then the optimal open-loop feedback control of problem $(M_a(\mu))$ at stage 0, $u^*(0)$, can be obtained.

Proceeding to stage k, we can view stage k as the initial stage and $\hat{x}(k)$ as an estimate of the initial state when we consider a truncated dual control problem from stage k to stage N. Based on the principle of optimality and the concept of a rolling horizon, the optimal value of r should be equal to the following using the same derivation scheme as in Theorem 3,

$$r^* = \Phi^{-1}(k)\Psi(k)\hat{X}(k). \quad (2.33)$$

Substitute (2.33) into the control law (2.23), an optimal open-loop feedback control of problem $(M_a(\mu))$ at stage k, $u^*(k)$, can be obtained.

We have derived in the above discussion an optimal open-loop feedback control for problem $(M_a(\mu))$ with a fixed value of μ. The next natural question to be answered is how to determine the value of μ which represents a degree of importance of active learning. Entropy is a measure of uncertainty. Saridis (1988) and Tsai et al. (1992) studied the entropy formulation of optimal and adaptive control problems. We propose in our solution algorithm to assign the value of μ on-line at stage k to be proportional to the entropy of the probability distribution of θ at stage k, i.e.,

$$\mu \propto -\sum_{i=1}^{s} q_i(k) \ln q_i(k). \quad (2.34)$$

Conceptually, at the first few stages, since there exist parameter uncertainties, more effort will be put in active learning. As time involves, the value of μ will decrease. When the true parameter is identified, the entropy will be equal to zero such that the optimal solution of $(M_a(\mu))$ will converge to the optimal control of problem (P). The proportional constant that relates the entropy to μ can be determined numerically. Notice that a too large proportional constant may result in a poor control performance due to too much effort was devoted to learning.

2.3.4 *Optimal Nominal Dual Control*

Minimizing a covariance term at the final stage provides a feature of active learning for the derived control law. The control law obtained, however, is not a closed-loop law but an optimal open-loop feedback control. Under this framework, the impact from the future learning has not been considered.

The key research issues are: (1) what is the best possible (partial) closed-loop control for (2.10), and (2) what is the active learning strategy to achieve this best possible outcome. A major difficulty in solving (2.10) is that the optimal control cannot be determined when the future posterior probabilities are unknown, while at the same time the future posterior probabilities depend on the control applied at the early stages. In order to break this loop, a possible solution scheme is to derive the relationship between the posterior probability and the control. A control which satisfies a deterministic version of this relationship is defined as the nominal control. The expected posterior probabilities when applying the nominal control are called nominal future posterior probabilities. Applying the nominal future posterior probabilities generated by the nominal control instead in (2.10), the effect of future learning can be taken into account. Since in this situation, all the achievable future information is used in terms of its expected value, the control law obtained can be considered to be the best possible closed-loop control law in this sense.

Assume that the current time is k and consider the truncated control problem from stage k to the end of the time horizon. For given $\lambda^t = [\lambda_1^t, \ldots, \lambda_s^t]' \in R_+^s$, $t = k, k+1, \ldots, N$, with $\lambda^k = [q_1(k, I^k), q_2(k, I^k), \ldots, q_s(k, I^k)]'$, consider the following optimal control problem,

$$
\begin{aligned}
(ONC(\lambda)) \min\ & E\left\{\sum_{t=k}^{N}(\sum_{i=1}^{s} \lambda_i^t J_i(t, I^t))\right\} \\
\text{s.t.}\ \ & x_i(t+1) = A_i(t)x_i(t) + B_i(t)u(t) + w(t), \\
& \qquad t = k, k+1, \cdots, N-1,\ i = 1, 2, \cdots, s \\
& y_i(t) = C_i(t)x_i(t) + v(t), \\
& \qquad t = k+1, k+2, \cdots, N,\ i = 1, 2, \cdots, s,
\end{aligned}
$$

where $A_i(t) = A(t, \theta_i)$, $B_i(t) = B(t, \theta_i)$, $C_i(t) = C(t, \theta_i)$, and $x_i(t)$ and $y_i(t)$ are the state and observation of the ith fictitious system, respectively, when assuming $\theta = \theta_i$.

Let

$$\begin{aligned}
X(t) &= [x_1'(t), x_2'(t), \ldots, x_s'(t)]' \\
Y(t) &= [y_1'(t), y_2'(t), \ldots, y_s'(t)]' \\
\bar{A}(t) &= diag(A_1(t), A_2(t), \ldots, A_s(t)) \\
\bar{B}(t) &= [B_1'(t), B_2'(t), \ldots, B_s'(t)]' \\
\bar{C}(t) &= diag(C_1(t), C_2(t), \ldots, C_s(t)) \\
\bar{Q}(t, \lambda) &= diag(\lambda_1^t Q(t), \lambda_2^t Q(t), \ldots, \lambda_s^t Q(t)) \\
D_1 &= [I_n, I_n, \ldots, I_n]' \\
D_2 &= [I_m, I_m, \ldots, I_m]',
\end{aligned}$$

where *diag* denotes a block diagonal matrix. We can obtain a compact form for the above multi-model formulation of $(ONC(\lambda))$ as follows:

$$\begin{aligned}
\min E &\left\{ X'(N)\bar{Q}(N, \lambda)X(N) + \sum_{t=k}^{N-1} [X'(t)\bar{Q}(t, \lambda)X(t) + u'(t)R(t)u(t)] \mid I^k \right\} \\
\text{s.t. } & X(t+1) = \bar{A}(t)X(t) + \bar{B}(t)u(t) + D_1 w(t), \quad t = k, k+1, \cdots, N-1 \\
& Y(t) = \bar{C}(t)X(t) + D_2 v(t), \quad t = k+1, k+2, \cdots, N.
\end{aligned}$$

Define

$$\hat{X}(t) = [\hat{x}_1'(t|t), \hat{x}_2'(t|t), \ldots, \hat{x}_s'(t|t)]'.$$

The optimal solution to $(ONC(\lambda))$ can be derived by using dynamic programming,

$$u^*(t) = -\Gamma(t, \lambda)\hat{X}(t), \tag{2.35}$$

where for $t = k, k+1, \ldots, N-1$

$$\Gamma(t, \lambda) = -G^{-1}(t, \lambda)\bar{B}'(t)S(t+1, \lambda)\bar{A}(t) \tag{2.36}$$

$$G(t, \lambda) = \bar{B}'(t)S(t+1, \lambda)\bar{B}(t) + R(t) \tag{2.37}$$

$$S(t, \lambda) = \bar{A}'(t)S(t+1, \lambda)\bar{A}(t) + \bar{Q}(t, \lambda) - \Gamma'(t, \lambda)G(t, \lambda)\Gamma(t, \lambda), \tag{2.38}$$

with the boundary condition $S(N,\lambda) = \bar{Q}(N,\lambda)$. Note that the optimal control, $\{u^*(t)\}_{t=k}^{N-1}$, is linear in the augmented state estimation $\hat{X}(t)$ and the feedback gain matrix Γ is nonlinear in λ.

At stage k, the true observation $y(k)$ is known, therefore $\hat{x}_i(k|k)$ can be obtained by the Kalman filter (1) to (5). Since future observations cannot be known in advance, a predicted nominal state trajectory $\{\hat{x}_i^*(t)\}_{t=k+1}^{N}$ and a predicted nominal observation trajectory $\{\hat{y}_i^*(t)\}_{t=k+1}^{N}$, can be calculated by setting all random variables at their expected values, i.e.

$$\hat{x}_i^*(t+1) = A_i(t)\hat{x}_i^*(t) + B_i(t)u^*(t),\ t = k, k+1, \dots, N-1, \tag{2.39}$$

$$\hat{y}_i^*(t) = C_i(t)\hat{x}_i^*(t),\ t = k+1, k+2, \dots, N, \tag{2.40}$$

with the initial condition $\hat{x}_i^*(k) = \hat{x}_i(k|k)$. For $t = k+1, k+2, \dots, N$, let

$$\hat{X}(t) = [\hat{x}_1^*(t)', \hat{x}_2^*(t)', \dots, \hat{x}_s^*(t)']'.$$

Substituting $\hat{X}(t)$ back into Eq. (2.35), we can close the loop and obtain a predicted nominal control.

Comparing problem $(ONC(\lambda))$ with the closed-loop control problem (2.10) at stage k, it is easy to recognize that if λ_i^t plays the same role as the posterior probabilities $q_i(t, I^t)$ at every stage, the optimal control of problem $(ONC(\lambda))$ is also optimal to problem (P) at stage k. However, those posterior probabilities at the later stages are unattainable. A feasible way is to use the nominal posterior probabilities generated by the nominal control instead. The control law achieved under this framework is referred to as the optimal nominal control to the original problem.

Define for $t = k+1, k+2, \dots, N$

$$\hat{y}^*(t) = \sum_{i=1}^{s} \lambda_i^t \hat{y}_i^*(t). \tag{2.41}$$

Using the Bayes formula, the predicted nominal posterior probability of mode i at stage k, $i = 1, 2, \dots, s$, satisfies the following recursive equation:

$$\tilde{q}_i(t) = \frac{L_i(t)}{\sum_{j=1}^{s} \tilde{q}_j(t-1)L_j(t)}\tilde{q}_i(t-1), \quad t = k+1, k+2, \dots, N, \tag{2.42}$$

with the initial condition $q_i(k, I^k)$, where $L_i(t)$ is still the same as given in (7) except

$$\tilde{y}(t|t-1, \theta_i) = \hat{y}^*(t) - \hat{y}_i^*(t). \tag{2.43}$$

It is clear that $\tilde{q}_i(t)$ is a function of $\lambda^k, \lambda^{k+1}, \dots \lambda^N$.

In order to force the weighting coefficients λ_i^t to be equal to the nominal posterior probability $\tilde{q}_i(t)$ for all $t = k+1, k+2, \ldots, N$, we construct the following optimization problem at stage k

$$\min \sum_{t=k+1}^{N} \sum_{i=1}^{s} (\lambda_i^t - \tilde{q}_i(t))^2 \tag{2.44}$$

$$\text{s.t.} \sum_{i=1}^{s} \lambda_i^t = 1, \ \textit{and all}\ \lambda_i^t \geq 0,\ t = k+1, \ldots, N.$$

This is a nonlinear programming problem and can be solved by using general nonlinear programming solvers.

2.4 Dual Control in Big Data Era

In the Big Data era, massive amounts of information are generated every day. The high volume, high velocity, and high variety features of Big Data make capturing, managing, analyzing, storing, and retrieving information extremely challenging. In addition, the large-scale interconnected systems such as economic systems, power systems, manufacturing systems, health systems, water distribution systems, biological systems, etc. are complex and rapidly changing. It is not realistic and possible to develop mathematical models precisely to describe the system dynamics. Dual controls with probing features are advantageous in regulating these stochastic systems, especially in two situations: (1) when the time horizon is short and the initial estimates are poor, it is essential to stimulate the systems and rapidly find good estimates before reaching the end of the control horizon; (2) when the parameters of the process are changing very rapidly (Wittenmark 1975a). Some successful applications of dual control are summarized as below.

2.4.1 Economic Systems

Most economic problems are stochastic. There is uncertainty about the present state of the system, uncertainty about the response of the system to policy measures, and uncertainty about future events. For example, in macroeconomics some time series are known to contain more noise than others. Also, policy makers are uncertain about the magnitude and timing of responses to changes in tax rates, government spending, and interest rates. In international commodity stabilization, there is uncertainty about the effects of price changes on consumption (Kendrick 1981). Because of the short time horizon and highly stochastic nature of the parameters in the economic processes, dual controls have been seen in solving economic systems (Bar-Shalom and Wall 1980; Kendrick 1981). Kendrick demonstrated examples of using dual control to solve MacRae problem and a macroeconometric model with measurement error (Kendrick 1981).

2.4.2 *Manufacturing Processes*

Dual control is also successfully applied in manufacturing processes. The grinding processes in the pulp industry (Allison 1994), where the parameters are changing fairly rapidly and the gain is also changing sign, is probably the first application of dual control to process control. The controller is an active adaptive controller, which consists of a constrained certainty equivalence approach coupled with an extended output horizon and a cost function modification to get probing (Wittenmark 1975a).

Another application of dual control in capital intensive semiconductor manufacturing processes has been seen in Arda Vanli et al. (2011). In such processes, it is often impractical to run large designed experiments and the amount of experimental data available is often not adequate to build sufficiently accurate statistical models or reliably estimating optimal conditions. A dual control approach that simultaneously considers model estimation and optimization objectives is adopted and an adaptive Bayesian response surface model is used. It is shown that by employing the proposed adaptive Bayesian approach one can simultaneously learn the process while not requiring excessive perturbations away from the target level and can achieve faster model estimation than central composite experimental designs.

2.4.3 *Automobile Systems*

A driver assistance system with a dual control scheme was developed in Saito et al. (2016), which can effectively identify drivers' drowsiness and prevent sleep-related vehicle accidents. The dual control has two purposes: (1) to effect the partial control initiated by the assistance system, preventing lane departure, and (2) enabling the assistance system to judge, through the interaction between the driver and the assistance system, whether the driver recognizes that the vehicle is going to deviate from the lane. The assistance system implements partial control in the event of lane departure and gives the driver the chance to voluntarily take the action needed. If the driver fails to implement the steering action needed within a limited time, the assistance system judges that "the driver's understanding of the given situation is incorrect" and executes the remaining control.

2.4.4 *Robotics*

Adaptive dual control using neural networks has also been extensively investigated. Neural networks have been used to approximate the unknown functions in the system dynamics of the nonlinear stochastic systems. Such dual control was successfully applied to kinematic control of nonholonomic mobile robots in which the robot dynamic functions are nonlinear with varying uncertain/unknown parameters

(Bugeja et al. 2009). Two schemes are developed in discrete time, and the robot's nonlinear dynamic functions are assumed to be unknown. The Gaussian radial basis function and sigmoidal multilayer perception neural networks are used for function approximation. In each scheme, the unknown network parameters are estimated stochastically in real time, and no preliminary offline neural network training is used. In contrast to other adaptive techniques hitherto proposed in the literature on mobile robots, the dual control laws do not rely on the heuristic certainty equivalence property but account for the uncertainty in the estimates. This results in a major improvement in tracking performance, despite the plant uncertainty and unmodeled dynamics.

2.4.5 Information Retrieval

An Information Retrieval (IR) system consists of a collection of documents and an engine that retrieves documents described by user queries. In large systems, such as the Web, queries are typically too vague, hence an iterative process in which the users refine their queries gradually has to take place. An active learning approach was proposed in Jaakkola and Siegelmann (2001) to reduce the IR users dissatisfactions due to long, tedious repetitive search sessions. The system responds to the initial user's query by successively probing the user for distinctions at multiple levels of abstraction. The system's initiated queries are optimized for speedy recovery and the user is permitted to respond with multiple selections or may reject the query. The information is in each case unambiguously incorporated by the system and the subsequent queries are adjusted to minimize the need for further exchanges. More applications in information retrieval and image retrieval can be seen in Zhang and Chen (2002) and Dagli et al. (2005).

2.5 Conclusions

This overview presents the dual control methods, elaborated from the Feldbaum's seminal work in the 1960s until present. The author and collaborators' research on dual control for a class of discrete-time linear quadratic Gaussian problems with parameter uncertainty in both state and observation equations is summarized to demonstrate different control laws. It is shown that minimizing a covariance term at the final stage introduces a feature of active learning for the derived control law. By exploring the future nominal posterior probabilities, the control law takes into account the function of future learning, thus the best possible closed-loop feedback control can be achieved. Successful applications of dual controls in various areas indicate although cautious, the controller with the probing/active learning feature can help reduce system uncertainties and hence it performs better than the controller with passive or without learning ability.

References

B.J. Allison, Dual adaptive control of a chip refiner motor load. Ph.D. Thesis, University of British Columbia, 1994

J. Alster, P.R. Belanger, A technique for dual adaptive control. Automatica **10**, 627 (1974)

J. Alster, P.R.B. Langer, A technique for dual adaptive control. Automatica, **10**, 627–634 (1974)

M. Aoki , *Optimization of Stochastic Systems* (Academic, New York 1967)

O. Arda Vanli, C. Zhang, B. Wang, An adaptive Bayesian method for semiconductor manufacturing process control with small experimental data sets. IEEE Trans. Semicond. Manuf. **24**(3), 418–431 (2011)

K.J. Astrom, *Introduction to Stochastic Control Theory* (Academic Press, New York, 1970)

K.J. Astrom, B. Wittenmark, Problems of identification and control. J. Math. Analysis Appl. **34**, 90 (1971)

K.J. Astrom, B. Wittenmark, *Adaptive Control* (Addison Wesley, Reading, 1989)

Y. Bar-Shalom, Stochastic dynamic programming: caution and probing. IEEE Trans. Autom. Control **AC-26**, 1184–1194 (1981)

Y. Bar-Shalom, R. Sivan, On the optimal control of discrete-time linear systems with random parameters. IEEE Trans. Autom. Control **AC-14**, 3–8 (1969)

Y. Bar-Shalom, E. Tse, Dual effect, certainty equivalence, and separation in stochastic control. IEEE Trans. Autom. Control **AC-19**(5), 494–500 (1974)

Y. Bar-Shalom, K.D. Wall, Dual adaptive control and uncertainty effects in macroeconomic systems optimization. Automatica **16**, 147–156 (1980)

M.K. Bugeja, S.G. Fabri, L. Camilleri, Dual adaptive dynamic control of mobile robots using neural networks. IEEE Trans. Syst. Man Cybern. **39**(1), 129–141 (2009)

F. Casiello, K.A. Loparo, Optimal policies for passive learning controllers. Automatica **25**, 757–763 (1989)

C.K. Dagli, S. Rajaram, T.S. Huang, Combining diversity-based active learning with disriminant analysis in image retrieval, in *Third International Conference on Information Technology and Applications*, vol. 2 (2005), pp. 173–178

J.G. Deshpande, T.N. Upadhyay, D.G. Lainiotis, Adaptive control of linear stochastic systems. Automatica **9**, 107–115 (1973)

A.A. Feldbaum, *Optimal Control Systems* (Academic, New York, 1965)

N.M. Filatov, H. Unbehauen, Survey of adaptive dual control methods. IEE Proc. Control Theory Appl. **147**, 118–128 (2000)

J.J. Florentine, Optimal probing, adaptive control of a simple Bayesian system. J. Electronics Control **13**, 165 (1962)

P. Fu, D. Li, F. Qian, Active dual control for linear-quadratic Gaussian system with unknown parameters, in *Proceedings of the 15th World Congress of IFAC*, Barcelona, Spain, July 21–16 (2002)

D. Gorman, J. Zaborszky, Stochastic optimal control of continous time systems with unknown gain. IEEE Trans. Autom. Control **AC-13**, 630–638 (1968)

A.J. Grammaticos, B.M. Horowitz, The optimal adaptive control law for a linear plant with unknown input gains. Int. J. Control **12**, 337 (1970)

B.E. Griffiths, K.A. Loparo, Optimal control of jump linear systems. Int. J. Control. **42**(4), 791–819 (1985)

R. Hasting-James, M.W. Sage, Recursive generalized least squares procedure for on-line identification of process parameters. IEE Proc. **116**, 2057 (1969)

D.J. Hughes, O.L.R. Jacobs, Turn-off, escape and probing in nonlinear stochastic control, in *Proceedings of IFAC Symposium Adaptive Control*, Budapest, Hungary, Sept. 1974

T. Jaakkola, H. Siegelmann, Active information retrieval. Adv. Neural Inf. Proces. Syst. **14**, 777–784 (2001)

D.H. Jacobson, Optimal stochastic linear systems with exponential performance criteria and their relation to deterministic differential games. IEEE Trans. Autom. Control **18**, 124–131 (1973)

O.L.R. Jacobs, S.M. Langdon, An optimal extremum control system. Automatica **6**, 297 (1970)
K. Jenkins, R. Roy, A design procedure for discrete adaptive control systems, Preprints JACC (1966), p. 624
R.E. Kalman, Design of self-optimizing control system, Trans. Am. Soc. Mech. Eng. **80**, 468 (1958)
D.A. Kendrick, *Stochastic Control for Economic Models* (McGraw-Hill, New York, 1981)
D. Lainiotis, J.G. Deshpande, T.N. Upadhyay, Optimal adaptive control: a nonlinear separation theorem. Int. J. Control **15**, 877–888 (1972)
D. Li, F. Qian, P. Fu, Variance minimization approach for a class of dual control problems. IEEE Trans. Autom. Control **AC-47** 2010–2020 (2002)
D. Li, F. Qian, P. Fu, Optimal nominal dual control for discrete-time linear-quadratic Gaussian problems with unknown parameters. Automatica **44**, 119–127 (2008)
B. Lindoff, J. Holst, B. Wittenmark, Analysis of approximations of dual control. Int. J. Adapt. Control Signal Process. **13**, 593–620 (1999)
J.C. Luxat, L.H. Lees, Suboptimal control of a class of nonlinear systems. Int. J. Control **17**, 965 (1973)
R. Milito, C. Padilla, R. Padilla, D. Cardorin, An innovations approach to dual control. IEEE Trans. Autom. Control **AC-27**, 132–137 (1982)
V. Panuska, A stochastic approximative method for identification of linear systems using adaptive filter, in *Preprint Joint Automatic Control Conference* (1968)
Y. Saito, M. Itoh, T. Inagaki, Driver assistance system with a dual control scheme: effectiveness of identifying driver drowsiness and preventing lane departure accidents. IEEE Trans. Hum. Mach. Syst. **46**(5), 660–671 (2016)
G.N. Saridis, Entropy formulation of optimal and adaptive control. IEEE Trans. Autom. Control **AC-33**(8), 713–721 (1988)
J. Sternby, A regulator for time-varying stochastic systems, in *Proceedings of 7th IFAC World Congress*, Helsinki, Finland, June 1978
Y.A. Tsai, F.A. Casiello, K.A. Loparo, Discrete-time entropy formulation of optimal and adaptive control problems. IEEE Trans. Autom. Control **AC-37**(7), 1083–1088 (1992)
E. Tse, M. Athans, Adaptive stochastic control for a class of linear systems. IEEE Trans. Autom. Control **AC-17**, 38–52 (1972)
E. Tse, Y. Bar-Shalom, L. Meier, Wide-sense adaptive dual control for nonlinear stochastic systems. IEEE Trans. Autom. Control **AC-18**, 98–108 (1973)
B. Wittenmark, Stochastic adaptive control methods: a survey. Int. J. Control **21**, 705–730 (1975a)
B. Wittenmark, An active suboptimal dual controller for systems with stochastic parameters. Autom. Control Theory Appl. **3**, 13–19 (1975b)
B. Wittenmark, Stochastic adaptive control methods: a survey. Int. J. Control **21**(5), 705–730 (1975c)
J. Wieslander, B. Wittenmark, An approach to adaptive control using real-time identification. Automatica **7**, 211 (1971)
P.C. Young, The use of linear regression and related procedure for the identification of dynamic processes. in *Proceedings of Seventh I.E.E.E Symposium on Adaptive Processes* (1968)
C. Zhang, T. Chen, An active learning framework for content-based information retrieval. IEEE Trans. Multimedia **4**(2), 260–268 (2002)

Chapter 3
Time Inconsistency and Self-Control Optimization Problems: Progress and Challenges

Yun Shi and Xiangyu Cui

Abstract Time inconsistency has been an important issue in many stochastic decision problems arisen in real life and financial decision making, especially in the dynamic investment area. When a stochastic decision problem is time inconsistent, the decision maker would be puzzled by his/her conflicting decisions "optimally" derived from his/her time-varying preferences at different time instants. In the literature, the time inconsistent problem is also called the self-control problem, as the decision maker needs to exert proper self-control to resist present temptation and then achieve a better long-term performance. Different approaches dealing with time inconsistency in the literature are reviewed in this paper. After that, the open questions and challenges are also discussed.

Keywords Time inconsistency • Self-control • Present bias • Non-separable problem • Quasi-hyperbolic discounting function • Dynamic portfolio optimization

3.1 Introduction

Time inconsistency, also called dynamic inconsistency, refers to the phenomena that the long-term optimal decision policy determined at time 0 is no longer optimal when reconsidering the truncated decision problem at time t. Many real-life decision

---

Y. Shi
School of Management, Shanghai University, Shanghai, People's Republic of China
e-mail: y_shi@shu.edu.cn

X. Cui (✉)
School of Statistics and Management, Shanghai University of Finance and Economics, Shanghai, People's Republic of China
e-mail: cui.xiangyu@mail.shufe.edu.cn

T.-M. Choi et al. (eds.), *Optimization and Control for Systems in the Big-Data Era*, International Series in Operations Research & Management Science 252,
DOI 10.1007/978-3-319-53518-0_3

problems are time inconsistent problems, such as whether to abstain from smoking.[1] Apparently, the long-term optimal decision is to insist abstaining from smoking. However, when coming into action, the smoker may prefer to smoke today and abstain tomorrow. The reason behind is that comparing to the present temptation (the pleasure of smoking), the decision maker evaluates the bad consequences received in the future (bad health) with a large discount factor, which is also called the present bias (see Thaler 1981; Loewenstein and Prelec 1992). In the literature, the time inconsistent problem is also called the self-control problem, as the decision maker needs to exert proper self-control in order to achieve a good long-term performance.

Essentially, when a decision maker faces a dynamic decision-making problem, his/her preferences at different time instants for the corresponding tail parts of the time horizon could be time-varying and/or state dependent. Actually, we all have faced occasions in which we change our minds, but usually we do not go to extraordinary steps to prevent ourselves from deviating from the original plan. The only circumstances in which we would want to commit ourselves to our planned course of action is when we have a good reason to believe that if we change our preferences later, this change of preferences will be a mistake. In the smoking case, the smoker definitely believes that abstaining from smoking is good for health. Thus, the crucial question is this: if I know I am going to change my mind about my preferences, when and how would I take some actions to restrict my future behaviour?

In this article, we start by summarizing some progress in dealing with the time inconsistency and self-control issue, especially in the dynamic investment area. We then describe some major open questions in this area.

3.2 Progress

3.2.1 Separable Problem Versus Non-separable Problem

The time inconsistent problems can be classified into two categories: separable ones and non-separable ones, which are caused by the present bias and the non-separable preference, respectively. Dynamic investment and consumption problems with quasi-hyperbolic discounting and dynamic mean-variance portfolio selection problems are two salient time inconsistent decision problems in the literature. The past few years have seen substantial progress in our understanding of the time consistency issue. Much of the progress concerns these two problems.

For the dynamic investment and consumption problem with quasi-hyperbolic discounting, the decision maker's time preference, which represents the value at time t of \$1 received at future time s, is described by

$$D_t(s) = \begin{cases} 1, & \text{if } s = t, \\ \beta\delta^{(s-t)}, & \text{if } s > t, \end{cases}$$

[1] More time inconsistency examples can be found in Gul and Pesendorfer (2001), Grenadier and Wang (2007), Björk and Murgoci (2010), Basak and Chabakauri (2010) and Cui et al. (2012).

where $0 < \beta < 1$ is the quasi-hyperbolic discounting parameter and represents the short-run discounting, and δ represents the long-run discounting (see Laibson 1997; O'Donoghue and Rabin 1999). The quasi-hyperbolic discounting is a typical and well-documented time preference with present bias, under which the decision maker tends to underestimate the value of payoff in the future. The decision maker's investment objective at time t is the expected sum of discounted utilities of the future consumptions and the terminal wealth,

$$\mathbb{E}_t\left[\sum_{s=t}^{T-1} D_t(s)U(c(s))\right] + D_t(T)\mathbb{E}_t[U(X(T))].$$

Due to the existence of the short-run discounting parameter β, the preferences at different time instants are inconsistent. The decision maker may switch his/her mind at a later time t and prefer to consume more than his/her original plan to maximize his/her short-term preference (see Thaler and Shefrin 1981). The smaller the parameter β, the larger the conflict between the long-term optimal consumption plan and the short-term optimal consumption plan.

For dynamic mean-variance portfolio selection problem, the decision maker's preference at time t is the weighted sum of the conditional expected value and the conditional variance of the terminal wealth,

$$\mathbb{E}_t[X(T)] - \lambda \mathrm{Var}_t(X(T)),$$

where $\lambda > 0$ is the risk aversion parameter. As the variance term does not satisfy the smooth property, i.e., $\mathrm{Var}_t(X(T)) \neq \mathrm{Var}_t(\mathrm{Var}_s(X(T)))$ for $s > t$, the preferences at different time instants would be definitely inconsistent. After a mean-variance investor derives his/her long-term optimal investment strategy at time 0, he/she could be tempted to adopt a different strategy at a later time t in order to achieve a short-term mean-variance efficiency (see Basak and Chabakauri 2010; Cui et al. 2012).

One significant difference between the above two problems is that the objectives of the dynamic investment and consumption problem with quasi-hyperbolic discounting at different time instants only involve the separable expectation operator (which can be represented as the expected sum of the future discounted performance measures), while the objectives of the dynamic mean-variance portfolio selection problem at different time instants involve a non-separable operator: variance. In addition, almost all widely adopted risk measures in static portfolio selection, including variance, semi-variance, value-at-risk (VaR) and conditional value-at-risk (CVaR), become time inconsistent in dynamic mean-risk framework (see Boda and Filar 2006). Moreover, all those risk measures are of a non-separable nature.

3.2.2 Approaches Dealing with Time Inconsistency

Before summarizing the approaches, we need to know the mathematical meaning of time inconsistency. When a decision maker faces a time inconsistent dynamic decision problem, the overall objective for the entire time horizon under consideration does not conform with the local objective for a tail part of the entire time horizon. In the language of dynamic programming, *Bellman's principle of optimality* is not applicable in such situations, as the global and local interests derived from their respective objectives are not consistent (see Artzner et al. 2007).

Apparently, one direct approach to overcome the time inconsistency issue is to construct a time consistent decision model. This approach is widely studied in the field of dynamic risk measures and dynamic risk management. As a basic requirement, all the suitable dynamic risk measures should necessarily possess certain functional structure, such as

$$\rho_t(X_T) = \rho_t(-\rho_s(X_T)), \quad t < s < T,$$

in order to satisfy time consistency (see Rosazza Gianin 2006; Artzner et al. 2007; Jobert and Rogers 2008). When a dynamic risk measure is time consistent, it not only justifies the mathematical formulation for risk management, but also facilitates the solution process in finding the optimal decision (see Cherny 2010). However, this approach cannot be applied to the time inconsistent decision problems caused by present bias or non-separable preferences.

In the literature, there are mainly three different solution schemes in dealing with the time inconsistency issue for the general dynamic decision problem. The first solution scheme is the so-called pre-committed policy approach. In this approach, the decision maker strictly adheres to the global long-term optimal decision policy over the entire time horizon. In other words, the decision maker only cares about the global objective and fully ignores local objectives. Such policy is called the *pre-committed policy.*[2] To adopt the pre-committed policy, the decision maker can make the pre-committed policy the only feasible policy or the only economically reasonable policy via a strict self-control commitment or external contractual commitment, which is not easy in reality. Investment plan 401(k) is one such example in reality that forces employees not to withdraw pensions before retirement through a contractual penalty scheme (see Madrian and Shea 2001).[3]

[2]This policy is also termed "strategy of pre commitment" in Strotz (1956).

[3]In the United States, a 401(k) plan is the tax-qualified, defined-contribution pension account defined in section 401(k) of the Internal Revenue Code. The Internal Revenue Code imposes severe restrictions on withdrawals of pre-tax or Roth contributions while a person remains in service with the company and is under the age of 59.5. Any withdrawal that is permitted before the age of 59.5 is subject to an excise tax equal to ten percent of the amount distributed (on top of the ordinary income tax that has to be paid).

The second solution scheme is the time consistent policy approach. In this approach, the decision maker is aware of the inconsistency between the global objective and local objectives, but is unable to adhere to the pre-committed policy. Thus, the decision maker totally bows to local objectives (i.e., local temptations). As the decision maker at current time instant has a decision advantage with respect to the ones at future time instants, the decision maker's problem can be modelled as an intrapersonal sequential game. In the game, the decision maker at any time instant acts as a Stackelberg leader and makes his/her "best" decision by taking into account his/her decisions in future periods. The corresponding subgame perfect Nash equilibrium decision policy is called the *time consistent policy*.[4] This approach is widely applied to separable and non-separable time inconsistent decision problems. Laibson (1997), O'Donoghue and Rabin (1999, 2001), and Grenadier and Wang (2007) studied the time consistent policies for different financial decision problems with quasi-hyperbolic discounting. Basak and Chabakauri (2010), Hu et al. (2012), Lioui (2013), Chen et al. (2014b), Björk et al. (2014) and Cui et al. (2016) studied time consistent policies for the mean-variance preference under different market settings.

The third solution scheme is the self-control policy approach developed in the literature recently. In this approach, the decision maker intends to resolve conflicts between the long-term and short-term interests by reconciling the global objective and local objectives. To achieve this goal, the decision maker is required to possess a degree of willpower to exert self-control and resist the local temptation in the future time instants (see Rachlin 2004). Several theoretical models with a self-control feature have been developed to guide decision makers in achieving such a balance. For example, O'Donoghue and Rabin (1999, 2001) proposed the partial naive decision maker assumption, which assumes that the decision maker can exert self-control to have a larger quasi-hyperbolic discounting parameter, i.e.,

$$\hat{D}_t(s) = \begin{cases} 1, & \text{if } s = t, \\ \hat{\beta}\delta^{(s-t)}, & \text{if } s > t, \end{cases}$$

where $\beta < \hat{\beta} \leq 1$. With the larger quasi-hyperbolic discounting parameter $\hat{\beta}$, the decision maker decreases the conflict between the long-term and short-term preferences, and thus achieves a better balance between the long-term and short-term interests. Gul and Pesendorfer (2001, 2004) proposed the axiomatic theory of self-control, under which the decision maker integrates the opportunity costs of deviating from the local optimal policies into the global objective. By doing so, a policy taking into account both the long-term and short-term interests is obtained. Thaler and Shefrin (1981) and Bénabou and Pycia (2002) proposed the planner-doer model, and Fudenberg and Levine (2006, 2012) proposed the dual-self model,

[4]This policy is also termed the "strategy of consistent planning" in Strotz (1956), and the decision maker who adopts this policy is called the *sophisticated* decision maker in O'Donoghue and Rabin (1999, 2001) and Grenadier and Wang (2007).

which both assume that the global self can influence the myopic preferences of local selves through different self-control schemes and then derive the equilibrium policy between global self and local selves. We call this type of policies the *self-control policy* in this review.[5] These models with self-control features have been successfully applied to decision problems, whose time inconsistency is caused by present bias (see O'Donoghue and Rabin 1999, 2001; Grenadier and Wang 2007; Chen et al. 2014a; Tian 2016). However, these models are not applicable for non-separable time inconsistent decision problems.

Besides the above approaches, Cui et al. (2012) and Cui et al. (2015a) proposed a new angle for dealing with the time inconsistency of dynamic mean-variance portfolio problem. As mean-variance problem is a multi-objective optimization problem, a multi-objective version of principle of optimality is applied. In other words, any tail part of an efficient policy is also efficient for any realizable state at an intermediate period (see Li and Haimes 1987; Li 1990). In spirit of this logic, Cui et al. (2012) extended the concept of time consistency to a relaxed version to incorporate efficiency, namely, *time consistency in efficiency*. Through showing that the dynamic mean-variance formulation is not time consistent in efficiency, they demonstrated that the investor may have irrational local preferences of minimizing the risk and the return at the same time. Cui et al. (2012) relaxed the self-financing restriction and allow withdrawal of positive dollar amounts out of the market during the investment process. Furthermore, they proposed a better revised policy which can achieve the original mean-variance pair but obtain some extra (positive) dollar amounts with a strictly positive probability under certain probability distribution assumptions. Moreover, Cui et al. (2015a) studied the effect of portfolio constraints on the time consistency in efficiency of convex cone-constrained markets and established a general procedure for constructing time consistent in efficiency dynamic mean-variance portfolio selection problems by introducing suitable portfolio constraints.

3.3 Challenges

3.3.1 Dynamic Mean-Risk Portfolio Optimization Problems

Mean-risk portfolio selection models are widely used in portfolio management practices. Although most of these models suffered the problem of time inconsistency, only the dynamic mean-variance model attracts enough attentions in the literature.

[5]In Gul and Pesendorfer (2001, 2004), the preference of such a type of decision maker is called "preference with self-control". In O'Donoghue and Rabin (1999, 2001) and DellaVigna and Malmendier (2004, 2006), the decision maker who takes this type of policy is classified as *partially naive*. In Fudenberg and Levine (2006, 2012), this policy is termed "SR-Perfect equilibrium strategy".

Moreover, the published works on time inconsistency issue of dynamic mean-variance models mainly concentrate on the time consistent policy and the revised policy. Thus, there are several open questions in this field.

The first question is how to derive the time consistent policies for the mean-risk models beyond mean-variance. As these models are non-separable, there does not exist analytical or semi-analytical form policy. Thus, suitable numerical methods can be developed. Furthermore, the properties of the time consistent polices under different risk aversion parameter settings are worth investigation. Cui and Shi (2014) made an attempt to analyse the time consistent policy for multi-period mean-CVaR model with finite states.

The second question is whether the mean-risk models beyond mean-variance satisfy time consistency in efficiency. If not, the decision maker may devote himself/herself to constructing the revised policies, which is better than the pre-committed policy.

The third question is how to construct the self-control policies for the mean-risk models. Although the existing theoretical models with a self-control feature are not directly applicable to the non-separable mean-risk models, the idea is very useful in constructing some new theoretical models under non-separable mean-risk framework. To our best knowledge, there are some preliminary work in this direction. By extending the planner-doer model of Thaler and Shefrin (1981) and Bénabou and Pycia (2002), Cui et al. (2017) developed a two-tier planner-doer game framework with self-coordination, which is theoretically applicable to discrete-time non-separable decision problems. They successfully applied the proposed framework to deal with dynamic mean-variance portfolio selection problem and a two-period mean-CVaR portfolio selection problem. Similarly, Cui et al. (2015b) extended the dual-self model of Fudenberg and Levine (2006, 2012) and proposed a two-tier dual-self game model, which is theoretically applicable to continuous-time non-separable decision problems. Although the above two new frameworks have an important theoretical value, how to apply them to construct suitable self-coordination schemes and compute the corresponding self-control policies for the mean-risk models beyond mean-variance are still unclear.

3.3.2 Time Inconsistency Generated by Probability Weighting

The probability weighting function, proposed by Tversky and Kahneman (1992), transforms objective probabilities into decision weights. The original motivation for this transformation function was the simultaneous demand many people had for both lotteries and insurance. Typically, people prefer a 0.001 chance of $50,000 to a certain $50 but meanwhile prefer to pay $50 rather than face a 0.001 chance of a $50,000 loss. This combination of behaviours is difficult to explain under the expected utility theory. However, under the probability weighting framework, the unlikely events—gaining or losing $50,000—are overweighted, thereby explaining these choices. The probability weighting is a key feature of many behaviour portfolio selection models, such as the rank-dependent utility model (see Schmeidler 1989; Abdellaoui 2002) and the cumulative prospect theory model (see Tversky and Kahneman 1992; He and Zhou 2011).

In a dynamic setting, probability weighting also generates time inconsistency (see Barberis 2012), which, once again, lies in the domain of non-separability. This inconsistency may be useful for understanding some real trading behaviours, for example, people sometimes hold on to losing investments longer than they were planning to, known as the disposition effect in the literature (see Odean 1998). However, there is relatively little research on it, especially when compared to the large literature on the inconsistency generated by present bias. Shi et al. (2015) suggested one possible approach to analyze the time inconsistency generated by probability weighting in dynamic setting, but other approaches are surely also possible and deserved more studies.

3.3.3 Data Challenge

As long as the data becomes more and more easy to collect, the decision makers begin to formulate their decision problems based on rich data. On the one hand, they can use the rich data to describe the dynamics of the uncertainties, which makes the constructions of dynamic decision problems possible. On the other hand, they may build data-driven decision problems by fully using the rich data (see Bertsimas and Thiele 2006; Delage and Ye 2010; Hou and Wang 2013; Huh et al. 2011 for data-driven decision-making examples). Based on these two developing directions, there will be more and more data-driven dynamic decision problems in research and practices.

In general, these data-driven dynamic decision problems are time inconsistent. Comparing to the dynamic decision problems with explicit assumptions on the uncertainties, the data-driven dynamic decision problems may introduce great computation challenges.

Acknowledgements This work is dedicated to the 65th birthday of the authors' supervisor, Professor Duan Li. Both the authors are in debt to Professor Duan Li for his invaluable guidances and advices during the authors' studies, career developments and lives. This work was partially supported by National Natural Science Foundation of China under Grants 71601107, 71671106, 71201094, by the State Key Program in the Major Research Plan of National Natural Science Foundation of China under Grant 91546202, by Shanghai Pujiang Program under Grant 15PJC051, by Program for Innovative Research Team of Shanghai University of Finance and Economics.

References

M. Abdellaoui, A genuine rank-dependent generalization of the Von Neumann-Morgenstern expected utility theorem. Econometrica **70**, 717–736 (2002)

P. Artzner, F. Delbaen, J.M. Eber, D. Heath, H. Ku, Coherent multiperiod risk adjusted values and Bellman's principle. Ann. Oper. Res. **152**, 5–22 (2007)

S. Basak, G. Chabakauri, Dynamic mean-variance asset allocation. Rev. Financ. Stud. **23**, 2970–3016 (2010)
N. Barberis, A model of casino gambling. Manag. Sci. **58**(1), 35–51 (2012)
D. Bertsimas, A. Thiele, Robust and data-driven optimization: modern decision-making under uncertainty. Tutor. Oper. Res. **4**, 95–122 (2006)
R. Bénabou, M. Pycia, Dynamic inconsistency and self-control: a planner-doer interpretation. Econ. Lett. **77**, 419–424 (2002)
T. Björk, A. Murgoci, A general theory of Markovian time inconsistent stochasitc control problem, working paper (2010). Available at SSRN: http://ssrn.com/abstract=1694759
T. Björk, A. Murgoci, X.Y. Zhou, Mean-variance portfolio optimization with state dependent risk aversion. Math. Financ. **24**, 1–24 (2014)
K. Boda, J.A. Filar, Time consistent dynamic risk measures. Math. Methods Oper. Res. **63**, 169–186 (2006)
S.M. Chen, Z.F. Li, Y. Zeng, Optimal dividend strategies with time-inconsistent preferences. J. Econ. Dyn. Control. **46**, 150–172 (2014a)
Z.P. Chen, G. Li, Y.G. Zhao, Time-consistent investment policies in Markovian markets: a case of mean-variance analysis. J. Econ. Dyn. Control **40**, 293–316 (2014b)
A.S. Cherny, Risk-reward optimization with discrete-time conherent risk. Math. Financ. **20**, 571–595 (2010)
X.Y. Cui, Y. Shi, Multiperiod mean-CVaR portfolio selection. in *Modelling, Computation and Optimization in Information Systems and Management Sciences*. (Springer, Berlin, 2014), pp. 293–304
X.Y. Cui, D. Li, S.Y. Wang, S.S. Zhu, Better than dynamic mean-variance: time inconsistency and free cash flow stream. Math. Financ. **22**, 346–378 (2012)
X.Y. Cui, D. Li, X. Li, Mean-variance policy for discrete-time cone-constrained markets: time consistency in efficiency and the minimum-variance signed supermartingale measure. Math. Financ. (2015a) doi:10.1111/mafi.12093
X.Y. Cui, X. Li, Y. Shi, Resolving time inconsistency of decision problem with non-expectation operator: from internal conflict to internal harmony by strategy of self-coordination. Working Paper (2015b)
X.Y. Cui, D. Li, Y. Shi, Self-coordination in time inconsistent stochastic decision problems: A planner-doer game framework. J. Econ. Dyn. Control **75**, 91–113 (2017)
X.Y. Cui, L. Xu, Y. Zeng, Continuous time mean-variance portfolio optimization with piecewise state-dependent risk aversion. Optim. Lett. **10**, 1681–1691 (2016)
E. Delage, Y.Y. Ye, Distributionally robust optimization under moment uncertainty with application to data-driven problems. Oper. Res. **58**, 595–612 (2010)
S. DellaVigna, U. Malmendier, Contract design and self-control: theory and evidence. Q. J. Econ. **119**, 353–402 (2004)
S. DellaVigna, U. Malmendier, Overestimating self-control: evidence from the health club industry. Am. Econ. Rev. **96**, 694–19 (2006)
D. Fudenberg, D.K. Levine, A dual-self model of impulse control. Am. Econ. Rev. **96**, 1449–1476 (2006)
D. Fudenberg, D.K. Levine, Timing and self-control. Econometrica **80**, 1–42 (2012)
S.R. Grenadier, N. Wang, Investment under uncertainty and time-inconsistent preferences. J. Financ. Econ. **84**, 2–39 (2007)
F. Gul, W. Pesendorfer, Temptation and self-control. Econometrica **69**, 1403–1435 (2001)
F. Gul, W. Pesendorfer, Self-control and the theory of consumption. Econometrica **71**, 119–158 (2004)
X.D. He, X.Y. Zhou, Portfolio choice under cumulative prospect theory: an analytical treatment. Manag. Sci. **57**, 315–331 (2011)
Z.S. Hou, Z. Wang, From model-based control to data-driven control: survey, classification and perspective. Inf. Sci. **235**, 3–35 (2012)
Y. Hu, H. Jin, X. Zhou, Time-inconsistent stochastic linear-quadratic control. SIAM J. Control Optim. **50**, 1548–1572 (2012)

W.T. Huh, R. Levi, P. Rusmevichientong, J.B. Orlin, Adaptive data-driven inventory control with censored demand based on Kaplan-Meier estimator. Oper. Res. **59**, 929–941 (2011)
A. Jobert, L.C. Rogers, Valuations and dynamic convex risk measures. Math. Financ. **18**, 1–22 (2008)
D. Laibson, Golden eggs and hyperbolic discounting. Q. J. Econ. **112**, 443–477 (1997)
D. Li, Multiple objectives and nonseparability in stochastic dynamic programming. Int. J. Syst. Sci. **21**, 933–950 (1990)
D. Li, Y.Y. Haimes, The envelope approach for multiobjective optimization problems. IEEE Trans. Syst. Man Cybern. **17**, 1026–1038 (1987)
A. Lioui, Time consistent vs. time inconsistent dynamic asset allocation: some utility cost calculations for mean variance preferences. J. Econ. Dyn. Control. **37**, 1066–1096 (2013)
G. Loewenstein, D. Prelec, Anomalies in intertemporal choice: evidence and an interpretation. Q. J. Econ. **107**, 573–598 (1992)
B.C. Madrian, D. Shea, The power of suggestion: inertia in 401(k) participation and savings behavior. Q. J. Econ. **116**, 1149–1187 (2001)
T. Odean, Are investors reluctant to realize their losses? J. Financ. **53**(5), 1775–1798 (1998)
T. O'Donoghue, M. Rabin, Doing it now or later. Am. Econ. Rev. **89**, 103–124 (1999)
T. O'Donoghue, M. Rabin, Choice and procrastination. Q. J. Econ. **116**, 121–160 (2001)
H. Rachlin, *The Science of Self-Control* (Harvard University Press, Cambridge, 2004)
E. Rosazza Gianin, Risk measures via g-expectations. Insur. Math. Econ. **39**, 19–34 (2006)
D. Schmeidler, Subject probability and expected utility without additivity. Econometrica **57**, 571–587 (1989)
Y. Shi, X.Y. Cui, D. Li, Discrete-time behavioral portfolio selection under cumulative prospect theory. J. Econ. Dyn. Control **61**, 283–302 (2015)
R.H. Strotz, Myopia and inconsistency in dynamic utility maximization. Rev. Econ. Stud. **23**, 165–180 (1956)
R.H. Thaler, Some empirical evidence on dynamic inconsistency. Econ. Lett. **8**, 201–207 (1981)
R.H. Thaler, H.M. Shefrin, An economic theory of self-control. J. Polit. Econ. **89**, 392–406 (1981)
Y. Tian, Optimal capital structure and investment decisions under time-inconsistent preferences. J. Econ. Dyn. Control **65**, 83–104 (2016)
A. Tversky, D. Kahneman, Advances in prospect theory: cumulative representation of uncertainty. J. Risk Uncertain. **5**(4), 297–323 (1992)

Chapter 4
Quadratic Convex Reformulations for Integer and Mixed-Integer Quadratic Programs

Baiyi Wu and Rujun Jiang

Abstract We review recent advances in the quadratic convex reformulation (QCR) approach that is employed to derive efficient equivalent reformulations for mixed-integer quadratically constrained quadratic programming (MIQCQP) problems. Although MIQCQP problems can be directly plugged into and solved by standard MIQP solvers that are based on branch-and-bound algorithms, it is not efficient because the continuous relaxation of the standard MIQCQP reformulation is very loose. The QCR approach is a systematic way to derive tight equivalent reformulations. We will explore the QCR technique on subclasses of MIQCQP problems with simpler structures first and then generalize it step by step such that it can be applied to general MIQCQP problems. We also cover the recent extension of QCR on semi-continuous quadratic programming problems.

Keywords Quadratic programming • Quadratic convex reformulation • Recent advances • Semi-continuous quadratic programming

4.1 Introduction

Mixed-integer quadratically constrained quadratic programming (MIQCQP) problems are mathematical programming problems with continuous and discrete variables and quadratic functions in the objective function and constraints. The use of MIQCQP is a natural approach of formulating nonlinear problems where it is necessary to simultaneously optimize the system structure (discrete) and parameters (continuous).

B. Wu
School of Finance, Guangdong University of Foreign Studies, Guangzhou, Guangdong, China
e-mail: baiyiwu@outlook.com

R. Jiang (✉)
Department of Systems Engineering and Engineering Management, The Chinese University of Hong Kong, Shatin, N.T., Hong Kong
e-mail: rjjiang@se.cuhk.edu.hk

T.-M. Choi et al. (eds.), *Optimization and Control for Systems in the Big-Data Era*,
International Series in Operations Research & Management Science 252,
DOI 10.1007/978-3-319-53518-0_4

MIQCQPs have been used in various applications, including the process industry and the financial, engineering, management science and operations research sectors. It includes problems such as the unit commitment problem (Frangioni and Gentile 2006; Frangioni et al. 2011), the Markowitz mean-variance mode with practical constraints (Mitra et al. 2007), the chaotic mapping of complete multipartite graphs (Fu et al. 2001), the material cutting problem (Cui 2005), the capacity planning problem (Hua and Banerjee 2000). More MIQCQP applications can be found in Grossmann and Sahinidis (2002). The MIQCQP problem is in general NP-hard. It also nests many NP-hard problems as its special cases, such as the binary quadratic programming problem, the integer quadratic programming problem, the semi-continuous quadratic programming problem, etc. All these problems are very difficult to solve and yet have many useful real-life applications such as the max-cut problem (Rendl et al. 2010), the portfolio lot sizing problem (Li et al. 2006), the cardinality constrained portfolio selection and control problems (Gao and Li 2011, 2013b; Zheng et al. 2014; Gao and Li 2013a). The needs in such diverse areas have motivated research and development in solving MIQCQP problems, as they become more and more challenging with larger problem sizes along with the "big data" era. Various methods to find local or global optimal solutions and good lower bounds for different classes of MIQCQP problems can be found in Li and Sun (2006), which is also an excellent survey for nonlinear integer programming.

The general form of an MIQCQP problem is:

$$
\begin{aligned}
(\mathrm{P}) \quad \min \quad & x^T Q_0 x + c_0^T x \\
\text{s.t.} \quad & x^T Q_i x + c_i^T x \le b_i, \ i = 1, \ldots, m, \\
& x \in \mathbb{R}^n, \\
& x_i \in \mathbb{Z}, \ 0 \le x_i \le u_i, \ i \in I \subset \{1, \ldots, n\},
\end{aligned}
$$

where we assume that, after fixing the values of $x_i, \ i \in I$, the remaining problem is convex. The standard continuous relaxation of problem (P) resulting from removing the integral constraint $x_i \in \mathbb{Z}$ is a convex problem. Thus, problem (P) can be directly plugged into and solved by many off-the-shelf solvers such as CPLEX and Gurobi, which use branch-and-bound schemes. The major issue is that the bound from the standard continuous relaxation is usually very loose, resulting in a large search tree in the branch-and-bound process.

One remedy is to find equivalent reformulations that have a tighter continuous relaxation. In this chapter, we review a general technique, the quadratic convex reformulation (QCR), that is used to find good equivalent reformulations for problem (P). When a better reformulation is solved by CPLEX or Gurobi, the computation time needed can be significantly reduced.

The QCR approach focuses on finding a tighter MIQCQP formulation that is equivalent to the original problem (P). It has the following characteristics:

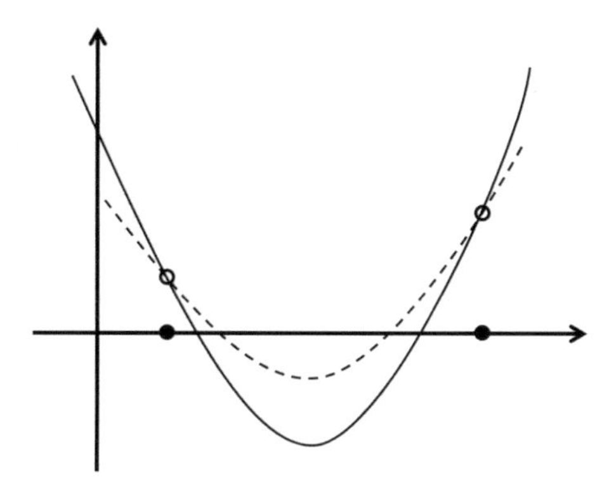

Fig. 4.1 Equivalent reformulation that has a tighter lower bound

(a) Quadratic functions that vanish in the feasible region can be added to the objective function and the constraints.
(b) These added quadratic functions can be characterized by a set of parameters.
(c) A convex problem is solved to get the best parameters such that the continuous relaxation of the reformulation is convex and at the same time it provides a lower bound that is as tight as possible.

The process of QCR can be demonstrated by a simple example. The solid curve of Fig. 4.1 is the original objective function and the two dots are the feasible solutions. We can find a new objective curve, the dashed curve, that passes through the corresponding objective points of the feasible solutions and at the same time provides a better lower bound from its continuous relaxation. It is essential that the two curves coincide on the feasible region. The curvature of the dashed curve is smaller than that of the solid curve. That is why the QCR process is also called a "flattening" process in Plateau (2006) and Ahlatçıoğlu et al. (2012).

We will explore the QCR technique on subclasses of problem (P) with simpler structures first and then generalize it step by step such that it can be applied to the general form of problem (P). In Sect. 4.2, we review QCR for binary quadratic programming problems. Two schemes are covered, with or without additional variables. In Sect. 4.3, we review QCR for linear equality constrained binary quadratic programming problems. In Sect. 4.4, we show step by step how we can apply QCR on the general from of problem (P). In Sect. 4.5, we extend the QCR on semi-continuous quadratic programming problems. We conclude the review in Sect. 4.6.

Notation In remaining sections, we denote by $\mathrm{v}(\cdot)$ the optimal value of problem $(\cdot)$, and $\mathbb{R}^n_+$ the nonnegative orthant of $\mathbb{R}^n$. For any $a \in \mathbb{R}^n$, we denote by $\mathrm{Diag}(a) = \mathrm{Diag}(a_1, \ldots, a_n)$ the diagonal matrix with a_i being its ith diagonal element. We denote by e the all-one vector. We denote by $\mathbb{S}^n$ the set of $n \times n$ symmetric real matrix..

4.2 QCR for Binary Quadratic Programming

QCR is firstly applied on the following binary quadratic programming problem,

$$\begin{aligned}(\text{BQP}) \quad \min \ & x^T Q_0 x + c_0^T x \\ \text{s.t.} \ & x \in \{0,1\}^n.\end{aligned}$$

There are two main schemes, one with no additional variables and the other with additional variables.

4.2.1 *QCR with No Additional Variables*

The QCR scheme with no additional variables was pioneered by Hammer and Rubin (1970) and later improved by Plateau (2006) and Billionnet et al. (2008, 2009). Li et al. (2012) explored the geometry of QCR from another angle.

For any $d \in \mathbb{R}^n$, the following problem is equivalent to problem (BQP):

$$\begin{aligned}(\text{BQP}(d)) \quad \min \ & x^T (Q_0 - \text{Diag}(d))x + (c_0 + d)^T x \\ \text{s.t.} \ & x \in \{0,1\}^n.\end{aligned}$$

By relaxing the binary constraint as continuous constraint and noting $0 \le x_i \le 1$ is equivalent to $x_i^2 - x_i \le 0$, we can represent problem $(\overline{\text{BQP}}(d))$ as

$$\begin{aligned}(\overline{\text{BQP}}(d)) \qquad \min_{x \in \mathbb{R}^n} \ & x^T (Q_0 - \text{Diag}(d))x + (c_0 + d)^T x \\ \text{s.t.} \ & x_i^2 - x_i \le 0, \ i = 1, \ldots, n.\end{aligned}$$

We want to choose d such that the continuous relaxation of (BQP(d)) is a convex problem and the lower bound from this relaxation is as tight as possible. This can be done by solving the following problem,

$$(\text{BQP\_MAX\_d}) \ \ \max\{\text{v}(\overline{\text{BQP}}(d)) \mid d \in \mathbb{R}^n, \ Q_0 - \text{Diag}(d) \succeq 0\}.$$

Note that problem $(\overline{\text{BQP}}(d))$ is convex if and only if $Q_0 - \text{Diag}(d) \succeq 0$.

Theorem 1 (Billionnet et al. 2009) *Problem* (BQP_MAX_d) *is equivalent to the following semi-definite programming (SDP) problem,*

$$\begin{aligned}(\text{BQP\_DSDP\_d}) \quad \max \ & \tau \\ \text{s.t.} \ & \begin{pmatrix} -\tau & 0.5(c_0 + d)^T \\ 0.5(c_0 + d) & Q_0 - \text{Diag}(d) \end{pmatrix} \succeq 0, \ d \in \mathbb{R}^n.\end{aligned}$$

Proof The Lagrangian dual of $(\overline{\text{BQP}}(d))$ is

$$(\overline{\text{LBQP}}(d)) \qquad \max_{\alpha \geq 0} \min_{x \in \mathbb{R}^n} x^T(Q_0 - \text{Diag}(d))x + (c_0 + d)^T x + x^T \text{Diag}(\alpha) x - \alpha^T x,$$

which is equivalent to

$$\max_{\alpha \geq 0} \min_{x \in \mathbb{R}^n} \tau + (1\ x^T) \begin{pmatrix} -\tau & 0.5(c_0 + d - \alpha)^T \\ 0.5(c_0 + d - \alpha) & Q_0 - \text{Diag}(d) + \text{Diag}(\alpha) \end{pmatrix} \begin{pmatrix} 1 \\ x \end{pmatrix},$$

for any $\tau \in \mathbb{R}$. By noting

$$\min_{x \in \mathbb{R}^n} \tau + (1\ x^T) \begin{pmatrix} -\tau & 0.5(c_0 + d - \alpha)^T \\ 0.5(c_0 + d - \alpha) & Q_0 - \text{Diag}(d) + \text{Diag}(\alpha) \end{pmatrix} \begin{pmatrix} 1 \\ x \end{pmatrix}$$

$$= \begin{cases} \tau & \text{if } \begin{pmatrix} -\tau & 0.5(c_0 + d - \alpha)^T \\ 0.5(c_0 + d - \alpha) & Q_0 - \text{Diag}(d) + \text{Diag}(\alpha) \end{pmatrix} \succeq 0, \\ -\infty & \text{otherwise,} \end{cases}$$

we obtain that $(\overline{\text{LBQP}}(d))$ is equivalent to

$$\begin{aligned} &\max \ \tau \\ &\text{s.t.} \ \begin{pmatrix} -\tau & 0.5(c_0 + d - \alpha)^T \\ 0.5(c_0 + d - \alpha) & Q_0 - \text{Diag}(d) + \text{Diag}(\alpha) \end{pmatrix} \succeq 0, \ \alpha \geq 0. \end{aligned}$$

Note that the constraints of $(\overline{\text{BQP}}(d))$ satisfy the Slater condition since $(0.5, \ldots, 0.5)^T$ is an interior point. If $Q_0 - \text{diag}(d) \succeq 0$, then problem $(\overline{\text{BQP}}(d))$ is convex and from the no duality gap theory of convex programming (see, e.g., Proposition 6.5.6 in Bertsekas et al. 2003), we have $\text{v}(\overline{\text{BQP}}(d)) = \text{v}(\overline{\text{LBQP}}(d))$. Thus (BQP_MAX_d) is equivalent to

$$(\text{LBQP\_MAX\_d}) \quad \max\{\text{v}(\overline{\text{LBQP}}(d)) \mid d \in \mathbb{R}^n, \ Q_0 - \text{Diag}(d) \succeq 0\},$$

which is further equivalent to problem (BQP_DSDP_d) by eliminating the variable α. □

In fact, by rewriting the binary constraint $x \in \{0, 1\}^n$ as $x_i^2 - x_i = 0, \ i = 1, \ldots, n$, one can check that problem (BQP_DSDP_d) is the Lagrangian dual of problem (BQP).

Convex optimization shows that the conic dual of problem (BQP_DSDP_d) is equivalent to the following SDP relaxation of problem (*BQP*), which is well known as Shor's relaxation (Shor 1987),

$$\begin{aligned}(\text{BQP\_PSDP\_d}) \quad \min \quad & Q_0 \bullet X + c_0^T x \\ \text{s.t.} \quad & X_{ii} = x_i, \ i = 1, \ldots, n, \\ & \begin{pmatrix} 1 & x \\ x^T & X \end{pmatrix} \succeq 0.\end{aligned}$$

4.2.2 QCR with Additional Variables

The QCR scheme with additional variables for binary quadratic programming problems can be found in Billionnet et al. (2012, 2013, 2015).

If we introduce new variables, the reformulation can be further strengthened. For any $S \in \mathbb{S}^n$, the following problem is equivalent to (BQP),

$$\begin{aligned}(\text{BQP}(S)) \quad \min \quad & x^T(Q_0 - S)x + c_0^T x + S \bullet Y \\ \text{s.t.} \quad & Y_{ij} \le x_i, \ Y_{ij} \le x_j, \ Y_{ij} \ge x_i + x_j - 1, \ Y_{ij} \ge 0, \ \forall\, 0 \le i < j \le n, \\ & Y_{ii} = x_i, \ i = 1, \ldots, n, \\ & x \in \{0,1\}^n, \ Y \in [0,1]^{n \times n}, \ Y \in \mathbb{S}^n.\end{aligned}$$

By relaxing $x \in \{0,1\}^n$, as $0 \le x_i \le 1, \ i = 1 \ldots, n$, we obtain the continuous relaxation of (BQP(S)) as follows:

$$\begin{aligned}\min \quad & x^T(Q_0 - S)x + c_0^T x + S \bullet Y \\ \text{s.t.} \quad & Y_{ij} \le x_i, \ Y_{ij} \le x_j, \ Y_{ij} \ge x_i + x_j - 1, \ Y_{ij} \ge 0, \ \forall\, 0 \le i < j \le n, \\ & Y_{ii} = x_i, \ i = 1, \ldots, n, \\ & x \in [0,1]^n, \ Y \in [0,1]^{n \times n}, \ Y \in \mathbb{S}^n,\end{aligned}$$

which is equivalent to

$$\begin{aligned}(\overline{\text{BQP}}(S)) \quad \min \quad & x^T(Q_0 - S)x + c_0^T x + S \bullet Y \\ \text{s.t.} \quad & Y_{ij} \le x_i, \ Y_{ij} \le x_j, \ Y_{ij} \ge x_i + x_j - 1, \ Y_{ij} \ge 0, \ \forall\, 0 \le i \le j \le n, \qquad (4.1) \\ & Y_{ii} = x_i, \ i = 1, \ldots, n, \\ & x \in \mathbb{R}^n, \ Y \in \mathbb{S}^n,\end{aligned}$$

where the constraint (4.1) includes the case of $i = j$. Note that the constraint $Y \le 1$ is implicit in the above problem since $Y_{ii} = x_i$ and $Y_{ii} \ge x_i + x_i - 1$ together imply $Y_{ii} \le 1$, thus $Y_{ij} \le x_i = Y_{ii} \le 1$. The constraint $0 \le x_i \le 1$ is also implicit in the above problem since $x_i = Y_{ii}$.

We want to choose S such that $(\overline{\text{BQP}}(S))$ is a convex problem and the lower bound from this relaxation is as tight as possible. This can be done by solving the following problem:

$$\text{(BQP\_MAX\_S)} \quad \max\{\text{v}(\overline{\text{BQP}}(S)) \mid S \in \mathbb{S}^n,\ Q_0 - S \succeq 0\}.$$

Theorem 2 (Billionnet et al. 2013) *Problem* (BQP_MAX_S) *is equivalent to the following semi-definite programming (SDP) problem,*

$$\begin{aligned}
\text{(BQP_DSDP_S)} \quad \max \quad & -\tau + Y_3 \bullet (ee^T) \\
\text{s.t.} \quad & \begin{pmatrix} \tau & 0.5y^T \\ 0.5y & Q_0 - Y \end{pmatrix} \succeq 0, \\
& y = c - Y^1 e - Y^2 e + 2Y^3 e - d, \\
& Y = S + Y^1 + Y^2 - Y^3 - Y^4 + Diag(d), \\
& Y^1, Y^2, Y^3, Y^4 \in \mathbb{N}^n \cap \mathbb{S}^n,\ d \in \mathbb{R}^n, Y \in \mathbb{S}^n, \\
& S \in \mathbb{S}^n, y \in \mathbb{R}^n, \tau \in \mathbb{R},
\end{aligned}$$

where e is a column vector with all entries being 1, and $\mathbb{N}^n$ represents the set of $n \times n$ nonnegative matrices.

Proof The Lagrangian dual of $(\overline{\text{BQP}}(S))$ is as follows:

$$\begin{aligned}
(\overline{\text{LBQP}}(S)) \quad \max \ & \tau + Y_3 \bullet (ee^T) \\
\text{s.t.} \ & \begin{pmatrix} -\tau & 0.5y^T \\ 0.5y & Q_0 - Y \end{pmatrix} \succeq 0 \\
& y = c_0 - Y^1 e - Y^2 e + 2Y^3 e - d, \\
& Y = S + Y^1 + Y^2 - Y^3 - Y^4 + \text{Diag}(d), \\
& Y^1, Y^2, Y^3, Y^4 \in \mathbb{N}^n \cap \mathbb{S}^n,\ d \in \mathbb{R}^n, Y \in \mathbb{S}^n, \\
& y \in \mathbb{R}^n, \tau \in \mathbb{R}.
\end{aligned}$$

Note that the constraints of $(\overline{\text{BQP}}(S))$ satisfy the Slater condition. If $Q_0 - S \succeq 0$, then problem $(\overline{\text{BQP}}(S))$ is convex and from the no duality gap theory of convex programming, we have $\text{v}(\overline{\text{BQP}}(S)) = \text{v}(\overline{\text{LBQP}}(S))$. Thus (BQP_MAX_S) is equivalent to

$$\text{(LBQP\_MAX\_S)} \quad \max\{\text{v}(\overline{\text{LBQP}}(S)) \mid S \in \mathbb{S}^n,\ Q_0 - S \succeq 0\}.$$

It is obvious that the above problem is further equivalent to (BQP_DSDP_S). □

The conic dual of problem (BQP_DSDP_S) is given as follows:

$$
\begin{aligned}
\text{(BQP\_PSDP\_S)} \ \min \quad & Q_0 \bullet X + c_0^T x \\
\text{s.t.} \quad & X_{ij} \le x_i,\ X_{ij} \le x_j,\ X_{ij} \ge x_i + x_j - 1, X_{ij} \ge 0, \forall\, 0 \le i \le j \le n, \\
& X_{ii} = x_i,\ i = 1, \ldots, n, \\
& \begin{pmatrix} 1 & x^T \\ x & X \end{pmatrix} \succeq 0,
\end{aligned}
$$

which is an SDP relaxation for problem (BQP) that is tighter than the Shor's relaxation.

4.3 QCR for Linear Equality Constrained Binary Quadratic Programming

In this section, we extend the QCR to the following binary quadratic programming problems with linear equability constraints,

$$
\begin{aligned}
\text{(EBQP)} \quad \min \quad & x^T Q_0 x + c_0^T x \\
\text{s.t.} \quad & Ax = b,\ x \in \{0, 1\}^n,
\end{aligned}
$$

where A is an $m \times n$ matrix and $b \in \mathbb{R}^n$. We consider only the scheme with no additional variables, which was proposed by Plateau (2006) and Billionnet et al. (2008, 2009).

For any $d \in \mathbb{R}^n, s \in \mathbb{R}^{(m+1)*m/2}$, the following problem is equivalent to problem (EBQP):

$$
\begin{aligned}
\text{(EBQP}(d, s)) \quad \min \quad & x^T (Q_0 - \text{Diag}(d))x + (c_0 + d)^T x \\
& + \sum_{1 \le i \le j \le m} s_{ij} (A_i x - b_i)^T (A_j x - b_j) \\
\text{s.t.} \quad & Ax = b,\ x \in \{0, 1\}^n.
\end{aligned}
$$

By relaxing the binary constraint as continuous constraint and noting $0 \le x_i \le 1$ is equivalent to $x_i^2 - x_i \le 0$, we can represent the continuous relaxation of (EBQP(d, s)) as

$$
\begin{aligned}
(\overline{\text{EBQP}}(d)) \quad \min \quad & x^T (Q_0 - \text{Diag}(d))x + (c_0 + d)^T x \\
& + \sum_{1 \le i \le j \le m} s_{ij} (A_i x - b_i)^T (A_j x - b_j) \\
\text{s.t.} \quad & Ax = b,\ x_i^2 - x_i \le 0.
\end{aligned}
$$

We want to choose d and s such that $(\overline{\text{EBQP}}(d, s))$ is a convex problem and the low bound from this relaxation is as tight as possible. This can be done by solving the following problem:

$$\text{(EBQP\_MAX\_ds)}\quad \max\{\text{v}(\overline{\text{EBQP}}(d,s)) \mid d \in \mathbb{R}^n,\ s \in \mathbb{R}^{(m+1)*m/2},\\ Q_0 - \text{Diag}(d) + \sum_{1\le i\le j\le m} s_{ij}A_i^T A_j \succeq 0\}.$$

Using similar proofs as in Sect. 4.2, we can show that (EBQP_MAXds) is equivalent to the following SDP problem,

$$\begin{aligned}
\text{(EBQP_PSDP_ds)}\quad \min\ & Q_0 \bullet X + c_0^T x \\
\text{s.t.}\ & Ax = b, \\
& X \bullet A_i^T A_j - (b_i^T A_j + b_j^T A_i)x + b_i^T b_j = 0,\ 1 \le i \le j \le n, \\
& X_{ii} = x_i,\ i = 1, \ldots, n, \\
& \begin{pmatrix} 1 & x^T \\ x & X \end{pmatrix} \succeq 0.
\end{aligned}$$

The formulation (EBQP(d, s)) is parameterized by d and s. The dimension of s is $(m + 1) * m/2$. As the number of linear equalities grows, the number of parameters increases quadratically. In fact, we can achieve the same good reformulation with only one parameter.

For any $d \in \mathbb{R}^n,\ w \in \mathbb{R}$, the following problem is equivalent to (EBQP):

$$\begin{aligned}
\text{(EBQP}(d,w))\quad \min\ & x^T(Q_0 - \text{Diag}(d))x + (c_0 + d)^T x + w(Ax - b)^T(Ax - b) \\
\text{s.t.}\ & Ax = b, \\
& x \in \{0, 1\}^n.
\end{aligned}$$

We want to choose d and w such that the continuous relaxation of (EBQP(d, w)) is a convex problem and the low bound from this relaxation is as tight as possible. This can be done by solving the following problem:

$$\text{(EBQP\_MAX\_dw)}\quad \max\{\text{v}(\overline{\text{EBQP}}(d,w)) \mid d \in \mathbb{R}^n,\ w \in \mathbb{R}\\ Q_0 - \text{Diag}(d) + wA^T A \succeq 0\}.$$

Similar to the last section, we have the following theorem.

Theorem 3 (Billionnet et al. 2012) *Problem* (EBQP_MAX_dw) *is equivalent to the following SDP problem:*

$$\begin{aligned} \text{(EBQP\_DSDP\_dw)} \quad \max \quad & -\tau + wb^T b \\ \text{s.t.} \quad & \begin{pmatrix} \tau & 0.5(c_0 + d - 2b^T A)^T \\ 0.5(c_0 + d - 2b^T A) & Q_0 - Diag(d) + wA^T A \end{pmatrix} \succeq 0, \\ & d \in \mathbb{R}^n, w \in \mathbb{R}. \end{aligned}$$

The conic dual of problem (EBQP_DSDP_dw) is given as follows:

$$\begin{aligned} \text{(EBQP\_PSDP\_dw)} \quad \min \quad & Q_0 \bullet X + c_0^T x \\ \text{s.t.} \quad & Ax = b, \\ & X \bullet (A^T A) - 2b^T Ax + b^T b = 0, \\ & X_{ii} = x_i, \ i = 1, \ldots, n, \\ & \begin{pmatrix} 1 & x^T \\ x & X \end{pmatrix} \succeq 0. \end{aligned}$$

which is also an SDP relaxation of (EBQP).

It is shown in Faye and Roupin (2007) that problem (EBQP_PSDP_dw) has the same optimal value with the SDP problem (EBQP_PSDP_ds). Thus the reformulation (EBQP(d, w)) with less parameters is as good as the reformulation (EBQP(d, s)).

4.4 Generalization of QCR to MIQCQP

In this section we extend QCR to general MIQCQP problems step by step using the techniques derived in Sects. 4.2 and 4.3.

4.4.1 QCR for Binary Quadratically Constrained Quadratic Programming

We first consider binary quadratic programming problems with quadratic constraints:

$$\begin{aligned} \text{(BQCQP)} \quad \min \quad & x^T Q_0 x + c_0^T x \\ \text{s.t.} \quad & x^T Q_i x + c_i^T x \le b_i, \ i = 1, \ldots, m, \\ & x \in \{0, 1\}^n. \end{aligned}$$

For any $d_0, d_1, \ldots, d_m \in \mathbb{R}^n$, the following problem is equivalent to problem (BQCQP):

$$\begin{aligned}
(\text{BQCQP}(d)) \quad \min \quad & x^T(Q_0 - \text{Diag}(d_0))x + (c_0 + d_0)^T x \\
\text{s.t.} \quad & x^T(Q_i - \text{Diag}(d_i))x + (c_i + d_i)^T x \le b_i, \ i = 1, \ldots, m, \\
& x \in \{0, 1\}^n.
\end{aligned}$$

For any $S_0, S_1, \ldots, S_m \in \mathbb{S}^n$, the following problem is equivalent to (BQCQP),

$$\begin{aligned}
(\text{BQCQP}(S)) \quad \min \quad & x^T(Q_0 - S_0)x + c_0^T x + S_0 \bullet Y \\
\text{s.t.} \quad & Y_{ij} \le x_i, \ Y_{ij} \le x_j, \ Y_{ij} \ge x_i + x_j - 1, \ Y_{ij} = Y_{ji}, \ \forall \, 0 \le i < j \le n, \\
& Y_{ii} = x_i, \ i = 1, \ldots, n, \\
& x^T(Q_i - S_i)x + c_i^T x + S_i \bullet Y \le b_i, \ i = 1, \ldots, m, \\
& x \in \{0, 1\}^n, \ Y \in \{0, 1\}^{n \times n}.
\end{aligned}$$

Using the same methods as in Sects. 4.2 and 4.3, one can derive SDP problems to solve for the best parameters d for problem (BQCQP(d)) and S for problem (BQCQP(S)). Note that in this case we also need to maintain the convexity in the constraints.

4.4.2 QCR for Mixed-Binary Quadratic Programming

In this section we consider the following problem with continuous and binary variables,

$$\begin{aligned}
(\text{MBQP}) \quad \min \quad & x^T Q_0 x + c_0^T x \\
\text{s.t.} \quad & x \in \mathbb{R}^n, \ x_i \in \{0, 1\}, \ \ i \in I \subset \{1, \ldots, n\},
\end{aligned}$$

where we assume that, after fixing the values of $x_i, \ i \in I$, the remaining problem is convex.

For any $s_{ij} \in \mathbb{R}, \ (i, j) \in I \times \{1, \ldots, n\}$, the following problem is equivalent to (MBQP),

$$\begin{aligned}
(\text{BQP}(s)) \quad \min \quad & x^T Q_0 x + c_0^T x - \sum_{(i,j) \in I \times \{1, \ldots, n\}} s_{ij}(x_i x_j - y_{ij}) \\
\text{s.t.} \quad & y_{ij} \le x_i, \ y_{ij} \le x_j, \ y_{ij} \ge x_i + x_j - 1, \ y_{ij} \ge 0, \ (i, j) \in I \times \{1, \ldots, n\}, \\
& y_{ii} = x_i, \ i = 1, \ldots, n, \\
& x \in \{0, 1\}^n, \ y_{ij} \in \mathbb{R}, \ (i, j) \in I \times \{1, \ldots, n\}.
\end{aligned}$$

Using the same methods as in Sects. 4.2 and 4.3, one can derive SDP problems to solve for the best parameters s for problem (MBQP(s)).

4.4.3 QCR for MIQCQP

The final step in extending QCR to the general form of problem (P) is to employ binary expansion of the integer variables. Each variable $x_i \in \mathbb{Z},\ 0 \leq x_i \leq u_i$, can be replaced by its unique binary decomposition:

$$x_i = \sum_{k=0}^{\lfloor \log(u_i) \rfloor} 2^k t_{ik},$$

$$t_{ik} \in \{0, 1\},\ k = 1, \ldots, \lfloor \log(u_i) \rfloor.$$

Then we can apply all the results in previous sections to the "binarized" problem.

4.4.4 Compact QCR for MIQCQP

Using simple binary expansion as in Sect. 4.4.3 could blow up the problem size very quickly. This is especially true for problems driven by a massive amount of "big data". Billionnet et al. (2012, 2013, 2015) proposed a relatively compact formulation. Consider the following integer quadratically constrained quadratic programming problem,

$$\begin{aligned}
\text{(IQCQP)}\quad \min\ & x^T Q_0 x + c_0^T x \\
\text{s.t.}\ & x^T Q_i x + c_i^T x \leq b_i,\ i = 1, \ldots, m, \\
& x_i \in \mathbb{Z},\ 0 \leq x_i \leq u_i,\ i = 1, \ldots, n.
\end{aligned}$$

For any $S_0, S_1, \ldots, S_m \in \mathbb{S}^n$, the following problem is equivalent to (IQCQP),

$$\begin{aligned}
\text{(IQCQP}(S)\text{)}\quad \min\ & x^T (Q_0 - S_0) x + c_0^T x + S_0 \bullet Y \\
\text{s.t.}\ & x^T (Q_i - S_i) x + c_i^T x + S_i \bullet Y \leq b_i,\ i = 1, \ldots, m, \\
& 0 \leq x_i \leq u_i.\ i = 1, \ldots, n, \\
& x_i = \sum_{k=0}^{\lfloor \log(u_i) \rfloor} 2^k t_{ik},\ t_{ik} \in \{0, 1\},\ k = 1, \ldots, \lfloor \log(u_i) \rfloor, \\
& i = 1, \ldots, m, \\
& Y_{ij} = \sum_{k=0}^{\lfloor \log(u_i) \rfloor} 2^k z_{ijk},\ i, j = 1, \ldots, n,
\end{aligned}$$

$$
\begin{aligned}
& z_{ijk} \leq u_j t_{jk},\ i,j = 1,\ldots,n,\ k = 1,\ldots,\lfloor \log(u_i) \rfloor, \\
& z_{ijk} \leq x_j,\ i,j = 1,\ldots,n,\ k = 1,\ldots,\lfloor \log(u_i) \rfloor, \\
& z_{ijk} \geq x_j - u_j(1 - t_{ik}),\ i,j = 1,\ldots,n,\ k = 1,\ldots,\lfloor \log(u_i) \rfloor, \\
& z_{ijk} \geq 0,\ i,j = 1,\ldots,n,\ k = 1,\ldots,\lfloor \log(u_i) \rfloor, \\
& Y_{ii} = x_i,\ i = 1,\ldots,n, \\
& Y_{ij} = Y_{ji},\ i,j = 1,\ldots,n,\ i < j \\
& Y_{ij} \geq u_j x_i + u_i x_j - u_i u_j,\ i,j = 1,\ldots,n,\ i \leq j, \\
& Y_{ij} \geq 0,\ i,j = 1,\ldots,n,\ i \leq j,
\end{aligned}
$$

The method for extending this formulation to mixed-integer problems can be found in Billionnet et al. (2015).

4.4.5 *With or Without Additional Variables*

In Sect. 4.2, we presented two schemes, one with additional variables and the other without additional variables. In fact, all the reformulations we considered can have the corresponding two schemes. A natural question is: which scheme is better? The scheme with additional variables has tighter reformulations but involves more variables. The trade-off could be problem-specific. Billionnet et al. (2013) showed that the scheme without additional variables could be more efficient in terms of overall computation time for general problem (P).

4.5 QCR for Semi-Continuous Quadratic Programming

The QCR in previous sections exploits the structures of binary variables and linear equality constraints to derive tighter reformulations. In this section, we look at another structure involving semi-continuous variables.

When modeling real-world optimization problems, due to some managerial and technological consideration, the decision variables are often required to exceed certain threshold if they are set to be nonzero. Such variables are termed semi-continuous variables. Mathematically, semi-continuous variables can be defined as $x_i \in \{0\} \cup [a_i, b_i]$, where $a_i < b_i$ for $i = 1,\ldots,n$. Using binary variables, semi-continuous variables can be expressed by a set of mixed-integer 0-1 linear constraints:

$$a_i y_i \leq x_i \leq b_i y_i,\ y_i \in \{0,1\},\ i = 1,\ldots,n.$$

Let us consider the following semi-continuous quadratic programming problem,

$$
\begin{aligned}
\text{(SQP)}\quad \min\ & x^T Q_0 x + c_0^T x + h_0^T y \\
\text{s.t.}\ & Ax + By \le d, \\
& a_i y_i \le x_i \le b_i y_i,\ y_i \in \{0,1\},\ i = 1, \dots, n,
\end{aligned}
$$

Wu et al. (2016) proposed the following equivalent reformulation,

$$
\begin{aligned}
(\text{SQP}(u,v))\quad \min\ & f_{u,v}(x,y) \triangleq x^T Q_0 x + c_0^T x + h_0^T y + \sum_{i=1}^{n} (u_i x_i y_i + v_i y_i^2 - u_i x_i - v_i y_i) \\
\text{s.t.}\ & Ax + By \le d, \\
& a_i y_i \le x_i \le b_i y_i,\ y_i \in \{0,1\},\ i = 1, \dots, n,
\end{aligned}
$$

for any $u, v \in \mathbb{R}^n$. They showed that the quadratic function

$$
\sum_{i=1}^{n} (u_i x_i y_i + v_i y_i^2 - u_i x_i - v_i y_i)
$$

is the most general quadratic function that can be added to the objective function.

We want to choose u, v such that $(\overline{\text{SQP}}(u, v))$ is a convex problem and the lower bound from this relaxation is as tight as possible. This can be done by solving the following problem:

$$
(\text{SQP\_MAX\_uv})\quad \max\{\text{v}(\overline{\text{SQP}}(u,v)) \mid u, v \in \mathbb{R}^n, f_{u,v}(x,y) \textit{ is convex}\}.
$$

Following the approaches in Sects. 4.2 and 4.3, we can derive an SDP problem to solve for the best parameters u, v for problem (SQP(u, v)).

Theorem 4 (Wu et al. 2016) *Problem (SQP_MAX_uv) is equivalent to the following SDP problem:*

$$
(\text{SDP}_q)\quad \max\ \tau
$$

$$
\text{s.t.}\ \begin{pmatrix} Q & \frac{1}{2}\text{diag}(u) & \frac{1}{2}\alpha(u,\eta,\mu,\sigma) \\ \frac{1}{2}\text{diag}(u) & \text{diag}(v) & \frac{1}{2}\beta(v,\eta,\mu,\sigma,\lambda,\pi) \\ \frac{1}{2}\alpha(u,\eta,\mu,\sigma)^T & \frac{1}{2}\beta(v,\eta,\mu,\sigma,\lambda,\pi)^T & -\eta^T d - e^T \pi - \tau \end{pmatrix} \succeq 0, \tag{4.2}
$$

$$
(\eta, \mu, \sigma, \lambda, \pi) \in \Re_+^m \times \Re_+^n \times \Re_+^n \times \Re_+^n \times \Re_+^n, \tag{4.3}
$$

$$
(u, v, \tau) \in \Re^n \times \Re^n \times \Re,
$$

where

$$\alpha(u, \eta, \mu, \sigma) = c - u + A^T\eta - \mu + \sigma, \tag{4.4}$$

$$\beta(v, \eta, \mu, \sigma, \lambda, \pi) = h - v + B^T\eta + \text{diag}(a)\mu - \text{diag}(b)\sigma - \lambda + \pi. \tag{4.5}$$

4.6 Concluding Remark

The quadratic convex reformulation (QCR) approach for solving mixed-integer quadratically constrained quadratic programming (MIQCQP) problems is very effective. Its goal is to find a tight and efficient equivalent reformulation, by adding quadratic functions that vanish in the feasible region to the objective function and the constraints. We have reviewed recent advances in the QCR approach. By exploring the QCR technique on subclasses of MIQCQP problems with simpler structures first, we are able to generalize the approach step by step such that it can be applied to general MIQCQP problems. We have also covered the recent extension of QCR on semi-continuous quadratic programming problems. In a broad picture, the QCR approach provides a very effective solution framework for solving MIQCQP problems.

As the problem size of the NP-hard MIQCQP problems increases along with the "big data" era, the QCR approach would be more and more important. This is because finding the best reformulation in the QCR approach reduces to an SDP problem, which is a convex problem that can be solved in polynomial time. Armed with the optimized reformulation, better approximation and heuristics can be developed.

In the coming future, the QCR approach can be further generalized according to different practical structures in the objective function and the constraints. Also, the integration of the QCR approach with other solution techniques for MIQCQP problems is also an interesting research direction.

References

A. Ahlatçıoğlu, M. Bussieck, M. Esen, M. Guignard, J. Jagla, A. Meeraus, Combining QCR and CHR for convex quadratic pure 0–1 programming problems with linear constraints. Ann. Oper. Res. **199**(1), 33–49 (2012)

D. Bertsekas, A. Nedić, A. Ozdaglar, *Convex Analysis and Optimization* (Athena Scientific, Belmont, 2003)

A. Billionnet, S. Elloumi, M. Plateau, Quadratic 0–1 programming: tightening linear or quadratic convex reformulation by use of relaxations. RAIRO Oper. Res. **42**(02), 103–121 (2008)

A. Billionnet, S. Elloumi, M. Plateau, Improving the performance of standard solvers for quadratic 0-1 programs by a tight convex reformulation: the QCR method. Discret. Appl. Math. **157**(6), 1185–1197 (2009)

A. Billionnet, S. Elloumi, A. Lambert, Extending the QCR method to general mixed-integer programs. Math. Program. **131**(1–2), 381–401 (2012)

A. Billionnet, S. Elloumi, A. Lambert, An efficient compact quadratic convex reformulation for general integer quadratic programs. Comput. Optim. Appl. **54**(1), 141–162 (2013)

A. Billionnet, S. Elloumi, A. Lambert, Exact quadratic convex reformulations of mixed-integer quadratically constrained problems. Math. Program. **158**(1), 235–266 (2015)

Y. Cui, Dynamic programming algorithms for the optimal cutting of equal rectangles. Appl. Math. Model. **29**(11), 1040–1053 (2005)

A. Faye, F. Roupin, Partial Lagrangian relaxation for general quadratic programming. 4OR **5**(1), 75–88 (2007)

A. Frangioni, C. Gentile, Perspective cuts for a class of convex 0–1 mixed integer programs. Math. Program. **106**, 225–236 (2006)

A. Frangioni, C. Gentile, E. Grande, A. Pacifici, Projected perspective reformulations with applications in design problems. Oper. Res.**59**, 1225–1232 (2011)

H.L. Fu, C.L. Shiue, X. Cheng, D.Z. Du, J.M. Kim, Quadratic integer programming with application to the chaotic mappings of complete multipartite graphs. J. Optim. Theory Appl.**110**(3), 545–556 (2001)

J. Gao, D. Li, Cardinality constrained linear-quadratic optimal control. IEEE Trans. Autom. Control **56**, 1936–1941 (2011)

J. Gao, D. Li, Optimal cardinality constrained portfolio selection. Oper. Res. **61**, 745–761 (2013)

J. Gao, D. Li, A polynomial case of the cardinality-constrained quadratic optimization problem. J. Glob. Optim. **56**, 1441–1455 (2013)

I. Grossmann, N. Sahinidis, Special issue on mixed integer programming and its application to engineering. Part I/II. Optim. Eng. 136–141 (2002)

P. Hammer, A. Rubin, Some remarks on quadratic programming with 0-1 variables. RAIRO Oper. Res. (Recherche Opérationnelle) **4**(V3), 67–79 (1970)

Z. Hua, P. Banerjee, Aggregate line capacity design for PWB assembly systems. Int. J. Prod. Res. **38**(11), 2417–2441 (2000)

D. Li, X. Sun, Nonlinear integer programming, vol. 84 (Springer, New York, 2006)

D. Li, X. Sun, J. Wang, Optimal lot solution to cardinality constrained mean-variance formulation for portfolio selection. Math. Financ. **16**, 83–101 (2006)

D. Li, X. Sun, C. Liu, An exact solution method for unconstrained quadratic 0-1 programming: a geometric approach. J. Glob. Optim. **52**(4), 797–829 (2012)

G. Mitra, F. Ellison, A. Scowcroft, Quadratic programming for portfolio planning: insights into algorithmic and computational issues. Part II: processing of portfolio planning models with discrete constraints. J. Asset Manage. **8**, 249–258 (2007)

M. Plateau, Reformulations quadratiques convexes pour la programmation quadratique en variables 0-1. Ph.D. Thesis, Conservatoire National des Arts et Métiers (2006)

F. Rendl, G. Rinaldi, A. Wiegele, Solving max-cut to optimality by intersecting semidefinite and polyhedral relaxations. Math. Program. **121**(2), 307–335 (2010)

N. Shor, Quadratic optimization problems. Sov. J. Comput. Syst. Sci. **25**, 1–11 (1987)

B. Wu, X. Sun, D. Li, X. Zheng, Quadratic convex reformulations for semi-continuous quadratic programming. Working Paper, The Chinese University of Hong Kong

X. Zheng, X. Sun, D. Li, Improving the performance of MIQP solvers for quadratic programs with cardinality and minimum threshold constraints: a semidefinite program approach. INFORMS J. Comput. **26**(4), 690–703 (2014)

Part II
Reviews on Optimization and Control Applications

Chapter 5
Measurements of Financial Contagion: A Primary Review from the Perspective of Structural Break

Xi Pei and Shushang Zhu

Abstract Financial contagion is an attractive topic in recent years, since it is one of the most important issues closely related to the financial systemic risk that could seriously hurt the economy. This review aims to summarize and clarify the different concepts and measurements of financial contagion investigated in the literatures and try to highlight their common feature and differences. Noting that "structural break" is the essential feature used to define financial contagion, most of the measurements of financial contagion proposed in the literature are along the line of modeling structural break according to different mechanisms. Although, a few measurements could be used to investigate financial contagion, there remain hardships in real applications. The emerging "Big Data" technology might be helpful to refine both the research and the practice of risk management relevant to financial contagion in model specification and information acquisition.

Keywords Financial contagion • Financial market • Interbank system • Structural break

5.1 Introduction

Financial contagion is the most typical manifestation and characteristic of financial systemic risk, which always exhibits as the spread of financial distress from one market, asset class, or geographical region to others continued unabatedly (see, e.g., Kolb 2011). For instance, the financial crisis of 2007–2009 initially originated in the USA in the subprime mortgage market, then it rapidly spread across real economic sectors, and to other both advanced and emerging countries. As a consequence, there were even many countries suffered sharper crashes than the USA.

X. Pei • S. Zhu (✉)
Department of Finance and Investment, Sun Yat-Sen Business School, Sun Yat-Sen University, Guangzhou 510275, China
e-mail: peixi1989@126.com; zhuss@mail.sysu.edu.cn

T.-M. Choi et al. (eds.), *Optimization and Control for Systems in the Big-Data Era*, International Series in Operations Research & Management Science 252,
DOI 10.1007/978-3-319-53518-0_5

As compared to the early representative theoretical literatures on financial contagion, such as Allen and Gale (2000), Calvo and Mendoza (2000), and Kodres and Pritsker (2002), the literatures on empirical tests of financial contagion occurred almost 10 years earlier, such as Engle et al. (1990), King and Wadhwani (1990), and Bekaert and Hodrick (1992), and the interest in empirically testing financial contagion continued to more recent days, for example, Bekaert et al. (2014) and Sahalia et al. (2015). Although, there are several measurements to analyze financial contagion, the common logic in testing contagion is by first setting benchmark model (diverse styles and names in different papers) to reflect the relationship of asset returns (or price) in non-crisis environment or tranquility time, and then to verify if there exists significant structural break (regarded as contagion after a shock) or not. Our paper is from the perspective of structural break and aims to unify a framework to highlight the similarities and differences of various approaches used in testing and measuring financial contagion of financial markets and interbank system. We hope this primary review can help people who are new to financial contagion to quickly capture the essence of it.

By revisiting the literature, we suggest three aspects that need more attentions before testing/measuring financial contagion. First, the essences of different definitions in literatures used to test contagion are almost consistent, although each definition may represent one aspect of financial contagion. Researchers usually chose the definition that can match their research objectives better. Second, we find that early studies (most papers before Forbes and Rigobon 2002) do not distinguish comovement and contagion well, as they mainly focused on the channels through which negative shocks propagate. After the work of Forbes and Rigobon (2002), researchers became to emphasis the differences between comovement caused by normal interdependence and contagion caused by structural change. Thus specifying a benchmark model to measure the normal interdependence in the tranquil period is the first key step to test contagion characterized by excessive comovement after a shock. Hence, tests for contagion should be based on the identification of structural breaks either in the data-generating process or in some of its statistics, such as volatility and correlation of asset returns. Third, in practical tests of contagion usually require a clearly recognizable initial shock in one market or a group of markets. Given the first shock, one can apply different methodologies to verify whether the spread of instability is just the "business as usual", or reflects something more than normal interdependence. However, splitting the samples between crisis periods and tranquil periods is often arbitrary. In the empirical literature, such a split depends on an arbitrary cutoff value for the crisis indicator. Notice that those arbitrary choices may result different, or even conflicting results.

Based on the perspective of structural break, the paper proceeds as follows. In Sect. 5.2, we reclassify the concepts of financial contagion as the foundations for further study. In Sect. 5.3, we present the key steps of most popular measurements used in financial markets contagion by comparing setting benchmark model with testing structure break. Along the same framework, we further review the methods to model the financial contagion in an interbank system in Sect. 5.4. Finally, we discuss the potential applications of Big Data technology in financial contagion in Sect. 5.5.

5.2 Concepts of Financial Contagion

Intuitively, financial contagion is a phenomenon usually happened during financial distress and in turn accelerates the distress or even gives rise to a financial crisis. While there is no consensus on what financial contagion is, a few definitions are used to ascertain financial contagion in the literature. For example, Pericoli and Sbracia (2003) proposed five definitions of financial contagion and argued that the empirical results of testing contagion by different definitions are similar and concluded that the discrepancies mainly stem from the differences among the data sets used to detect the contagion. Noting that there would be no uniform definition on financial contagion, we summarize in this paper the following three typical definitions of financial contagion in a broad sense.

I. **Volatility spillover.** A stylized fact in international financial markets is the rise in volatility of asset return during periods of financial turmoil (Pericoli and Sbracia 2003). This definition exploits a volatility spillover from one market to another as empirical evidence of financial contagion, where the volatilities are usually estimated by GARCH models (see, e.g., Connolly and Wang 2003; Beirne et al. 2013; Karmann 2014). Volatility of asset return is generally regarded as a good approximation of market uncertainty. Hence, an interpretation of this definition of contagion also refers to the spread of uncertainty across international financial markets.

   Note that a simultaneous rise in volatility in different markets might be due to the unchanged normal interdependence between these markets or a structural change of cross-market linkages. But this distinction was neglected by early works, which merely focused on the occurrence of a volatility spillover but not on its causes. As a remedy, some researchers suggested to measure the contagion by the excessive volatilities of asset returns caused by structural change of cross-market linkages (see, e.g., Corsetti et al. 2001; Rigobon 2003; Bekaert et al. 2005). We refer to the definition of financial contagion as volatility spillover in this sense.

II. **Extra comovements of asset returns conditional on a crisis/events.** This definition is characterized by the significantly increased comovements of asset returns during the spread of financial distress after a crisis event (see, e.g., Dungey and Zhumabekova 2001; Romero-Meza et al. 2015), such as the subprime crisis from the USA to global markets in 2008 (see, e.g., Longstaff 2010), the Russian crisis in the summer of 1998 (see, e.g., Forbes and Rigobon 2002), the Hong Kong stock market crash in October 1997 (see, e.g., Basu 2002). By emphasizing on "significant increase", this definition conveys the notion of contagion as "excessive comovements" relative to some benchmark. The essence of this definition is thus to draw a distinction between normal comovements in return due to normal interdependence and excessive comovements in return due to some kind of structural break.

III. **Severe systemic instability induced by domino effect of a shock.** In the complex financial market, the crisis of one individual participant, especially the bankruptcy of a big company (usually commercial bank, investment bank, or other systemic important financial institution) would rapidly affect others through their close linkages and release danger signal to market, which will trigger domino effect of default in the whole system (see, e.g., Eisenberg and Noe 2001; Elsinger et al. 2006a; Caccioli et al. 2014). Take interbank system as an example, banks hold assets of other banks and engage in interbank lending. If one bank defaults, it will increase bad debts of its cooperative banks, which may worsen their operations and even causes others to bankruptcies (Elsinger et al. 2006a). In order to deal with their problem rapidly, banks in distress have to fire sale their assets that always held by multi-banks, which will lead to further defaults (Caccioli et al. 2014) and rise instability of the whole interbank system.

It deserves mention that the above three definitions emphasize the same point, "structural break", although it may be measured in different ways in different definitions.

5.3 Contagion of Financial Markets

In this section we review the literature on contagion of financial market, more specifically, stock market, bond market, and currency market. Since the methods applied to different markets are similar, we just try to introduce the commonly used typical methods, without reviewing each paper in details.

For an overview of the literature on measuring financial contagion, we classify relevant papers into three broad types. Figure 5.1 showed the representative researches according to this classification.

The first type is to measure the extra increase of volatility of asset returns via GARCH type models, since extra increase of volatility is regarded as the evidence of contagion during the crisis period (see, e.g., Connolly and Wang 2003; Beirne et al. 2013).

The second type is to measure the significant changes in the dependence of asset returns, where Pearson correlation (see, e.g., Forbes and Rigobon 2002) or tail dependence modeled by copula (see, e.g., Rodriguez 2007; Durante and Jaworski 2010) are used as the indicators.

The third type employs typical factor model (see, e.g., Rigobon 2003; Bekaert et al. 2014, etc.) to test structural break in crisis period. Another merit of factor model is that it can help to verify which factor is significantly changed to represent the contagion source (see, e.g., Corsetti et al. (2001), Bekaert et al. (2005), Corsetti et al. (2005), etc.).

In addition to the above three type methods, there are some other methods used to measure financial contagion. For example, Probit/Logit models used to detect the probability change on occurrence of crisis (see, e.g., Bae et al. 2003; Caramazza et al. 2004; Markwat et al. 2009). Since we are unable to produce a complete review, we omit these literature in this paper, which does not mean that they are unimportant.

	Measurements		Subprime crisis	European debt crisis	LTCM	Russia crisis	Asia crisis
Review		Pericoli & Sbracia (2003); Dungey et al. (2005)					
Stock market	Volatility Spillover	Connolly & Wang (2003); Beirne et al. (2009) ; Sahalia et al. (2015)			Dungey et al. (2007)		Markwat et al. (2009); Karmann & Herrera (2014)
	Excess Correlation	Dungey & Zhumabekova (2001); Meza et al. (2015)					Forbes & Rigobon (2002)
	Copula	Rodriguez & Carlos (2007); Durante & Jaworski (2010)					Rodriguez (2007)
	Factor Model	Engle et al. (1990); Bekaert (1992)	Bekaert et al. (2014)			Rigobon (2003)	
	Factor Model + Excess Correlation + Volatility Spillover				Corsetti et al. (2001); Rigobon (2003); Bekaert et al. (2005);		Corsetti et al. (2005);
Bond market Bond Market	Excess Correlation		Longstaff (2010)	Longstaff et al. (2010); Blatt et al. (2015)			Basu (2002)
	Factor Model + Excess Correlation				McKibbin et al. (2005); Dungey et al. (2006)		
Currency market	Factor Model						Dungey & Martin (2004); Cipollini & Kapetanios (2009)
	Excess Correlation					Favero &Giavazzi (2002);	
Bond & stock markets	Excess Correlation	Brière et al. (2012)					
Stock, bond & currency markets	Volatility spillover	Beirne & Gieck (2014)					

Fig. 5.1 Literature of financial contagion among financial markets

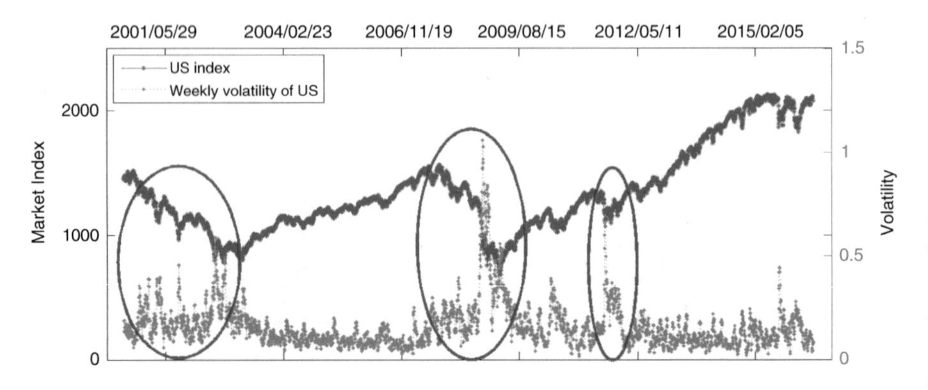

Fig. 5.2 Volatility of US financial market

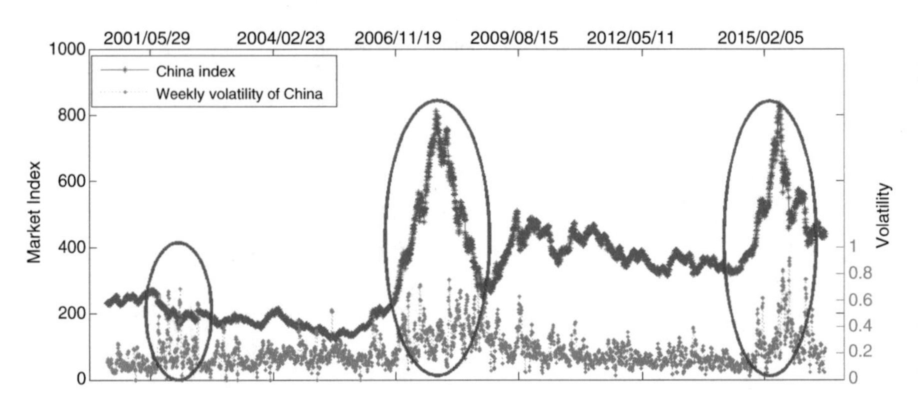

Fig. 5.3 Volatility of China financial market

5.3.1 Volatility Analysis

The first definition of financial contagion is according to the typical feature that financial returns during crises exhibit high volatility. Figures 5.2 and 5.3 show the S&P 500 Index and its volatility, and Shanghai Composite Index and its volatility, respectively. Intuitively, we can see both volatilities of US market and China market increase significantly in crash time as compared to the normal time.

To depict the volatility in crisis time, GARCH type models might be the most popular model used in calculating the conditional variances to test volatility spillovers. As a work relatively easy to understand, Beirne et al. (2013) applied GARCH-BEKK models to study volatility spillovers from mature stock market to emerging stock market.

Setting Benchmark Model

Beirne et al. (2013) set a tri-variate VAR-GARCH(1,1) process as the benchmark model that takes the following form:

$$\boldsymbol{r}_t = \boldsymbol{\alpha} + \boldsymbol{\beta}\boldsymbol{r}_{t-1} + \boldsymbol{\varepsilon}_t \tag{5.1}$$

where $\boldsymbol{r}_t = (r_{1,t}, r_{2,t}, r_{3,t})'$ denote returns of the local emerging market, regional emerging market, and mature market at time t, respectively, $\boldsymbol{\alpha}$ and $\boldsymbol{\beta}$ are autoregression coefficients, and $\boldsymbol{\varepsilon}_t = (\varepsilon_{1,t}, \varepsilon_{2,t}, \varepsilon_{3,t})'$ is the residual vector following a normal distribution $\boldsymbol{\varepsilon}_t | I_{t-1} \sim N(0, H_t)$ with its conditional covariance matrix:

$$H_t = \begin{pmatrix} h_{11,t} & h_{12,t} & h_{13,t} \\ h_{21,t} & h_{22,t} & h_{23,t} \\ h_{31,t} & h_{32,t} & h_{33,t} \end{pmatrix}. \tag{5.2}$$

According to the multivariate GARCH(1,1)-BEKK representation proposed by Engle and Kroner (1995), H_t can be decomposed as:

$$H_t = C'C + A' \begin{pmatrix} \varepsilon^2_{1,t-1} & \varepsilon_{1,t_1}\varepsilon_{2,t-1} & \varepsilon_{1,t-1}\varepsilon_{3,t-1} \\ \varepsilon_{2,t-1}\varepsilon_{1,t-1} & \varepsilon^2_{2,t-1} & \varepsilon_{2,t-1}\varepsilon_{3,t-1} \\ \varepsilon_{3,t-1}\varepsilon_{1,t-1} & \varepsilon_{3,t-1}\varepsilon_{2,t-1} & \varepsilon^2_{3,t-1} \end{pmatrix} A + G'H_{t-1}G \tag{5.3}$$

with parameters

$$C = \begin{pmatrix} c_{11} & 0 & 0 \\ c_{21} & c_{22} & 0 \\ c_{31} & c_{32} & c_{33} \end{pmatrix}, A = \begin{pmatrix} a_{11} & 0 & 0 \\ a_{21} & a_{22} & 0 \\ a_{31} & a_{32} & a_{33} \end{pmatrix} \text{ and } G = \begin{pmatrix} g_{11} & 0 & 0 \\ g_{21} & g_{22} & 0 \\ g_{31} & g_{32} & g_{33} \end{pmatrix}.$$

Equation (5.3) models the dynamic process of H_t as a linear function of its own past values H_{t-1} and past values of innovations $(\varepsilon_{1,t-1}, \varepsilon_{2,t-1}, \varepsilon_{3,t-1})$, allowing for own-market and cross-market influences in the conditional variances.

Testing Structure Break

Beirne et al. (2013) defined volatility contagion as a shift in the transmission of volatility from mature stock market to emerging stock market during episodes of turbulence occurred in the former. In order to test for such shifts, they include a dummy d in Eq. (5.3) that allows the parameters governing volatility spillovers from mature market to change in these episodes. In such a case, the equation for the conditional variance of return of local emerging market becomes:

$$\begin{aligned} h_{11,t} = {} & c^2_{11} + a^2_{11}\varepsilon^2_{1,t-1} + a^2_{22}\varepsilon^2_{2,t-1} + (a_{33} + a_{33d}d)^2\varepsilon^2_{3,t-1} + 2a_{11}a_{21}\varepsilon_{1,t-1}\varepsilon_{2,t-1} \\ & + 2a_{11}(a_{31} + a_{31d}d)\varepsilon_{1,t-1}\varepsilon_{3,t-1} + 2a_{21}(a_{32} + a_{32d}d)\varepsilon_{2,t-1}\varepsilon_{3,t-1} \\ & + g^2_{11}h_{11,t-1} + g^2_{22}h_{22,t-1} + (g_{33} + g_{33d}d)^2h_{33,t-1} + 2g_{11}g_{21}h_{12,t-1} \\ & + 2g_{11}(g_{31} + g_{31d}d)h_{13,t-1} + 2g_{21}(g_{32} + g_{32d}d)h_{23,t-1}. \end{aligned} \tag{5.4}$$

In (5.4), parameters a_{31}, a_{32}, a_{33} and g_{31}, g_{32}, g_{33} can reflect the volatility spillover from mature stock market to local emerging market caused by a normal relationship, and $a_{31d}, a_{32d}, a_{33d}$ and $g_{31d}, g_{32d}, g_{33d}$ can capture shifts in these parameters, which imply contagion. Thus contagion can be detected by testing whether those parameters are equal to zero or not.

While GARCH model allows to incorporate volatility spillovers in the model, it is hard to estimate the model involving multiway simultaneous spillovers due to the issue of endogeneity. Karmann (2014) suggested the SVAR model as a remedy when encountering such a difficulty.

5.3.2 Correlation Analysis

The comovements of asset prices rise abruptly in crisis period and contagion can be characterized as the excessive increase in correlation between two asset returns during a crisis period. Figure 5.4 shows the correlation of daily returns of S&P 500 index and Hang Seng Composite Index during different time periods. One can see that correlation during crash becomes larger.

The correlation based approach for analyzing contagion is popularized by Forbes and Rigobon (2002), although their test based on the conditional correlation is biased upwards and might result in evidence of spurious contagion (Boyer et al. 1997; Loretan and English 2000; Corsetti et al. 2005). Here, we introduce the approach proposed by Corsetti et al. (2001), which can be regarded as an extension of correlation analysis framework of Forbes and Rigobon (2002). The way to implement test of contagion is to perform a regression on scaled asset return samples.

Setting Benchmark Model

Consider an example with two assets. Firstly, define the following regression equation during the non-crisis period where the returns are scaled by their respective standard deviations:

$$\left(\frac{r_{2,x,t}}{\sigma_{2,x}}\right) = \alpha_0 + \alpha_1 \left(\frac{r_{1,x,t}}{\sigma_{1,x}}\right) + \varepsilon_{x,t} \tag{5.5}$$

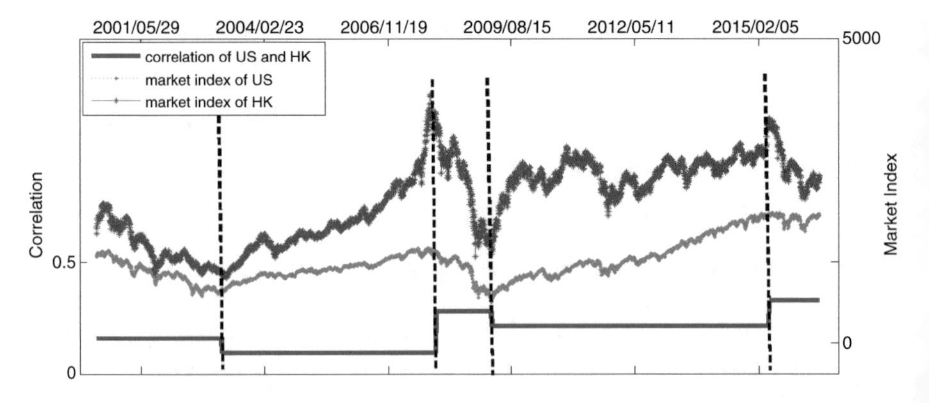

Fig. 5.4 Correlation of US market and HK market

where, $r_{i,x,t}$ and $\sigma_{i,x}$, $i = 1, 2$ denote the returns (at time t) and the standard deviations of returns during non-crisis period, respectively, and $\varepsilon_{x,t}$ is the residual term. Then the regression parameter α_1 is exactly the correlation coefficient of assets 1 and 2 during the non-crisis period.

Secondly, the regression equation for the crisis period is given as follows:

$$\left(\frac{r_{2,y,t}}{\sigma_{2,x}}\right) = \beta_0 + \beta_1 \left(\frac{r_{1,y,t}}{\sigma_{1,x}}\right) + \varepsilon_{y,t} \tag{5.6}$$

where y means samples are from crisis period, and the scaling of asset returns is still by the respective standard deviations of the non-crisis period. The regression parameter β_1 is exactly the adjusted correlation coefficient given by Forbes and Rigobon (2002).

Testing Structural Break

By (5.5) and (5.6), we see that verifying the contagion can be transformed to test the equality of the regression slope parameters estimated by ordinary least squares (OLS). This test is equivalent to a Chow test for a structural break of the regression slope. Implementation of the test can be based on the following pooled regression equation over the entire sample.

$$\left(\frac{r_{2,z,t}}{\sigma_{2,x}}\right) = \gamma_0 + \gamma_1 d_t + \gamma_2 \left(\frac{r_{1,z,t}}{\sigma_{1,x}}\right) + \gamma_3 \left(\frac{r_{1,z,t}}{\sigma_{1,x}}\right) d_t + \varepsilon_{z,t} \tag{5.7}$$

where $r_{z,i} = (r_{i,x,1}, r_{i,x,2}, \ldots, r_{i,x,T_x}, r_{i,y,1}, r_{i,y,2} \ldots, r_{i,y,T_y})$ represents the $(T_x + T_y) \times 1$ pooled data set by stacking the non-crisis and crisis data. The slope dummy, d_t, is defined as

$$d_t = \begin{cases} 1: & t > T_x, \\ 0: & \text{else} \end{cases} \tag{5.8}$$

The parameter $\gamma_3 = \beta_1 - \alpha_1$ in (5.7) captures the effect of contagion. If the dummy variable provides no new additional information during the crisis period, then $\gamma_3 = 0$. Forbes and Rigobon (2002) test of contagion can be implemented by estimating (5.7) by OLS and performing a one-sided t-test of $H_0 : \gamma_3 = 0$, which is equivalent to testing $H_0 : \alpha_1 = \beta_1$.

The difference between the regression approach to correlation testing for contagion based on (5.7) and the approach implemented by Forbes and Rigobon (2002) is that the standard errors used in the test statistics are different with small samples.

The approaches proposed by Corsetti et al. (2001) and Forbes and Rigobon (2002) are bi-variate analysis. Rigobon (2003) suggested an alternative multivariate test of contagion. This test is based on comparing the covariance matrices estimated with two different data samples (non-crisis and crisis).

Tail dependence is another special type of correlation. Structural breaks in tail dependence also implies financial contagions. For example, Rodriguez (2007) uses copula (Nelsen 2007) based tail dependence measure to test the financial contagions of stock markets during the Asian crisis and the Mexican crisis.

5.3.3 *Factor Model Based Approaches*

Factor model is commonly used to model asset returns in pricing and investment (see, e.g., Sharpe 1963; Sun et al. 2009; Zhu et al. 2011). In a factor model, asset returns are determined by a set of common factors and an idiosyncratic factor.

The measurements reviewed in previous sections can only used to check the existence of phenomenon of financial contagion, but cannot be adopted to model and explain its reason and mechanism. Factor model is a natural choice to remedy these weaknesses since contagion can be regarded as the structure change of the factor model (see, e.g., Masson 1998, 1999a,b; Corsetti et al. 2001; Forbes and Rigobon 2002; Bekaert et al. 2005; Dungey and Zhumabekova 2001; Dungey et al. 2002, 2005).

5.3.3.1 Contagion of Individual Shocks

In this part, we consider the contagion of individual shocks among markets using factor models. In the next part, we will further introduce the contagion of common shocks and the transmission channels modeled by factor models. Figure 5.5 shows the mechanism of contagion of individual shocks between two markets. By factor model, the return of each market is driven by some common factors and an idiosyncratic factor in tranquility time. In the crisis time, we can test individual contagion by verifying whether or not the idiosyncratic event of a market significantly influences another market.

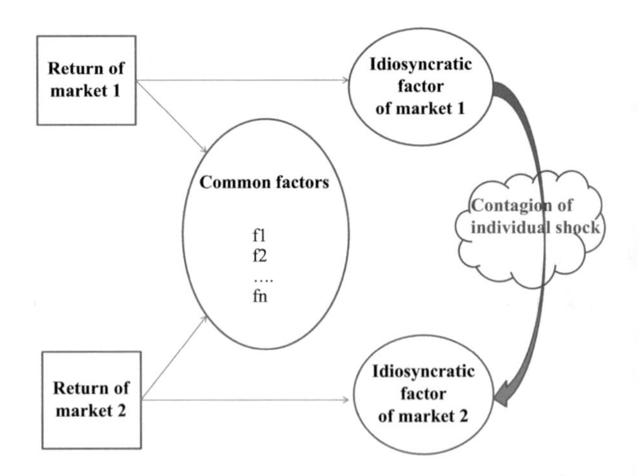

Fig. 5.5 Mechanism contagion of individual shocks

Setting Benchmark Model

Following Dungey et al. (2002), assuming the following factor model summarizes the dynamics of the demeaned return processes during a tranquility period:

$$r_{i,x,t} = \lambda_i f_t + \delta_i \varepsilon_{i,x,t}, \quad i = 1, 2 \tag{5.9}$$

satisfying

$$E(\varepsilon_{1,x,t}\varepsilon_{2,x,t}) = 0 \text{ and } E(\varepsilon_{i,x,t} f_t) = 0, \quad i = 1, 2$$

where $f_t \sim (0, 1)$ is common factor representing market fundamentals which determine the average level of asset returns across international markets during normal times.[1] $\varepsilon_{i,x,t} \sim (0, 1)$, $i = 1, 2$ are idiosyncratic factors with respect to different markets.

The interrelationship between these two market returns in (5.9) during a non-crisis period can be characterized by their covariance

$$E[r_{1,x,t} r_{2,x,t}] = \lambda_1 \lambda_2 \tag{5.10}$$

and variances

$$E[r_{i,x,t}^2] = \lambda_i^2 + \delta_i^2, \quad i = 1, 2. \tag{5.11}$$

Testing Structural Break

Consider the case of contagion from market 1 to market 2. The factor model in (5.9) is now augmented as follows:

$$\begin{aligned} r_{1,y,t} &= \lambda_1 f_t + \delta_1 \varepsilon_{1,y,t}, \\ r_{2,y,t} &= \lambda_2 f_t + \delta_2 \varepsilon_{2,y,t} + \gamma \varepsilon_{1,y,t} \end{aligned}$$

where the $r_{i,x,t}$ in (5.9) are replaced by $r_{i,y,t}$ to signify demeaned asset returns during the crisis period. The expression for $r_{2,y,t}$ now contains a contagious transmission channel as represented by local shocks from the market 1, with its impact measured by the parameter γ. The fundamental aim of all empirical models of contagion is to test the statistical significance of parameter γ.

The volatility of $r_{2,y,t}$ is now become

$$E[r_{2,y,t}^2] = \lambda_2^2 + \delta_2^2 + \gamma^2 \tag{5.12}$$

[1]The model can be extended to allow for a richer set of factors, including observed fundamentals (Eichengreen et al. 1996), trade linkages (Glick and Rose 1999; Pesaran and Pick 2007), financial flows (Van-Rijckeghem and Weder 2001), geographical distance (Bayoumi et al. 2003), and Fama–French factors (Flood and Rose 2004)

and the covariance of $r_{1,y,t}$ and $r_{2,y,t}$ is

$$E[r_{1,y,t}r_{2,y,t}] = \lambda_1\lambda_2 + \gamma\delta_1 \tag{5.13}$$

Dungey et al. (2005) suggested to decompose the effects of shocks into common, idiosyncratic, and contagion, respectively, as follows:

$$\frac{\lambda_2^2}{\lambda_2^2+\delta_2^2+\gamma_2^2}, \quad \frac{\delta_2^2}{\lambda_2^2+\delta_2^2+\gamma_2^2}, \quad \frac{\gamma_2^2}{\lambda_2^2+\delta_2^2+\gamma_2^2}. \tag{5.14}$$

This decomposition provides a descriptive measure of the relative strength of contagion in contributing to the volatility of returns during a crisis period. When extending model (5.9) to test contagion among multiple markets, generalized method of moments (GMM) can be used to estimate the unknown parameters (See Dungey et al. 2005).

The above method can be generalized to allow for time varying volatility. Suppose f_t is governed by the following GARCH process:

$$f_t \sim (0, h_t) \tag{5.15}$$

with conditional volatility h_t, given by the following GARCH structure (Diebold and Nerlove 1989):

$$h_t = (1-m-n) + mf_{t-1}^2 + nh_{t-1}. \tag{5.16}$$

The choice of the normalization, $(1-m-n)$, constrains the unconditional volatility to equal unity and is adopted for identification. Now, the volatility of return of market 2 during the crisis period is

$$\begin{aligned} E_{t-1}(r_{2,y,t}^2) &= E_{t-1}(\lambda_2 f_t + \delta_2\varepsilon_{2,y,t} + \gamma\varepsilon_{1,y,t}) \\ &= \lambda_2^2 h_t + \delta_2^2 + \gamma^2, \end{aligned} \tag{5.17}$$

and the conditional covariance between $r_{1,y,t}$ and $r_{2,y,t}$ is

$$\begin{aligned} E_{t-1}(r_{1,y,t}, r_{2,y,t}) &= E_{t-1}[(\lambda_1 f_t + \delta_1\varepsilon_{1,y,t})(\lambda_2 f_t + \delta_2\varepsilon_{1,y,t} + \gamma\varepsilon_{1,y,t})] \\ &= \lambda_1\lambda_2 h_t + \gamma\delta_1. \end{aligned} \tag{5.18}$$

For both constant volatility model and time varying volatility model, contagion has the same effect of causing a structural shift during the crisis period in the covariance by $\gamma\delta_1$ and the variance by γ^2, although the variance and the covariance change from $\lambda_2^2 + \delta_2^2$ and $\lambda_1\lambda_2$ under the constant volatility model to $\lambda_2^2 h_t + \delta_2^2$ and $\lambda_1\lambda_2 h_t$ under the time varying volatility model.

Clearly, with a factor model, we see how a financial contagion can be exhibited as increase of volatility and comovement of returns.

5.3.3.2 Contagion of Common Shocks and Transmission Channels

Although, contagions exhibit some common phenomena, such as volatility spillovers and extra comovements of asset returns, there is no consensus on the transmission mechanisms and channels. Crises may be triggered by shock via international transmission through trade links, competitive devaluations, financial links, or policy game among national governments (see, e.g., Schinasi and Smith 2001; Dungey and Tambakis 2003). For domestic contagion, fragile real economy and irregular financial market might be the main reasons of contagion (see, e.g., Lucas 1982; Dornbusch et al. 2000; Pritsker 2001). More recently, researchers pay more attention to market reaction mechanism of financial contagion, including investor behavior affected sentiment (see, e.g., Masson 1999a; Karolyi 2004), information asymmetry (see, e.g., Kodres and Pritsker 2002; Trevino 2014), and liquidity shortage (see, e.g., Brunnermeier and Pedersen 2009; Gai and Kapadia 2010). All these phenomena have been labeled as channels of contagion.

Bekaert et al. (2014) developed a factor model to set a benchmark for what the global equity market comovements should be, and define the unexplained increases in factor loadings and residual correlations as indicator of contagion in 2007–2009 financial crisis. They further disentangled the channels of contagion, and explained the heterogeneity in contagion across portfolios by testing whether and how the dependence of factor exposures on various instruments changed during the crisis.

Setting Benchmark Model

As in Bekaert et al. (2014), during non-crisis period, the excess return of portfolio i at time t denoted by $r_{i,t}$ is modeled as follows:

$$r_{i,t} = \alpha_{i,0} + \alpha_{i,1} r_{i,t-1} + \alpha_{i,2} dy_{i,t-1} + \boldsymbol{\beta}'_{i,t} \boldsymbol{f}_t + \varepsilon_{i,t}, \quad i = 1, \cdots, m \tag{5.19}$$

where dy_t is the dividend yield of the portfolio, $\boldsymbol{f}_t$ is the vector of three common factors that drive the returns.

Testing Structural Break

To test contagion and identify channels of contagion, Bekaert et al. (2014) extended (5.19) to the full model as follows:

$$r_{i,t} = \alpha_{i,0} + \alpha_{i,1} r_{i,t-1} + \alpha_{i,2} dy_{i,t-1} + \boldsymbol{\beta}'_{i,t} \boldsymbol{f}_t + \eta_{i,t} CR_t + \varepsilon_{i,t}, \tag{5.20}$$

$$\beta^j_{i,t} = \beta^j_{i,0} + \left(\boldsymbol{\beta}^j_{i,1}\right)' z_{t-k} + \gamma^j_{i,t} CR_t, \tag{5.21}$$

$$\gamma^j_{i,t} = \gamma^j_{i,0} + \left(\boldsymbol{\gamma}^j_{i,1}\right)' z_{t-k}, \quad j = U, G, D \tag{5.22}$$

$$\eta_{i,t} = \eta_{i,0} + \boldsymbol{\eta}'_{i,1} z_{t-k} \tag{5.23}$$

where CR_t is a crisis dummy (the model is referred to as the "interdependence model" when it is eliminated) and z_t is a vector of control variables designed to capture time and cross-sectional variation in factor exposures. When the model includes control variables z, the expected return also depends automatically on these lagged $z's$.

Equations (5.20) and (5.21) allow to uncover the sources of contagion through coefficients γ or η if they are significant unequal to zero. γ in Eq. (5.21) measures contagion via the factors $\boldsymbol{f}_t$, that is, changes in interdependence during the crisis, while η in Eq. (5.20) captures contagion unrelated to the observable factors $\boldsymbol{f}_t$ of the model.

Further to explore possible channels of contagion, Bekaert et al. (2014) recommended a series of instruments to model time variation in exposures. Equations (5.21), (5.22), and (5.23) contain a set of lagged instruments, z_{t-k}, which are used to model the time variation in the exposures β, γ, and η.

By choosing various instruments for z, they mainly examined six groups channels of contagion, more specifically, interbank linkage, financial policies to protect the domestic financial sector during the crisis, globalization, information asymmetries, and herding behavior (See Bekaert et al. 2014 for details).

In addition to volatility, correlation, and factor model based methods, there exist some other methods used in measuring financial contagion along the perspectives other than increase of volatility and correlation. Based on the fact that shock in one market may increase the probability of occurrence of crisis in other markets, Eichengreen et al. (1996) and Bae et al. (2003) used multinomial logistic regression models to evaluate contagion in financial markets.

5.4 Contagion of Interbank System

The recent literature on interbank system tried to find whether there exist optimal financial networks that can promote financial stability. Most papers find that contagion is not only a purely random phenomenon, but also depends on the structure of the financial system. More and more literatures on financial networks and contagion are contributing to analyze contagion between financial intuitions and/or real economy sectors.

5.4.1 *Network Model of Interbank Contagion*

Network model is widely applied to many different areas, which can be described by a set of nodes and a set of links between them. The complex financial linkages (interbank exposures) among many financial institutions can be modeled as a network which can be further used to address issues of system stability and risk contagion according to the third definition of financial contagion. Although there is not yet a unified framework that is able to fully embed the literature on financial networks into the wider field of economic theory, there seem to be some important economic intuitions.

Allen and Gale (2000) claimed that financial contagion could be modeled as an equilibrium phenomenon. Because liquidity preference shocks are imperfectly correlated across regions, banks hold interregional claims on other banks to provide insurance against liquidity preference shocks. When uncertainties in a fragile financial system aggregate to certain degree, a small liquidity preference shock in one region can spread by contagion throughout the economy. They also argued that the possibility of contagion depends strongly on the completeness of the structure of interregional claims, and concluded that complete claim structures seem to be more robust than incomplete structures. Eisenberg and Noe (2001), Elsinger et al. (2006a), Elsinger et al. (2006b), Upper (2011), and Summer (2013) set network model to simulate the exposure linkages of a given banking system, and explained how shocks are potentially amplified through the network of exposures. Hasman (2013) compared the existing theoretical and empirical literature on contagion through the banking system, and concluded that the structure of the interbank market, the bank size, the linkages among them, the level of correlation of investments, and the transparency of the regulator are main five factors in determining the possibility of contagion. The author also claimed that financial linkages promote stability for small shocks, but increase instability for big shocks.

Setting Benchmark Model

Notice that it is popular that banks hold assets of each other and engage in interbank lending. Following Eisenberg and Noe (2001) and Summer (2013), suppose there exist n banks in the interbank system, which have for each institution i non-interbank-related asset a_i^{NIB} and interbank asset a_i^{IB} on their balance sheet. On the liability side, there are interbank liability d_i^{IB}, as well as liability to creditors outside the network d_i^{NIB}. The equity is denoted by e_i. The value of the non-interbank assets a_i^{NIB} can be interpreted as an exogenous random variable. The values of all the other parts of the balance sheet are determined endogenously within the network conditional on a particular draw of $a^{NIB} = (a_1^{NIB}, \cdots, a_n^{NIB})$. The structure of the interbank liabilities is represented by a matrix $L = (l_{ij})_{n \times n}$, where l_{ij} represents the nominal obligation of bank i to bank j. These liabilities are nonnegative, and the diagonal elements of L are zero as banks are not allowed to hold liabilities against themselves.

Figure 5.6 illustrates the simplified network model of interbank system in normal time. In this model, the relationships between interbank assets and liabilities can be modeled by some linear equations. Take bank1 as instance, the interbank asset of bank1 a_i^{IB} equals to the sum of its liabilities to other three banks, mathematically, $a_1^{IB} = l_{21} + l_{31} + l_{41}$, and interbank liability d_i^{IB} equals to the sum of its obligations to other three banks, i.e., $d_1^{IB} = l_{12} + l_{13} + l_{14}$.

In the case of default, the nominal values of assets and liabilities of banks are usually different to their market value. Denote

$$\bar{d}_i^{IB} = \sum_{j=1}^{n} l_{ij} \quad \text{and} \quad \bar{a}_i^{IB} = \sum_{j=1}^{n} l_{ji} \tag{5.24}$$

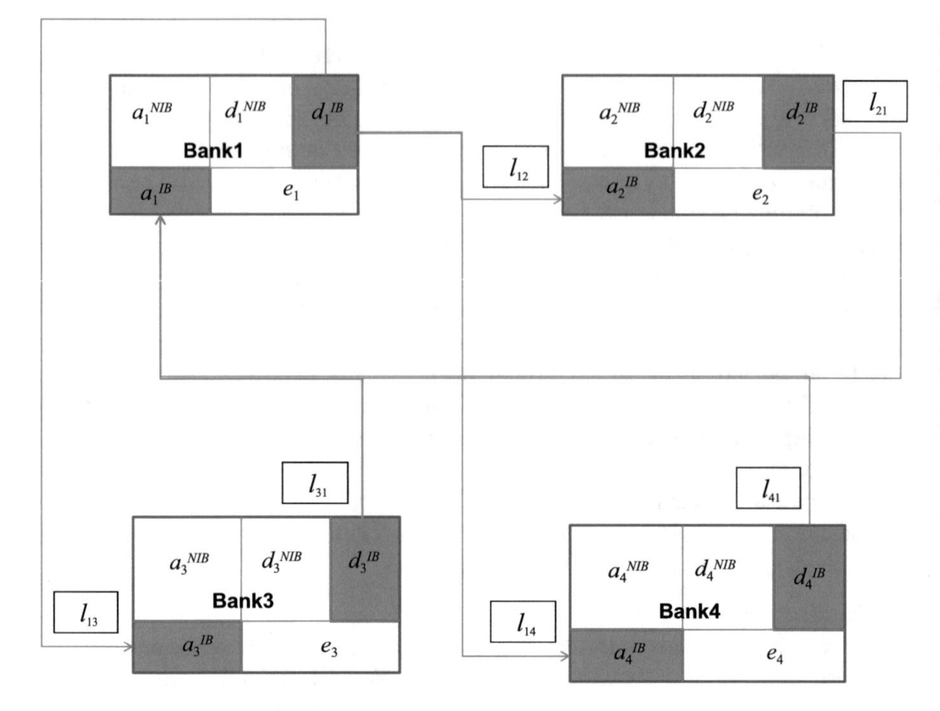

Fig. 5.6 Network of interbank system

the nominal values of the total interbank liability of bank i and claim of bank j in contrast to the market values d_i^{IB} and a_j^{IB} determined endogenously by the interbank system.

In the case of default, three rules have to be respected for clearing: *Limited liability*, *Priority of debt claims*, and *Proportionality paid off* (See Summer 2013 for details). To operationalize proportionality, let $\bar{d}_i^I$ be the total nominal obligations of node i, i.e.,

$$\bar{d}_i^I = \bar{d}_i^{IB} + \bar{d}_i^{NIB} = \sum_{j=1}^{n} l_{ij} + \bar{d}_i^{NIB} \tag{5.25}$$

and define the proportionality matrix $\Pi = (\pi_{ij})$ as

$$\pi_{ij} = \begin{cases} \dfrac{l_{ij}}{\bar{d}_i} : & \text{if } \bar{d}_i > 0, \\ 0 : & \text{otherwise} \end{cases} \tag{5.26}$$

The amount available for bank i to pay off its debt equals $a_i^{NIB} + \sum_{j=1}^{n} \pi_{ji} d_j$. Thus, the interbank system will remain stable if the actual payments made by all the banks can be able to completely pay off their debts, i.e.,

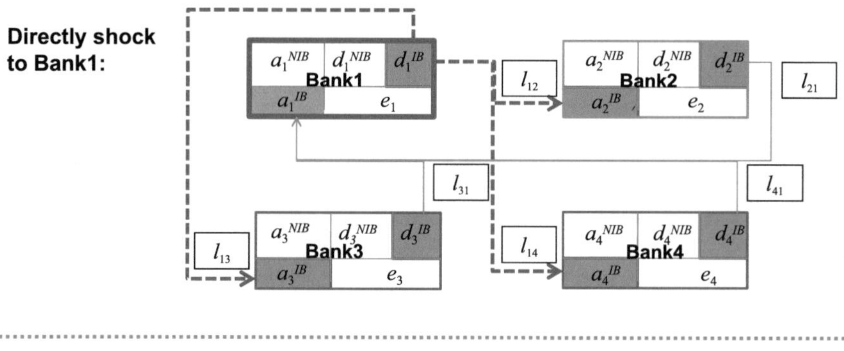

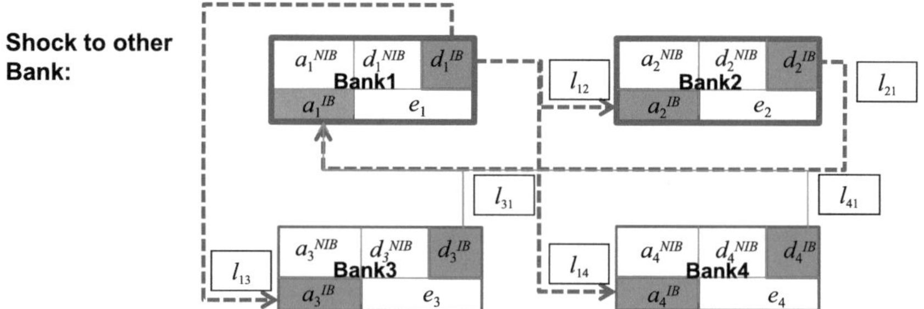

Fig. 5.7 Mechanism of contagion of interbank system

$$a_i^{NIB} + \sum_{j=1}^{n} \pi_{ji} d_j \geq \bar{d}_i, \quad i = 1, \cdots, n. \tag{5.27}$$

Testing Structural Break

In the case when a individual bank encountering default event, it cannot completely pay off its debts to other banks, thus may cause a "domino effect" of contagion in the interbank system. Figure 5.7 illustrates the contagion mechanism of interbank system. In the first situation, there is a big shock directly to Bank1 that induces default of Bank1 to other banks. For the bank holding large amount assets of Bank1, the bad debt rises suddenly and further results in bankruptcy that may start a new round defaults in this interbank system. In the second situation, there is a big shock to other bank, like Bank2, which defaults to other banks connected to it. If Bank1 is affected so much by the default of Bank2, then the system is going to run the same as the first situation.

Now we discuss the conditions which will lead to network structure break that causes contagion in system. More specifically, for bank i, there are two different situations considering the payment rules described in previous, i.e., "pay nothing" if

$$a_i^{NIB} + \sum_{j=1}^{n} \pi_{ji} d_j < 0 \tag{5.28}$$

and "proportionally pay off" characterized by

$$\bar{d}_i > a_i^{NIB} + \sum_{j=1}^{n} \pi_{ji} d_j \geq 0. \tag{5.29}$$

Noting that (5.27), (5.28), and (5.29), the clearing payment vector $\boldsymbol{d}^*$ satisfying $0 \leq \boldsymbol{d}^* \leq \bar{\boldsymbol{d}}$ is a payment vector determined by the following system:

$$\boldsymbol{d}^* = \{[\boldsymbol{a}^{NIB} + \Pi' \boldsymbol{d}^*] \vee 0\} \wedge \bar{\boldsymbol{d}}$$

where

$$\boldsymbol{x} \vee \boldsymbol{y} = (\min\{x_1, y_1\}, \cdots, \min\{x_n, y_n\})$$
$$\boldsymbol{x} \wedge \boldsymbol{y} = (\max\{x_1, y_1\}, \cdots, \max\{x_n, y_n\})$$

Clearing vectors can be calculated in different ways using relatively simple and fast algorithms. One can refer to Elsinger et al. (2013) for technical details. Notice that Elsinger et al. (2006a) further distinguished default of bank as fundamental default (if $a_i^{NIB} + \sum_{j=1}^{n} \pi_{ji} d_j < \bar{d}_i$) and contagious default (if $a_i^{NIB} + \sum_{j=1}^{n} \pi_{ji} \bar{d}_j \geq \bar{d}_i$, but $a_i^{NIB} + \sum_{j=1}^{n} \pi_{ji} d_j^* \leq \bar{d}_i$).

5.4.2 *Contagion via Portfolio Overlapping*

In the previous subsection, we focus on the endogenous contagion within interbank system. Generally, banks hold not only assets of other banks, but also non-interbank-related assets a_i^{NIB}. Now, we further address how the overlapping of non-interbank-related assets induces contagion of interbank system.

Caccioli et al. (2014) developed a network approach to measure the amplification of financial contagion of interbank system due to portfolio overlapping and leverage.

Setting Benchmark Model

First, they set a basic model of financial system with links in the network connecting banks to assets. Suppose there are two groups of nodes, n banks and k assets, of a financial network. The number of assets in the portfolio of bank i, i.e., the number of links of the corresponding node, is its degree k_i. Figure 5.8 shows the simplified network of portfolio overlapping among interbank system, including three assets and four banks. It deserves mention that the spirit implied by Fig. 5.8 goes beyond the essential idea of Caccioli et al. (2014) itself, which was partly investigated by Caccioli et al. (2015).

The average diversification, i.e., the average degree of banks in the network, is then:

$$\mu_b = \frac{1}{n} \sum_{i=1}^{n} k_i \tag{5.30}$$

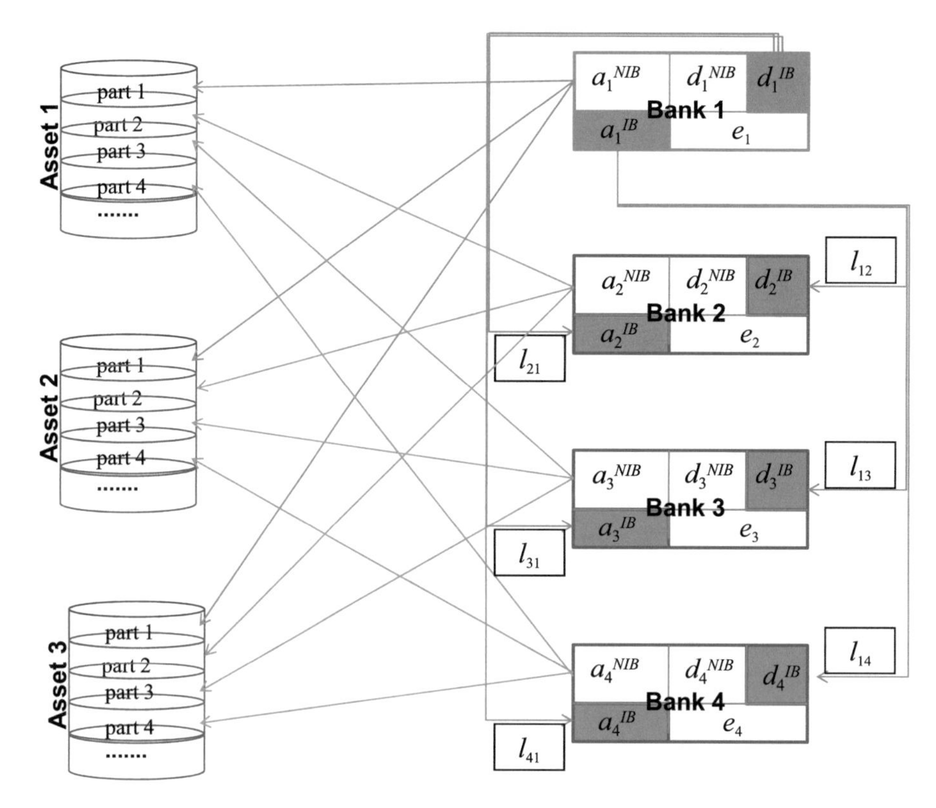

Fig. 5.8 Network of interbank system with portfolio overlapping

where the sum runs over all n banks. Conversely, the number of banks that hold asset j in their portfolio is its degree l_j, and the average degree of the assets is

$$\mu_a = \frac{1}{m}\sum_{j=1}^{m} l_i. \tag{5.31}$$

Since each link connects a bank to an asset, the total degree of the banks must equal the total degree of the assets, $\mu_b n = \mu_a m$. Thus, a rough characterization of the network topology can be given in terms of two parameters μ_b and $\nu = n/m$.

Second, they model the solvency condition. Denote the portfolio value of bank i at time t as

$$a_{i,t} = \sum_{j=1}^{m} q_{ij} p_{j,t} \tag{5.32}$$

where q_{ij} is the number of shares of asset j held by bank i and $p_{j,t}$ is the price of asset j at time t. Suppose bank i holds cash c_i, and denote its liability as l_i, neither of them are time dependent. $a_{i,0}$ is the initial portfolio value of bank i, its initial equity (or capital) is therefore $e_{i,0} = a_{i,0} + c_i - l_i$. The condition for bank i to be solvent at time t is

$$\sum_{j=1}^{m} q_{ij} p_{j,t} + c_i \geq l_i. \tag{5.33}$$

Finally, they set up a model for market impact. Whenever a bank does not satisfy the solvency condition, portfolio undergoes a fire sale which causes the price of assets in the bank portfolios to drop. If $x_{j,t}$ is the fraction of asset j that has been liquidated till time t, the price is updated as

$$p_j \rightarrow p_j f_j(x_{j,t}) \tag{5.34}$$

where $f_j(x_{j,t})$ is a market impact function set as $e^{-\alpha x_{j,t}}$ in Caccioli et al. (2014).

Testing Structural Break

Suppose that the initial shock occurred at time $t = 0$. Then at each time $t = 1, 2, \ldots$, the solvency condition is checked for every bank, the portfolios of newly insolvent or bankrupted banks are liquidated, and new prices are computed for each asset according to (5.34). The dynamics stops when no new bankruptcies occur between two consecutive time steps. This can be summarized as the following iteration:

(1) *Introduce the initial shock in the system;*
(2) *Liquidate the portfolio of insolvent banks;*
(3) *Recompute prices of assets;*
(4) *If new banks are insolvent go to step (2), otherwise end.*

Caccioli et al. (2014) further investigated the circumstances under which systemic instability is likely to occur as a function of parameters that representing leverage, market crowding, diversification, and market impact. By comprehensive simulation analysis, they concluded that (1) There is a critical threshold for leverage; below it financial networks are always stable, and above it the unstable region grows as leverage increases; (2) Although diversification may be good for individual institutions, it can create dangerous systemic effects; (3) Dynamic deleveraging during a crisis can amplify instabilities. The financial system exhibits "robust yet fragile" behavior, where contagion is rare but catastrophic whenever it occurs.

As illustrated in Fig. 5.9, suppose a bank default, then the financial distress spreads not only through network of interbank assets in the manner that we discussed in previous part, but also via the overlapping of non-interbank assets. Caccioli et al. (2015) claimed that neither channel of contagion results in large effects on its own, but bankruptcies are much more common and have large systemic effects when both channels are active at the same time. We guess that this framework might deserve further investigation.

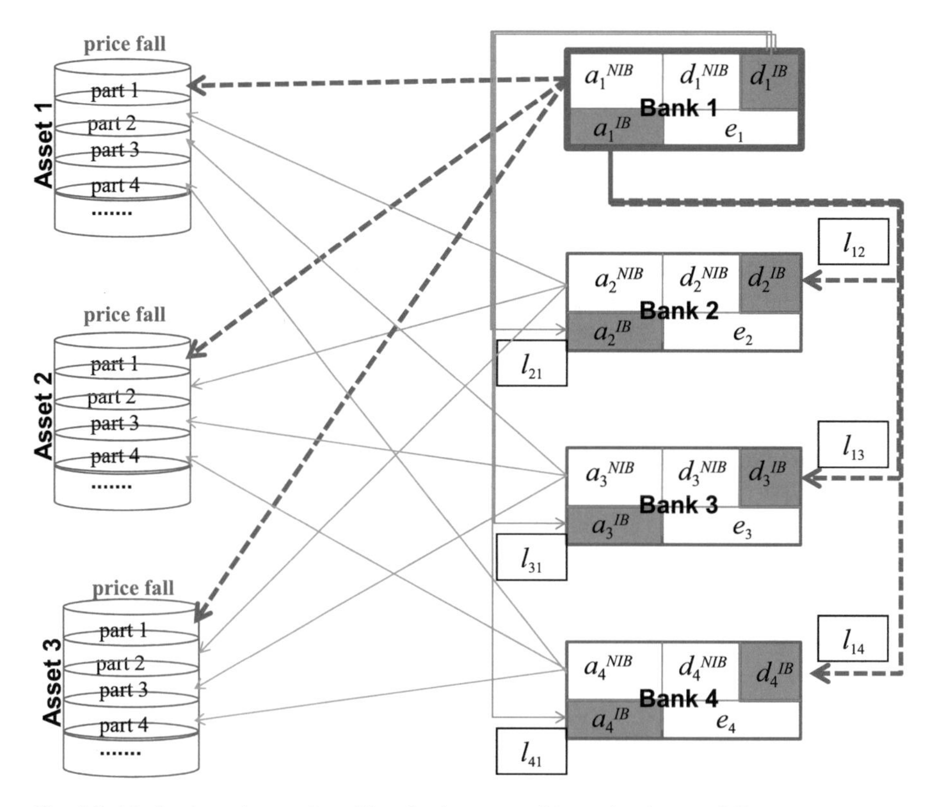

Fig. 5.9 Mechanism of contagion of interbank system with overlapping portfolios

5.5 Potential Applications of Big Data to Financial Contagion

Financial system is an uncertain and complex system interacting with human behavior. Up to date, there is no consensus on "what financial contagion is", "why it happens", and "how to cope with it". Although some typical methods and models are developed to simulate financial contagion, there remain many difficulties in measuring and coping with it.

First, it is a hard work to understand multiple financial contagion mechanisms. Many researchers had contributed their wisdoms to analyze contagion through various perspectives. However, more measurements and information are needed to disentangle internal relations of all contagion mechanisms and to develop a generalized way to comprehensively and properly measure multi-channel contagion.

Second, lack of information often forces researchers to alter their most suitable research scheme. For instance, to simplify model specification, investor behaviors are always supposed to be homogeneous. However, there exist significant differences in investor behaviors due to the difference in their capabilities in information acquisition and risk attitudes. By allowing for information asymmetry and inhomogeneous investor behaviors, the researches may become more close to vivid financial contagion in reality.

Third, sufficient and timely information is needed to make better financial risk managements in practice. However, the cost is too much for ordinary individual investors to get information and deal with it properly with the traditional methodologies. Furthermore, in order to gain competitiveness and keep business secrets, the needed data for financial risk management are often unavailable publicly. Most of individual investors in the financial system have to adopt "following strategy" that often manifests as "herd behavior" in the financial market and in turn amplifies financial contagion.

No matter for regulation, forecasting, or forewarning, the methodologies that can timely and frequently collect, update, and integrate information emerged in the market through different forms are necessary. Big data characterized by "4V", i.e., Volume, Variety, Value, and Velocity, might be exactly the right thing to improve and conquer these hardships in measuring financial contagion by providing diversiform information in cheaper, faster, and more efficient way.

Acknowledgements This research is partially supported by NSF of China, under project number 71471180.

References

F. Allen, D. Gale, Financial contagion. J. Polit. Econ. **108**(1), 1–33 (2000)

K.H. Bae, G.A. Karolyi, R.M. Stulz, A new approach to measuring financial contagion. Rev. Financ. Stud. **16**(3), 717–763 (2003)

R. Basu, Financial contagion and investor "learning": an empirical investigation. Working Paper, No. WP/02/218, International Monetary Fund (2002)

T. Bayoumi, G. Fazio, M. Kumar, R. MacDonald, Fatal attraction: A new measure of contagion. Working Paper, No. WP/03/80, International Monetary Fund (2003)

J. Beirne, G.M. Caporale, M. Schulze-Ghattas, N. Spagnolo, Volatility spillovers and contagion from mature to emerging stock markets. Rev. Int. Econ. **5**(21), 1060–1075 (2013)

G. Bekaert, R.J. Hodrick, Characterizing predictable components in excess returns on equity and foreign exchange markets. J. Financ. **47**(2), 467–509 (1992)

G. Bekaert, M. Ehrmann, M. Fratzscher, A. Mehl, The global crisis and equity market contagion. J. Financ. **69**(6), 2597–2649 (2014)

G. Bekaert, C.R. Harvey, A. Ng, Market integration and contagion. J. Bus. **78**(1), 39–69 (2005)

B.H. Boyer, M.S. Gibson, M. Loretan, Pitfalls in tests for changes in correlations. International Finance Discussion Papers, No. 597, Board of Governors of the Federal Reserve System (1997)

M.K. Brunnermeier, L.H. Pedersen. Market liquidity and funding liquidity. Rev. Financ. Stud. **22**(6), 2201–2238 (2009)

F. Caccioli, M. Shrestha, C. Moore, J.D. Farmer, Stability analysis of financial contagion due to overlapping portfolios. J. Bank. Financ. **46**, 233–245 (2014)

F. Caccioli, J.D. Farmer, N. Foti, D. Rockmore, Overlapping portfolios, contagion and financial stability. J. Econ. Dyn. Control **51**, 50–63 (2015)

G.A. Calvo, E.G. Mendoza, Rational contagion and the globalization of securities markets. J. Int. Econ. **51**(1), 79–113 (2000)

F. Caramazza, L. Ricci, R. Salgado, International financial contagion in currency crises. J. Int. Money Financ. **23**(1), 51–70 (2004)
R.A. Connolly, F.A. Wang, International equity market comovements: economic fundamentals or contagion? Pac. Basin Financ. J. **11**(1), 23–43 (2003)
G. Corsetti, M. Pericoli, M. Sbracia, Correlation analysis of financial contagion: what one should know before running a test. Center Discussion Paper, No.822, Economic Growth Center of Yale University (2001)
G. Corsetti, M. Pericoli, M. Sbracia, Some contagion, some interdependence: More pitfalls in tests of financial contagion. J. Int. Money Financ. **24**(8), 1177–1199 (2005)
F.X. Diebold, M. Nerlove, The dynamics of exchange rate volatility: A multivariate latent factor arch model. J. Appl. Economet. **4**(1), 1–21 (1989)
R. Dornbusch, Y.C. Park, S. Claessens, Contagion: understanding how it spreads. World Bank Res. Obs. **15**(2), 177–197 (2000)
M. Dungey, D.N. Tambakis, International financial contagion: what do we know? Working Paper, No.9 (2003), http://rspas.anu.edu.au/economics/staff/dungey//
M. Dungey, D. Zhumabekova, Testing for contagion using correlations: some words of caution. Working Paper, No. PB01-09, Center for Pacific Basin Monetary and Economic Studies and Federal Reserve Bank of San Francisco (2001)
M. Dungey, R. Fry, B. González-Hermosillo, V. Martin, The transmission of contagion in developed and developing international bond markets. in *Proceedings of the Third Joint Central Bank Research Conference* (2002)
M. Dungey, R. Fry, B. González-Hermosillo, V.L. Martin, Empirical modelling of contagion: A review of methodologies. Quant. Financ. **5**(1), 9–24 (2005)
F. Durante, P. Jaworski, Spatial contagion between financial markets: a copula based approach. Appl. Stoch. Model. Bus. Ind. **26**(5), 551–564 (2010)
L. Eisenberg, T.H. Noe, Systemic risk in financial systems. Manag. Sci. **47**(2), 236–249 (2001)
B. Eichengreen, A.K. Rose, C. Wyplosz, Contagious currency crises. Working Paper, No.5681, National Bureau of Economic Research (1996)
H. Elsinger, A. Lehar, M. Summer, Risk assessment for banking systems. Manag. Sci. **52**(9), 1301–1314 (2006a).
H. Elsinger, A. Lehar, M. Summer, Using market information for banking system risk assessment. Int. J. Cent. Bank. **2**(1), 137–165 (2006b)
H. Elsinger, A. Lehar, M. Summer, Network models and systemic risk assessment. *Handbook on Systemic Risk*, vol. 1 (Cambridge University Press, Cambridge, 2013), 287–305
R.F., Engle, K.F. Kroner, Multivariate simultaneous generalized arch. Econ. Theory **11**(01), 122–150 (1995)
R.F. Engle, T. Ito, W.L. Lin, Meteor showers or heat waves? Heteroskedastic intra-daily volatility in the foreign exchange market. Econometrica **58**(3), 525–542 (1990)
R.P. Flood, A.K. Rose, Equity integration in times of crisis (2004), https://www.researchgate.net/publication/229021532/
K.J. Forbes, R. Rigobon, No contagion, only interdependence: measuring stock market comovements. J. Financ. **57**(5), 2223–2261 (2002)
P. Gai, S. Kapadia, Contagion in financial networks. Proc. R. Soc. **466**, 2401–2423 (2010)
R. Glick, A.K. Rose, Contagion and trade: Why are currency crises regional? J. Int. Money Financ. **18**(4), 603–617 (1999)
A. Hasman, A critical review of contagion risk in banking. J. Econ. Surv. **27**(5), 978–995 (2013)
A. Karmann, R. Herrera, Volatility contagion in the Asian crisis: new evidence of volatility tail dependence. Rev. Dev. Econ. **18**(2), 354–371 (2014)
G.A. Karolyi, Does international financial contagion really exist. J. Appl. Corp. Financ. **16**(2), 136–146 (2004)
M.A. King, S. Wadhwani, Transmission of volatility between stock markets. Rev. Financ. Stud. **3**(1), 5–33 (1990)
L.E. Kodres, M. Pritsker, A rational expectations model of financial contagion. J. Financ. **57**(2), 769–799 (2002)

R.W. Kolb, *Financial Contagion: The Viral Threat to the Wealth of Nations* (Wiley, Hoboken, 2011)
F.A. Longstaff, The subprime credit crisis and contagion in financial markets. J. Financ. Econ. **97**(3), 436–450 (2010)
M. Loretan, W.B. English, Evaluating correlation breakdowns during periods of market volatility. Working Paper, No. 658, Board of Governors of the Federal Reserve System International Finance (2000)
R.E. Lucas Interest rates and currency prices in a two-country world. J. Monet. Econ. **10**(3), 335–359 (1982)
T. Markwat, E. Kole, V.D. Dick, Contagion as a domino effect in global stock markets. J. Bank. Financ. **33**(11), 1996–2012 (2009)
P.R. Masson, Contagion: Monsoonal effects, spillovers, and jumps between multiple equilibria. Working Paper, No. WP/98/142, International Monetary Fund (1998)
P.R. Masson, Contagion: macroeconomic models with multiple equilibria. J. Int. Money Financ. **18**(4), 587–602 (1999a)
P.R. Masson, Multiple equilibria, contagion, and the emerging market crises. Working Paper, No. WP/99/164, International Monetary Fund (1999b)
R.B. Nelsen, *An Introduction to Copulas* (Springer, New York, 2007)
M. Pericoli, M. Sbracia, A primer on financial contagion. J. Econ. Surv. **17**(4), 571–608 (2003)
M.H. Pesaran, A. Pick, Econometric issues in the analysis of contagion. J. Econ. Dyn. Control **31**(4), 1245–1277 (2007)
M. Pritsker, The channels for financial contagion. Working paper, Board of Governors of the Federal Reserve System, pp. 67–95 (2001)
R. Rigobon, On the measurement of the international propagation of shocks: Is the transmission stable? J. Int. Econ. **61**(2), 261–283 (2003)
J.C. Rodriguez, Measuring financial contagion: A copula approach. J. Empir. Financ. **14**(3), 401–423 (2007)
R. Romero-Meza, C. Bonilla, H. Benedetti, A. Serletis, Nonlinearities and financial contagion in Latin American stock markets. Econ. Model. **51**, 653–656 (2015)
Y. Sahalia, J. Cacho-Diaz, R.J. Laeven, Modeling financial contagion using mutually exciting jump processes. J. Financ. Econ. **117**(3), 585–606 (2015)
G.J. Schinasi, R.T. Smith, Portfolio diversification, leverage, and financial contagion. Working Paper, No. WP/99/136, International Monetary Fund (2001)
W.F. Sharpe, A simplified model for portfolio analysis. Manag. Sci. **9**(2), 277–293 (1963)
M. Summer, Financial contagion and network analysis. Annu. Rev. Financ. Econ. **5**(1), 277–297 (2013)
X.L. Sun, S.F. Niu, D. Li, An exact algorithm for factor model in portfolio selection with roundlot constraints. Optimization **58**(2), 305–318 (2009)
I. Trevino, Informational channels of financial contagion. Working Paper (2014), https://site.stanford.edu/sites/default/files/contagion_may2014.pdf/
C. Upper, Simulation methods to assess the danger of contagion in interbank markets. J. Financ. Stab. **7**(3), 111–125 (2011)
C. Van-Rijckeghem, B. Weder (2001) Sources of contagion: is it finance or trade? J. Int. Econ. **54**(2), 293–308
S.S. Zhu, X.T. Cui, X.L. Sun, D. Li, Factor-risk constrained mean-variance portfolio selection: Formulation and global optimization solution approach. J. Risk **14**(2), 51–89 (2011)

Chapter 6
Asset-Liability Management in Continuous-Time: Cointegration and Exponential Utility

Mei Choi Chiu

Abstract Using the technique of dynamic portfolio optimization, Chiu and Li (Insur. Math. Econ. 39:330–355, 2006) pioneered the optimal asset-liability management (ALM) framework for investors and insurers in a continuous-time economy. Their approach has been generalized to different objective functions under different stochastic models for the assets and the liabilities. This paper briefly summarizes recent advances along this research direction based on the author's personal interest and the required quantitative tools from stochastic optimal control theory. A new ALM solution is then derived for constant absolute risk averse insurers subject to cointegrated assets and compound Poisson-type insurance liabilities.

Keywords Asset-liability management • Cointegration • Utility theory

6.1 Introduction

The use of optimization techniques in portfolio management originated in the seminal work of Markowitz (1952), in which the mean-variance (MV) criterion is proposed. Although the static MV portfolio approach was important enough for Markowitz to receive the Nobel prize in Economics in 1990, its extension to dynamic portfolio choice was solved until the works by Li and Ng (2000) and Zhou and Li (2000) were published. Chiu and Wong (2011) generalize it to the cointegration economy. Cui et al. (2012) are among the first to study the MV portfolio in a time-consistency manner. Their work not only stimulates active research in time-consistent MV portfolio problems but is also extended to the concept of time-consistency in efficiency using cone constrained approach (Cui et al. 2015) and a unified framework using mean-field formulation (Cui et al. 2014). To

M.C. Chiu (✉)
Department of Mathematics and Information Technology,
The Education University of Hong Kong, Tai Po, Hong Kong
e-mail: mcchiu@eduhk.hk

T.-M. Choi et al. (eds.), *Optimization and Control for Systems in the Big-Data Era*,
International Series in Operations Research & Management Science 252,
DOI 10.1007/978-3-319-53518-0_6

address practical concerns, Gao et al. (2015a) investigate the optimal market timing via the time cardinality constrained MV portfolio selection.

The success of the dynamic portfolio choice stimulates the application of a stochastic optimal control framework to finance problems. Among many applications associated with the dynamic MV criteria, the present paper concentrates on the use of the optimal portfolio technique for asset-liability management (ALM) problems in continuous time proposed by Chiu and Li (2006). The major difficulty in an optimal ALM problem is that it contains an uncontrollable stochastic variable: liability. After subtracting liability from the investor's wealth, the objective function is based on the surplus process in place of the wealth process in the classical portfolio problems. Therefore, the ALM problem is inevitably formulated in an incomplete market. In the case of insurance liability or insurance claims, the surplus process becomes a jump-diffusion model in which the Poisson shock is uncontrollable. Therefore, the ALM problem requires further treatment to remove the difficulty generated by the uncontrollable liability. Chiu and Li (2006) lay the theoretical ground to overcome such a difficulty through a series of transformation of variables. An alternative way to look at ALM problem considers the asset-to-liability ratio or the funding ratio. Chiu et al. (2012) discover that the practical constraint on the bounded funding ratio removes the ill-posedness of some ALM problems. Recently, Gao et al. (2015b) use the bounded funding ratio constraint to solve mean-LPM and mean-CVaR portfolio optimization problems.

In fact, the optimal ALM approach has been further generalized to many practical stochastic models. Yi et al. (2008) extend this approach to discrete-time framework with uncertain investment horizon. Using the MV criteria, Chiu and Wong (2012, 2013a) consider the cointegrated assets with a compound Poisson liability and a diffusive liability, respectively. Chiu and Wong (2014) study the ALM problem in which risky assets have a stochastic variance–covariance matrix and the liability follows a compound Poisson process. Wong et al. (2014) derive optimal longevity risk management solutions for time-consistent and pre-commitment MV problems.

An alternative generalization considers other objective functions. Chiu and Li (2009) investigate the connection between the ALM problems under the MV criterion and the surrogate safety-first principle. Chiu et al. (2012) investigate the genuine safety-first ALM problems. Chiu and Wong (2013b) solve the ALM problem for constant relative risk aversion (CRRA) insurers.

This paper has two objectives. The first is to briefly review the dynamic ALM problem. The presentation is based on a rather general class of objective functions and a general stochastic differential equation (SDE) for the risky assets. The purpose is to highlight the major difference between the classical portfolio optimization and ALM problems.

The second is to offer a new research result. Using the established framework, I derive a closed-form solution to the optimal ALM strategy for a constant absolute risk averse (CARA) insurer subject to a high-dimensional cointegration system of risky assets and a compound Poisson insurance liability. This new finding not only generalizes the pair-trading strategy of Tourin and Yan (2013) from the two cointegrated risky assets case to a high-dimensional cointegration system but also from the portfolio management perspective to ALMs with a Poisson shock associated with the insurance claims. In addition, the model setting here also supplements that of Chiu and Wong (2013b) because their liability process is specially constructed to ensure a positive surplus. My consideration includes the possibility of ruin.

The theoretical optimal strategy in a financial market with a huge number of risky assets is useful for big data analysis as well. When the underlying financial problem involves many risky assets, traditional parameter estimation becomes unstable and degeneracy occurs by using the "optimized" strategy (Pun and Wong 2016). Existing big data methods can improve the implementation for static optimization problems but not for stochastic optimal control problems. The challenge is twofold. The first challenge is to estimate model parameters effectively under a big data environment. Even although the huge number of parameters are estimated satisfactorily, the error could be magnified exponentially when the parameters are plugged into the optimal portfolio policy. Therefore, the second challenge is to stabilize the aggregated estimation error through some additional big data techniques.

The cointegration model is closely related to the vector autoregression. When the vector autoregression system is stationary, Han et al. (2015) propose a high dimensional estimation scheme. Unfortunately, their method is not directly applicable for cointegration model because cointegration system contains a stationary sub-system but the whole system itself could be non-stationary.

The analytical solution obtained from the present paper is useful for the second step. Chiu et al. (2017) show that the analytical solution can serve as the base to construct a big data method to improve dynamic portfolio strategy. Therefore, the result presented in this paper has a great potential to correct estimation errors from high dimensionality. Potential candidate methods include shrinkage estimates and the constrained ℓ_1−minimization considered in Chiu et al. (2017). However, I leave this for a separate future research. An alternative approach considers robust optimization to reduce the effect from estimation errors. Corresponding references include Zhu et al. (2014, 2015). However, the extension of these latter approaches to incorporate cointegration is a very challenging mathematical task.

I concentrate on the cointegration system of risky assets because it is probably the most well-known, sufficiently general and empirically testable stochastic model in finance so far. Granger (1981) discover that a linear combination of two or more non-stationary time series can be stationary. Engle and Granger (1987) further formalize the idea of integrated variables sharing an equilibrium relation that turns out to be either stationary or to have a lower degree of integration than the original series. They denote this property as cointegration, signifying co-movements among trending variables that can be exploited to test for the existence of equilibrium relationships within a fully dynamic specification framework. Granger was thus awarded the Nobel Prize in Economics in 2003.

The remainder of this paper is organized as follows. Section 6.2 reviews the notion of ALM, cointegration, insurance liability and the problem formulation. Section 6.3 derives the optimal solution for CARA insurers within a cointegration system of risky assets and draws economic inferences from it. Section 6.4 concludes the paper.

6.2 Optimal ALM Formulation

Let $\{X(t) \in \mathbb{R}^n\}_{t\geq 0}$ be a stochastic vector process which is the strong solution of the following (SDE):

$$dX(t) = \mu(t, X(t))\, dt + \sigma(t)\, dW_t, \tag{6.1}$$

where $\mu(t, \cdot) \in \mathbb{R}^n$ is a deterministic vector-valued function, $\sigma(t) \in \mathbb{R}^{n\times n}$ is a deterministic matrix-valued function, and $\{W_t \in \mathbb{R}^n\}_{t\geq 0}$ is the vector of independent Weiner processes. Let $S_i(t) = e^{X_i(t)}$ be the price of risky asset i in the economy. Then, $\{X(t)\}_{t\geq 0}$ represents the vector of log-asset value processes. Consider the economy with one risk-free asset that

$$S_0(t) = S_0(0)e^{\int_0^t r(\tau)\, d\tau},$$

where $r(t)$ is the time-deterministic interest rate.

In classical dynamic portfolio problems, the investor determines the optimal allocation of her wealth, $A(t)$, to all of the available assets. Therefore,

$$A(t) = \sum_{j=0}^{n} N_j(t)S_j(t) = \sum_{j=0}^{n} u_j(t), \tag{6.2}$$

where $N_j(t)$ is the number of holdings of asset j at time t, and $u_j(t)$ is the investment amount in asset j at time t. An application of Itô's lemma shows that

$$dA(t) = [r(t)A(t) + u(t)'\beta(t)]\, dt + u(t)'\sigma(t)\, dW_t, \tag{6.3}$$

where $u(t) = [u_1(t) \cdots u_n(t)]'$ is the vector of risky asset investment amounts and

$$\beta(t) = \mu(t, X(t)) + \frac{1}{2}\mathcal{D}\mathbf{1} - r(t)\mathbf{1},$$

with $\mathcal{D}$ being the diagonal matrix sharing the same diagonal with $\sigma(t)\sigma(t)'$, and $\mathbf{1} \in \mathbb{R}^n$ being the vector of all elements equal to 1.

The classical portfolio problem can be expressed as

$$\sup_{u(t)\in\Pi} \mathrm{E}[U(A(T))|\mathcal{F}_t], \text{ s.t. (6.1) and (6.3)}, \tag{6.4}$$

where $U(\cdot)$ is a non-decreasing utility function, $\mathcal{F}_t$ is the information available to the investor at time t, and Π is the admissible set such that

$$\Pi = \left\{ u(t) \in \mathbb{R}^n \middle| \mathrm{E}\left[\int_0^T \|u(t)\|^2 \, dt \right] < \infty \right\}.$$

When the number of risky assets equals the number of driving processes, the market is called a complete market. Then, there exists a unique equivalent martingale measure so that the martingale approach, a kind of duality approach, can be applied. Otherwise, the solvability of the martingale approach could be a challenging technical problem.

6.2.1 Asset-Liability Management

The major difference between the portfolio problem and ALM is that the latter contains the uncontrollable liability. Hence, the investor's utility is based on the surplus rather than the wealth. The surplus is defined as

$$Y(t) = A(t) - L(t), \tag{6.5}$$

where $A(t)$ is defined in (6.2), and $L(t)$ is a stochastic liability. For instance, Chiu and Li (2006) consider a diffusive liability such that

$$dL(t) = \mu_L \, dt + \sigma_L \, dW_L(t),$$

where μ_L and σ_L are constant values, and $\{W_L(t)\}_{t\geq 0}$ is the real-valued Wiener process possibly correlated with the Wiener vector $\{W_t\}_{t\geq 0}$. For an insurer's ALM problem, Chiu and Wong (2012) use a compound Poisson process for the liability,

$$L(t) = \sum_{i=1}^{\mathcal{N}(t)} z_i, \tag{6.6}$$

where $\{\mathcal{N}(t)\}_{t\geq 0}$ is a Poisson process with intensity λ, and z_i for $i = 1, 2, \ldots$, are independent identically distributed random variables with a bounded moment-generating function.

Hence, the optimal ALM problem becomes

$$\sup_{u(t)\in\Pi} \mathrm{E}[U(Y(T))|\mathcal{F}_t], \text{ s.t. } (6.1), (6.3), \text{ and } (6.5). \tag{6.7}$$

In ALM problems, the utility is evaluated at the terminal surplus, and the optimal strategy $u(t)$ only affects the asset allocation. In practice, the payment schedule and amount of the liability can seldom be intervened by the investor. This is the situation of uncontrollable liability. Take the diffusive liability model as an example. The number of Weiner processes driving the surplus process is $n + 1$ while the number of controls is n so that the ALM problem falls into an incomplete market situation. The market incompleteness requires additional conditions for the solvability of the martingale approach (Chiu et al. 2012). When the liability is associated with jumps like (6.6), Chiu and Wong (2012) demonstrate the sophistication required to prove the existence and uniqueness of the optimal ALM strategy for a linear-quadratic objective function. More specifically, the duality approach requires the study of the existence and uniqueness of the corresponding forward–backward SDE (Chiu and Wong 2012).

When the martingale approach is too complicated to apply, the Hamilton–Jacobi–Bellman (HJB) framework is a useful alternative as shown in Chiu and Li (2006). For example, Chiu and Wong (2013b) adopt the HJB approach to the ALM problem with cointegrated assets and insurance liability for a CRRA insurer. Although (6.6) is the most popular model in the insurance literature, it makes the surplus possibly go negative, and hence, the CRRA utility becomes undefined. Chiu and Wong (2013b) then postulate the liability to be

$$L(t) = \int_0^t (1 - e^{Z_\tau}) Y(\tau)\, d\mathcal{N}(\tau)$$

so that the surplus always stays positive. However, insurers are primarily concerned with the probability of ruin or a negative surplus.

Nothing is better than giving a concrete example to demonstrate the solution process. This paper considers the cointegration system of risky assets and the liability of form (6.6). I demonstrate the solution process with the HJB framework below.

6.2.2 The Financial Market with Cointegration

Assumption 2.1 The vector of log prices of risky assets, $X(t)$, satisfies the SDE:

$$dX(t) = [\theta(t) - \mathcal{A}X(t)]\, dt + \sigma(t)dW_t,\ t \in [0, T], \tag{6.8}$$

where $W_t = (W_t^1, \ldots, W_t^n)'$ is a standard $\mathcal{F}_{t\geq 0}$-adapted n-dimensional Wiener process on a fixed filtered complete probability space $(\Omega, \mathcal{F}, \mathcal{P}, \mathcal{F}_{t\geq 0})$, W_t^i and W_t^j are mutually independent for all $i \neq j$, $\mathbb{F} := \{\mathcal{F}_t\}_{t\geq 0}$ is the filtration generated by W_t augmented by the null sets of $\mathcal{P}$, $\mathcal{A} \in \mathbb{R}^{m\times m}$ is a constant matrix, and $\Sigma(t) = \sigma(t)\sigma(t)'$ is the variance–covariance matrix of assets defined in the Banach space of the $\mathbb{R}^{n\times n}$-valued continuous function on $[0, T]$ such that the non-degeneracy condition of $\Sigma(t) \geq \delta I_n$ holds for all $t \in [0, T]$ and for some $\delta > 0$. Here, $\theta(t)$ and $\sigma(t)$ are time-deterministic functions. In addition, the real part of each eigenvalue of $\mathcal{A}$ is non-negative and $\mathcal{A}$ is non-zero. Note that $\mathcal{A}$ can be singular.

In other words, Assumption 2.1 postulates that $\mu(t, X) = \theta - \mathcal{A}X$ in (6.1), and it refers to the cointegration system of risky assets. Economically, there exist common equilibria among the risky assets in this financial market. Let us begin with the single risky asset case:

$$dX_1(t) = [\theta_1 - a\,X_1(t)]\,dt + \sigma_1\,dW_1(t).$$

By Itô's lemma, it is clear that

$$X_1(t) = X_1(0)e^{-at} + \theta_1 \frac{1 - e^{-at}}{a} + \sigma_1 \int_0^t e^{a(\tau - t)}\,dW_\tau$$
$$\Rightarrow X_1(t) \sim N\left(X_1(0)e^{-at} + \theta_1 \frac{1 - e^{-at}}{a}, \frac{\sigma_1^2}{2a}\left[1 - e^{-2at}\right]\right).$$

If $a > 0$, then $X_1(t)$ has a known distribution for all $t > 0$. In particular, its invariant distribution exists by taking t tends to infinity. In economic terms, the long-term equilibrium of X_1 exists and $X_1(\infty) \sim N(\theta_1/a, \sigma_1^2/(2a))$. The parameter θ_1/a is hence called the long-term mean of the equilibrium, and $\{X_1(t)\}_{t \geq 0}$ is called a (weakly) stationary process.

For $n > 1$, suppose $\mathcal{A}$ has k positive eigenvalues and the remainings are zero, where $k < n$. Suppose further that $\mathcal{A}$ is diagonalizable. Hence, $\mathcal{A} = PDP^{-1}$, where D is a diagonal matrix with the first k diagonal elements being positive. The system of SDE (6.8) is transformed into the following system of SDE:

$$dX^*(t) = (\theta^* - DX^*(t))\,dt + \sigma^*\,dW_t, \tag{6.9}$$

where $X^*(t) = P^{-1}X(t)$, $\theta^* = P^{-1}\theta$, and $\sigma^* = P^{-1}\sigma$. Let $Z(t) = [X_1^*(t) \cdots X_k^*(t)]'$. Then, there exist $\theta_z \in \mathbb{R}^k$ and invertible $D_z, \sigma_z \in \mathbb{R}^{k \times k}$ such that

$$dZ(t) = (\theta_z - D_z Z(t))\,dt + \sigma_z\,dW_t. \tag{6.10}$$

It is then easy to show that

$$Z(\infty) \sim N\left(D^{-1}\theta_z, \frac{1}{2}D^{-1}\sigma_z\sigma_z'\right).$$

In other words, there exists a linear transformation of log-asset prices that admits a stationary subsystem.

6.2.3 *The Surplus Process*

After specifying the risky asset dynamics, I consider the surplus of (6.5) subject to Assumption 2.1 and the liability process of (6.6). Notice that the ALM strategy is not a self-financing ones because the uncontrollable insurance liability is a random payment of cash outflow. This makes this ALM problem different from classical portfolio selection problems. Applying Itô's lemma to $Y(t)$ with respect to the cointegration system (6.8), the wealth process is given by

$$\begin{aligned} dY(t) &= \left[r(t)Y(t) + u(t)'\beta(t)\right]dt + u(t)'\sigma(t)dW_t - zd\mathcal{N}_t, \\ Y(0) &= Y_0, \end{aligned} \tag{6.11}$$

where z has the same distribution with z_j, and

$$\beta(t) = \theta(t) - AX(t) + \frac{1}{2}\mathcal{D}\mathbf{1} - r(t)\mathbf{1}, \tag{6.12}$$

in which $\mathcal{D}$ is the diagonal matrix with all diagonal elements equal to those of $\Sigma(t)$; $\mathbf{1}$ is the column vector with all elements being 1; $\mathcal{N}_t$ is a doubly stochastic Poisson process with $\mathbb{F}$-predictable non-negative intensity $\mu(t)$; and the parameters $\sigma_{ij}(t)$, θ, r, z and $\mu(t)$ are uniformly bounded and $\mathbb{F}$-predictable on $[0, T]$, for $i = 1, \cdots, m$ and $j = 1, \cdots, n$. Define $\mathbb{H} := \{\mathcal{H}_t\}_{t\geq 0}$, which is the filtration generated by $\mathcal{N}(t)$ augmented by the $\mathcal{P}$-null sets. Let $\mathbb{G}$ be the filtration $\{\mathcal{G}_t\}_{t\geq 0}$, where $\mathcal{G}_t := \mathcal{F}_t \vee \mathcal{H}_t$, the smallest filtration containing $\mathbb{F}$ and $\mathbb{H}$. Note that $\mathcal{G}_t$ can be regarded as the information available to the investor at time t. Define the compensated Poisson process $\mathcal{M}_t := \mathcal{N}_t - \int_0^t \mu(s)ds$, which is a $\mathbb{G}$-martingale.

6.2.4 The Optimization Problem

In economics, one school of thought on optimal investment decisions suggests maximizing the expected utility of an investor's future wealth, $\mathrm{E}[U(Y_T)]$. The standard approach assumes the utility to be positive, strictly increasing and concave. If the utility function is twice differentiable, then $U'(y) > 0$ and $U''(y) < 0$ for all y. One classical utility is an exponential utility:

$$U(y) = 1 - e^{-ay}, \quad a > 0. \tag{6.13}$$

The exponential utility implies CARA, with the coefficient of absolute risk aversion equal to a constant:

$$-\frac{U''(y)}{U'(y)} = a. \tag{6.14}$$

Given this background, I formally lay down the research problem.

Research Problem $\sup_{u(\cdot)} \mathrm{E}\left[1 - e^{-aY(T)}\right]$ s.t. (6.8), (6.11) and $u(\cdot) \in \Pi$.

(6.15)

If $\beta(t)$ is a time-deterministic function and the Poisson process is absent in (6.11), then the corresponding utility portfolio problem is reduced to the standard utility portfolio optimization problem. Unfortunately, $\beta(t)$ is essentially a stochastic vector depending on $X(t)$ through (6.12), and the insurance liability follows a compound Poisson process. Therefore, the wealth process (6.11) resembles a jump-diffusion model with a random drift.

Remarks

- When insurance liability is absent, i.e., $z \equiv 0$, Chiu and Wong (2015) used the dynamic time-consistent MV optimization to show that cointegration, i.e., Assumption 2.1, implies statistical arbitrage in the sense of Hogan et al. (2004). Although their results are useful for a trader, the present research problem is specifically useful for insurers. The major differences are as follows. 1. I permit different levels of risk aversion by selecting the constant risk aversion coefficient a; 2. I incorporate insurance liabilities into the optimization problem so that the insurer's wealth contains Poisson shocks; and 3. The investment strategy observes different budget equations because I use non-self-financing-type strategies.
- When $n = 2$ and $z \equiv 0$, the expected CARA utility maximization problem has been solved by Tourin and Yan (2013). Mine extends theirs by allowing a general $n > 1$ and by including an uncontrollable liability of the compound Poisson type.

6.3 The Optimal ALM Strategy

In the research problem of interest, the optimal decision is not affected by adding or subtracting a constant to the objective function. The CARA utility maximization problem can then be reduced to

$$\inf \mathrm{E}\left[e^{-aY(T)}\right], \tag{6.16}$$

where the insurer's wealth follows the SDE in (6.11).

Theorem 3.1 *Under Assumption 2.1, the research problem* (6.15) *with the exponential utility* (6.16) *has the optimal solution (investment policy)*

$$u^*(t, \beta(t)) = \frac{1}{ae^{\int_t^T r(s)ds}} \left[\left(\Sigma(t)^{-1} - \mathcal{A}'K(t,T)\right)\beta(t) - \mathcal{A}'N(t,T)\right] \tag{6.17}$$

and the optimal value of the objective function

$$\begin{aligned} \mathrm{E}\left[1 - e^{-aY(T)} \middle| \mathcal{G}_0\right]\Big|_{u=u^*} &= 1 - \exp\left\{-aY_0 e^{\int_0^T r(s)ds} \right. \\ &\quad \left. + \int_0^T \mathrm{E}\left[\left(\exp\left\{zae^{\int_s^T r(\tau)d\tau}\right\} - 1\right)\mu\right] ds\right\} \\ &\quad \times \exp\left[\frac{1}{2}\beta(0)'K(0,T)\beta(0) + N'(0,T)\beta(0) + M(0,T)\right], \end{aligned} \tag{6.18}$$

where $\beta(t)$ *is defined in (6.12);*

$$K(t,T) = -\int_t^T \Sigma^{-1}(s)ds; \tag{6.19}$$

$$N'(t,T) = \int_t^T \Theta'(s)K(s,T)ds; \tag{6.20}$$

$$M(t,T) = \int_t^T N'(s,T)\Theta(s) + \frac{1}{2}tr\left(K(s,T)\mathcal{A}\Sigma(s)\mathcal{A}'\right)ds; \tag{6.21}$$

and

$$\Theta(t) = \dot{\theta}(t) + \left[\frac{1}{2}\left(\dot{\mathcal{D}}(t) + \mathcal{A}\mathcal{D}(t)\right) - \dot{r}(t)I_m - \mathcal{A}r(t)\right]\mathbf{1}. \tag{6.22}$$

Proof The proof is based on the classic HJB framework. The dynamic of β can be obtained by applying Itô's lemma to (6.12) with respect to the dynamic of $X(t)$. Hence, we have

$$d\beta(t) = [\Theta(t) - \mathcal{A}\beta(t)]\,dt - \mathcal{A}\sigma(t)dW_t, \tag{6.23}$$

where

$$\Theta = \dot{\theta} + \left[\frac{1}{2}\left(\dot{\mathcal{D}} + \mathcal{A}\mathcal{D}\right) - \dot{r}I_m - \mathcal{A}r\right]\mathbf{1}.$$

Let

$$V(t,y,\beta) = \inf_{u\in\Pi} \mathrm{E}\left[e^{-aY^u(T)}\middle|\,\mathcal{G}_t\right].$$

Hence, the optimal function value of the research problem (6.15) with the exponential utility (6.16) is $1 - V(0,y,\beta)$. For a fixed terminal time T, the corresponding HJB equation is

$$\begin{aligned} &V_t + V_\beta'\left(\Theta - \mathcal{A}\beta\right) + \frac{1}{2}tr\left(V_{\beta\beta}\mathcal{A}\Sigma\mathcal{A}'\right) + \mathrm{E}\left[\left(V(t,y-z,\beta) - V(t,y,\beta)\right)\mu\right] \\ &+ \inf_u\left\{V_y\left(ry + u'\beta\right) + \frac{1}{2}V_{yy}u\Sigma u - u'\Sigma\mathcal{A}'V_{\beta y}\right\} = 0, \end{aligned} \tag{6.24}$$

with $V(T,y,\beta) = e^{-ay}$. Thus, the optimal feedback control, u^*, minimizes

$$V_y\left(ry + u'\beta\right) + \frac{1}{2}V_{yy}u\Sigma u - u'\Sigma\mathcal{A}'V_{\beta y}. \tag{6.25}$$

If $V_{yy} > 0$, differentiating (6.25) with respect to u and setting the differential to zero results in

$$u^* = -\Sigma^{-1}\frac{V_y}{V_{yy}}\beta + \mathcal{A}'\frac{V_{\beta y}}{V_{yy}}.$$

Otherwise, if $V_{yy} \le 0$, then the optimization has no solution. Substituting the u^* into the HJB equation (6.24), the partial integral differential equation (PIDE) of V becomes

$$V_t + V_\beta'(\Theta - \mathcal{A}\beta) + \frac{1}{2}tr\left(V_{\beta\beta}\mathcal{A}\Sigma\mathcal{A}'\right) + V_y r y + \mathrm{E}\left[\left(V(t, y-z, \beta) - V\right)\mu\right]$$
$$-\frac{1}{2}V_{yy}\left(\Sigma\mathcal{A}'\frac{V_{\beta y}}{V_{yy}} - \frac{V_y}{V_{yy}}\beta\right)'\Sigma^{-1}\left(\Sigma\mathcal{A}'\frac{V_{\beta y}}{V_{yy}} - \frac{V_y}{V_{yy}}\beta\right) = 0, \qquad (6.26)$$

with terminal condition $V(T, y, \beta) = e^{-ay}$. As $V(T, y, \beta) = e^{-ay}$, consider an exponential affine form for V:

$$V(t, y, \beta) = \exp\left\{-aye^{\int_t^T r(s)ds} + \int_t^T \mathrm{E}\left[\left(\exp\left\{zae^{\int_s^T r(\tau)d\tau}\right\} - 1\right)\mu\right]ds\right\}$$
$$\times \exp\left[\frac{1}{2}\beta' K(t, T)\beta + N'(t, T)\beta + M(t, T)\right], \qquad (6.27)$$

where K, N are M deterministic (matrix) functions of t and are defined as (6.19), (6.20) and (6.21) in order. Note that $K(t)$ is a symmetric matrix function of t. Clearly, the terminal value of the function in (6.27) satisfies the terminal condition in (6.26) and $V_{yy} > 0$. Taking partial derivatives to the affine form V with respect to t, y and β, we have

$$V_t = \left(raye^{\int_t^T r(s)ds} - \mathrm{E}\left[\left(\exp\left\{zae^{\int_t^T r(s)ds}\right\} - 1\right)\mu\right] + \frac{1}{2}\beta'\dot{K}\beta + \dot{N}'\beta + \dot{M}\right)V$$
$$-\left(raye^{\int_t^T r(s)ds} \quad \mathrm{E}\left[\left(\exp\left\{zae^{\int_t^T r(s)ds}\right\} - 1\right)\mu\right] + \frac{1}{2}\beta'\dot{K}'\beta + \dot{N}'\beta + \dot{M}\right)V;$$

$$V_\beta = \left[\frac{1}{2}\left(K' + K\right)\beta + N\right]V;$$

$$V_{\beta\beta} = \frac{K + K'}{2}V + \left[\frac{1}{2}\left(K' + K\right)\beta + N\right]\left[\frac{1}{2}\left(K' + K\right)\beta + N\right]'V;$$

$$V_y = -ae^{\int_t^T r(s)ds}V;$$

$$V_{yy} = a^2e^{2\int_t^T r(s)ds}V;$$

$$V_{\beta y} = -ae^{\int_t^T r(s)ds}\left[\frac{1}{2}\left(K' + K\right)\beta + N\right]V;$$

$$\mathrm{E}\left[\left(V(t, y - z), \beta\right) - V)\mu\right] = \mathrm{E}\left[\left(\exp\left\{zae^{\int_t^T r(s)ds}\right\} - 1\right)\mu\right]V.$$

After substituting these expressions into the left-hand side of (6.26), simple but tedious calculations easily verify that the proposed solution form satisfies the PIDE in (6.26). Thus, the solution form in (6.27) is actually a solution of the PIDE in (6.26). As the value function is twice continuously differentiable and all of the parameters are uniformly bounded and predictable, the classical verification theorems of Fleming and Soner (1993) (III, Theorem 8.1) confirm that the proposed affine form of the value function in (6.27) and the control in (6.17) are the optimal value function and optimal feedback control, respectively.

6.3.1 *Effect of Mean Reversion*

An interesting question is whether or not the statistical arbitrage index increases with the mean-reverting speed of the cointegrating factor. Is it beneficial for an insurance company to search for highly mean-reverting pairs? I investigate this by perturbing the cointegration coefficient matrix $A_\epsilon = \epsilon A$ and consider constant parameter settings.

Lemma 3.1 *Suppose that all parameters are constant values and the cointegration coefficient matrix is perturbed by* $\mathcal{A}_\epsilon = \epsilon\mathcal{A}$. *The research problem* (6.15) *with the exponential utility* (6.16) *has the optimal solution (investment policy)*

$$u^*(t, \beta(t)) = \frac{1}{ae^{\int_t^T r(s)ds}}\left[\left(\Sigma(t)^{-1} - \epsilon\mathcal{A}'K_\epsilon(t, T)\right)\beta(t) - \epsilon\mathcal{A}'N_\epsilon(t, T)\right] \quad (6.28)$$

and the optimal value of the objective function

$$\begin{aligned}\mathrm{E}\left[1 - e^{-aY(T)}\middle|\mathcal{G}_0\right]\Big|_{u=u^*} &= 1 - \exp\Big\{-aY_0e^{\int_0^T r(s)ds} \\ &\quad + \int_0^T \mathrm{E}\left[\left(\exp\left\{zae^{\int_s^T r(\tau)d\tau}\right\} - 1\right)\mu\right]ds\Big\} \\ &\quad \times \exp\Big[\frac{1}{2}\beta(0)'K_\epsilon(0, T)\beta(0) \\ &\quad + N_\epsilon{}'(0, T)\beta(0) + M_\epsilon(0, T)\Big], \end{aligned} \quad (6.29)$$

where

$$K_\epsilon(t,T) = K(t,T) = -\Sigma^{-1}(T-t); \tag{6.30}$$

$$N_\epsilon{}'(t,T) = \epsilon N'(t,T) = -\frac{\epsilon(T-t)^2}{2}\Theta'\Sigma^{-1}; \tag{6.31}$$

$$M_\epsilon(t,T) = \epsilon^2 M(t,T) = -\epsilon^2\left[\frac{(T-t)^2}{4}tr\left(\sigma'\mathcal{A}'\Sigma^{-1}\mathcal{A}\sigma\right) + \frac{(T-t)^3}{6}\Theta'\Sigma^{-1}\Theta\right] \leq 0; \tag{6.32}$$

and

$$\Theta_\epsilon = \epsilon\Theta = \epsilon\mathcal{A}\left[\frac{\mathcal{D}}{2} - rI_m\right]\mathbf{1}. \tag{6.33}$$

Proof The proof is similar to Theorem 3.1. □

From the above expression, we can conclude that the values $\frac{1}{2}\beta' K_\epsilon(t,T)\beta + N_\epsilon{}'(t,T)\beta + M_\epsilon(t,T)$ and $M_\epsilon(t,T)$ are not positive for all $t \in [0,T]$, β and $\epsilon \geq 0$. The details are as follows. Consider,

$$\begin{aligned}
&\frac{1}{2}\beta' K_\epsilon(t,T)\beta + N_\epsilon{}'(t,T)\beta + M_\epsilon(t,T)\\
&= \frac{1}{2}\beta' K_\epsilon(t,T)\beta + N_\epsilon{}'(t,T)\beta + M_\epsilon(t,T)\\
&\quad + \frac{\epsilon^2(T-t)^3}{8}\Theta'\Sigma^{-1}\Theta - \frac{\epsilon^2(T-t)^3}{8}\Theta'\Sigma^{-1}\Theta\\
&= \frac{1}{2}\beta' K_\epsilon(t,T)\beta + N_\epsilon{}'(t,T)\beta + M_\epsilon(t,T) + \frac{1}{2}N_\epsilon{}'(t,T)K_\epsilon^{-1}(t,T)N_\epsilon(t,T)\\
&\quad - \frac{1}{2}N_\epsilon{}'(t,T)K_\epsilon^{-1}(t,T)N_\epsilon(t,T)\\
&= \frac{1}{2}\left[\beta + K_\epsilon^{-1}(t,T)N_\epsilon(t,T)\right]' K_\epsilon^{-1}(t,T)\left[\beta + K_\epsilon^{-1}(t,T)N_\epsilon(t,T)\right]\\
&\quad + M_\epsilon(t,T) - \frac{1}{2}N_\epsilon{}'(t,T)K_\epsilon^{-1}(t,T)N_\epsilon(t,T) \leq 0.
\end{aligned}$$

Besides, the smaller the value of "$\frac{1}{2}\beta' K_\epsilon(t,T)\beta + N_\epsilon{}'(t,T)\beta + M_\epsilon(t,T)$" the greater the expected utility value. Hence, I would like to investigate the range of values of mean-reverting speed, ϵ, which can maximize the expected utility value.

6.3.2 *Optimal ϵ to Maximize the Expected Utility*

For the convenience, define

$$Q(\epsilon) = \frac{1}{2}\beta' K_\epsilon(t,T)\beta + N_\epsilon{}'(t,T)\beta + M_\epsilon(t,T) = \frac{1}{2}\beta' K(t,T)\beta$$
$$+\epsilon N'(t,T)\beta + \epsilon^2 M(t,T),$$

under constant parameters settings. Hence, $V^\epsilon(t,y,\beta)$, which is the optimal function value of the research problem (6.15) with specific mean-reverting speed ϵ, can be written as

$$V^\epsilon(t,y,\beta) = \exp\left\{-aye^{\int_t^T r(s)ds} + \int_t^T \mathrm{E}\left[\left(\exp\left\{zae^{\int_s^T r(\tau)d\tau}\right\} - 1\right)\mu\right] ds + Q(\epsilon)\right\}.$$

Taking partial derivatives to V^ϵ with respect to ϵ, we have

$$\frac{\partial Q}{\partial \epsilon} = N'\beta + 2\epsilon M; \quad \frac{\partial V^\epsilon}{\partial \epsilon} = \frac{\partial Q}{\partial \epsilon} V^\epsilon;$$
$$\frac{\partial V^\epsilon}{\partial \epsilon} = 0 \text{ when} \epsilon = \epsilon^* := -\frac{N'\beta}{2M};$$
$$\frac{\partial^2 Q}{\partial \epsilon^2} = 2M; \frac{\partial^2 V^\epsilon}{\partial \epsilon^2} = \left[\frac{\partial^2 Q}{\partial \epsilon^2} + \left(\frac{\partial Q}{\partial \epsilon}\right)^2\right] V^\epsilon;$$
$$\left.\frac{\partial^2 V^\epsilon}{\partial \epsilon^2}\right|_{\epsilon=\epsilon^*} = 2MV^\epsilon < 0.$$

Therefore, ϵ^* is the local maximum of V^ϵ and the local minimum of $\mathrm{E}[U^\epsilon(Y^*(T))]$, which is the optimal expected utility value with the mean-reverting speed ϵ.

The calculation above answers the question if it is beneficial for an insurer to search for highly mean-reverting pairs. It is shown that there exists the best target mean-reverting rate ϵ^*. In other words, it is not necessarily optimal for insurers to search for an extremely high mean-reverting pairs, not to mention the possible cost. An appropriate amount of mean reversion is beneficial to insurers. This is consistent with the conclusion of Chiu and Wong (2013b) that insurers better concentrate on their own insurance business rather than looking for additional profit by participating into risky arbitrage opportunities.

6.4 Conclusion

This paper gives a brief overview of asset-liability management with an optimal control framework. Then, I derive the optimal ALM strategy for a cointegration system of risky assets and an insurance liability that follows a compound Poisson process for a CARA insurer.

Acknowledgements The authors thank the Editor and an anonymous referee for their constructive comments. MC Chiu acknowledges the support by Research Grant Council of Hong Kong with ECS Project Number: 809913 and GRF Project Number: 18200114.

References

M.C. Chiu, D. Li, Continuous-time mean-variance optimization of assets and liabilities. Insur. Math. Econ. **39**, 330–355 (2006)

M.C. Chiu, D. Li, Asset-liability management under the safety-first principle. J. Optim. Theory Appl. **143**, 455–478 (2009)

M.C. Chiu, H.Y. Wong, Mean-variance portfolio selection of cointegrated assets. J. Econ. Dyn. Control **25**, 1369–1385 (2011)

M.C. Chiu, H.Y. Wong, Mean-variance asset-liability management: cointegrated assets and insurance liabilities. Eur. J. Oper. Res. **223**, 785–793 (2012)

M.C. Chiu, H.Y. Wong, Mean-variance principle of managing cointegrated risky assets and random liabilities. Oper. Res. Lett. **41**, 98–106 (2013a)

M.C. Chiu, H.Y. Wong, Optimal investment for an insurer with cointegrated assets: CRRA utility. Insur. Math. Econ. **52**, 52–64 (2013b)

M.C. Chiu, H.Y. Wong, Mean-variance asset-liability management with asset correlation risk and insurance liabilities. Insur. Math. Econ. **59**, 300–310 (2014)

M.C. Chiu, H.Y. Wong, Dynamic cointegrated pairs trading: Mean-variance time-consistent strategies. J. Comput. Appl. Math. **290**, 516–534 (2015)

M.C. Chiu, H.Y. Wong, D. Li, Roy's safety-first portfolio principle in financial risk management of disastrous events. Risk Anal. **32**, 1856–1872 (2012)

M.C. Chiu, C.S. Pun, H.Y. Wong, Big data challenges of high-dimensional continuous-time mean-variance portfolio selection and a remedy. Risk Anal. DOI: 10.1111/risa.12801, (2017)

X. Cui, D. Li, S.Y. Wang, S.S. Zhu, Better than dynamic mean-variance: time inconsistency and free cash flow stream. Math. Financ. **22**, 346–378 (2012)

X. Cui, D. Li, X. Li, Unified framework of mean-field formulations for optimal multi-period mean-variance portfolio selection. IEEE Trans. Autom. Control **59**, 1833–1844 (2014)

X. Cui, D. Li, X. Li, Mean-variance policy for discrete-time cone-constrained markets: time consistency in efficiency and the minimum-variance signed supermartingale measure. Math. Financ.. doi: 10.1111/mafi.12093 (2015)

R. Engle, C. Granger, Co-integration and error correction: representation, estimation and testing. Econometrica **55**, 251–276 (1987)

W.H. Fleming, H.M. Soner, *Controlled Markov Processes and Viscosity Solutions* (Springer, New York, 1993)

J.J. Gao, D. Li, X.Y. Cui, S.Y. Wang, Time cardinality constrained mean-variance dynamic portfolio selection and market timing: a stochastic control approach. Automatica **54**, 91–99 (2015a)

J.J. Gao, K. Zhou, D. Li, X.R. Cao, Dynamic mean-LPM and mean-CVaR portfolio optimization in continuous-time. SIAM J. Control Optim. (to appear) (2015b). Available at: http://arxiv.org/abs/1402.3464

C. Granger, Some properties of time series data and their use in econometric model specification. J. Econ. **23**, 121–130 (1981)

F. Han, H. Lu, H, Liu, A direct estimation of high dimensional stationary vector autoregressions. J. Mach. Learn. Res. **16**, 3115–3150 (2015)

S. Hogan, R. Jarrow, M. Teo, M. Warachkac, Testing market efficiency using statistical arbitrage with applications to momentum and value strategies. J. Financ. Econ. **73**, 525–565 (2004)

D. Li, W.L. Ng, Optimal dynamic portfolio selection: multiperiod mean-variance formulation. Math. Financ. **10**, 387–406 (2000)

H. Markowitz, Portfolio selection. J. Financ. **7**, 77–91 (1952)

C.S. Pun, H.Y. Wong, Resolution of degeneracy in Merton's portfolio problem. SIAM J. Financial Math. **7**, 786–811, (2016)

A. Tourin, R. Yan, Dynamic pairs trading using the stochastic control approach. J. Econ. Dyn. Control **37**, 1972–1981 (2013)

T.W. Wong, M.C. Chiu, H.Y. Wong, Time-consistent mean-variance hedging of longevity risk: effect of cointegration. Insur. Math. Econ. **56**, 56–67 (2014)

L. Yi, Z.F. Li, D. Li, Multi-period portfolio selection for asset-liability management with uncertain investment horizon. J. Ind. Manag. Optim. **4**, 535–552 (2008)

X.Y. Zhou, D. Li, Continuous time mean-variance portfolio selection: a stochastic LQ framework. Appl. Math. Optim. **42**, 19–33 (2000)

S.S. Zhu, M.J. Fan, D. Li, Portfolio management with robustness in both prediction and decision: a mixture model based learning approach. J. Econ. Dyn. Control **48**, 1–25 (2014)

S.S. Zhu, X.D. Ji, D. Li, A robust set-valued scenario approach for handling modeling risk in portfolio optimization. J. Comput. Financ. **19**, 11–40 (2015)

Chapter 7
A Review of Modern Cryptography: From the World War II Era to the Big-Data Era

Bojun Lu

Abstract This chapter briefly surveys the rapid development of *Modern Cryptography* from World War II (WW-II) to the prevailing Big-Data Era. Cryptography is the art and science of secret communication, which concerns about C.I.A., i.e., *Confidentiality*, *Integrity*, and *Authentication* of information, so as to guarantee the safety during information transmission. Meanwhile *Authentication* is the key step in information security, where an excellent example is online payment systems, which belongs to the field of Financial Technology (Fin-Tech) and is booming on multiple markets in recent years. The concept "Quantum" is popular in the recent decade, and the possibilities of inventing Quantum Cryptosystems are also raised in the literature, which is a promising direction in Modern Cryptosystem. We also select two classical cryptosystems, i.e., the Merkle–Hellman knapsack cryptosystem, and the subset sum problem (SSP)-based cryptosystem to present the mechanisms in encryption and decryption processes. Apart from being a brief survey, this chapter is also intended as an entry point to guide readers to this interesting and important field.

Keywords Big-Data • Cryptography • Cryptosystem • Information Security • Optimization • Financial Technology (Fin-Tech) • Quantitative Finance

7.1 Introduction

Cryptography is the art and science of secret communication (Singh 1999). In the recent decade, several brilliant research works in the field of *Modern Cryptography*, have successfully attracted the attention of *Turing Award*, which represents the highest honor to reward the achievements in the computing community, and is also stipulated that "The contributions should be of lasting and major technical importance to the computer field." For example, in 2015, winners are

B. Lu (✉)
Quantitative Researcher, Portfolio Management Department, Foresea Life Insurance Co., Ltd., Shenzhen, PRC
e-mail: bjlu.eva@foxmail.com

T.-M. Choi et al. (eds.), *Optimization and Control for Systems in the Big-Data Era*, International Series in Operations Research & Management Science 252,
DOI 10.1007/978-3-319-53518-0_7

Martin E. Hellman and Whitfield Diffie, who described and predicted the new directions of cryptography in their celebrated paper (Diffie and Hellman 1976) published in 1976, and the citation from Turing Award is shown as follows:

"*For fundamental contributions to* ***modern cryptography****. Diffie and Hellman's groundbreaking 1976 paper, 'New Directions in Cryptography' (Diffie and Hellman 1976) introduced the ideas of public-key cryptography and digital signatures, which are the foundation for most regularly-used security protocols on the internet today.*"

In 2002, winners are Ronald L. Rivest, Adi Shamir, and Leonard M. Adleman (please refer to Rivest et al. (1978) for their paper published in 1978), and the citation from Turing Award is: "*For their ingenious contribution for making* ***public-key cryptography*** *useful in practice.*" In 2000, winner is Andrew Chi-Chih Yao with citation from Turing Award as: "*In recognition of his fundamental contributions to the theory of computation, including the complexity-based theory of pseudorandom number generation,* ***cryptography****, and communication complexity.*"

There is no doubt about the importance of cryptography in its nature, and if we try to explain the importance of cryptography in more detail, we would like to emphasize that cryptography concerns about C.I.A., i.e., *Confidentiality*, *Integrity*, and *Authentication* of information, so as to guarantee the safety during information transmission. Please notice that the C.I.A. we defined in this review paper does not refer to the *Central Intelligence Agency* (CIA) of the United States, although the CIA of the United States does also have close relationship with highly confidential information.

If we try to seek the starting point of *Modern Cryptography Era*, we could trace back to the dates of World War II (WW-II), and several important and interesting questions could also be proposed, for instance,

1. *What* invention/technique invented/proposed by *whom* demonstrates that *Vintage Cryptography Era* begins to migrate to *Modern Cryptography Era*?
2. What event could be counted as the blasting fuse that boosts this migration?

To answer the first question, please let us use the electromechanical rotor based cipher system *Enigma Machine* invented by Arthur Scherbius at the end of World War I (WW-I), around 1918 [please refer to Jennifer (2006)], to be the representative invention/technique that represents the beginning of *Modern Cryptography Era*. Actually, before WW-II, mechanical and electromechanical cryptographic cipher machines were already in wide use, although almost all were *impractical manual systems*. Later, during WW-II, great advances on practical and theoretical cryptography were developed all in secrecy. Moreover, before and during WW-II, several models were developed based on Enigma Machine, and these models were specially adopted by military and government services of some countries, such as Germany, Japan, Russia, France, and Italy. during WW-II. In recent years, some of the WW-II cryptography related information has begun to be declassified, which partly owes to (1) the official 50-year (British) secrecy period has come to an end, (2) relevant US archives have been opened gradually, and (3) assorted memoirs and articles have been published, etc. Besides Enigma Machine, *Purple Machine* also deserves our attention, which was invented and improved by the Japanese during WW-II with

Fig. 7.1 Enigma in use 1943

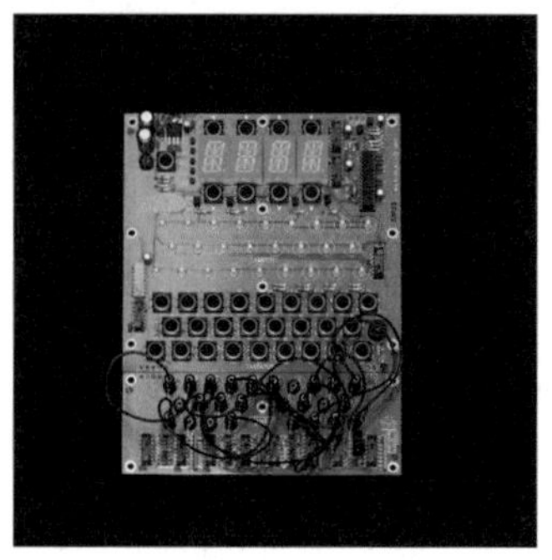

Fig. 7.2 Electronic implementation of an Enigma machine

inspiration from the mechanism of Enigma Machine used by Nazis; and which was used to transform the top level military secrets of Japanese Navy in the *Pearl Harbor War* (Figs. 7.1 and 7.2).

To answer the second question, one possible answer that we conjecture is that WW-II plays an important role as the blasting fuse that boosts the practical and theoretical development of modern cryptography. Meanwhile, because techniques become more mature, especially because the first computer has been invented around WW-II compared with scientific techniques in WW-I. All these enable cryptography to be used more widely in modern wars, for instance, in WW-II.

When we discuss cryptography, there are two angles of views, just like a coin has two sides, i.e., encryption technology and decryption technology. A good example is that by WW-II, there were unbreakable codes and then by the end, there was technology to break them. For example, Japan's Purple Machine was broken by US Army cryptographers (cryptologists) William F. Friedman, Frank Rowlett, and their subordinates in 1940, which enables America to hold a vantage position in the Pearl Harbor War during WW-II. It's worth to note that William F. Friedman is identified as the "Dean of American Cryptology" by the U.S. National Cryptologic Museum, and also the godfather of cryptology of the USA.

In the Big-Data Era now, cryptography continues to play an important role both in practical and theoretical aspects. Before listing the applications and emphasizing the importance of cryptography systems in Big-Data Era, we first briefly go through the evolvement of the concept of "big data." In 2001, Doug Laney, from META Group which is now re-named Gartner, defined big data in 3 dimensions, i.e., *Volume*, *Velocity*, and *Variety* with abbreviation "3Vs" [please see Gartner (2011)], which has been expanded to the following 5 dimensions in 2016 by Martin Hilbert (please refer to Hillbert (2015) and Wikipedia (https://en.wikipedia.org/wiki/Big_data) for more information):

- Volume: big data doesn't sample, but it observes and tracks what happens;
- Velocity: big data is often available in real time;
- Variety: big data draws from text, images, audio, video; plus it completes the missing pieces through data fusion;
- Machine Learning: big data often doesn't ask why and simply detects patterns;
- Digital footprint: big data is often a cost-free by-product of digital interaction.

Actually, based on our understanding, Big-Data Era precisely captures the trend of *information explosion*, since people interact so actively and share information so frequently through the internet, thus a huge amount of data is generated as "by-product." At the ACM Turing Centenary Celebration in 2012, Cerf et al. (2012) discussed the topic on "information, data, security in a networked future" to emphasize the importance of security of information and data in the modern real world. Then referring to our definition of C.I.A., in which the 3-dimensions depiction of information is provided, and as information is exploding, thus the importance of C.I.A. of information is manifested. Consequently, techniques and theories in cryptography, developed to protect information becomes even more vital nowadays. Real-world applications of cryptography can be evidenced in many industrial fields, such as modern financial systems, telecommunications, the newly emerging field called "Financial Technology (Fin-Tech)," etc.

In this paragraph, we would like to mention that the online payment systems is a good example to illustrate the crucial role that cryptography plays in modern finance with "big data." An online payment system called *WeChat Pay*, invented and run by Tencent Holdings, and another online payment system called *Alipay*, invented and run by Alibaba Group, are now the two biggest and most popular online payment platforms in Mainland China. Also as evidenced in the report from Credit Suisse, the online payment market grows rapidly, and the total value of online transactions in China grows from an insignificant size in 2008 to around RMB 4 trillion (US$660 billion) in 2012 (see Watling 2014). We also would like to mention that on February 18, 2016, the online payment platform called *Apply pay* developed by Apply Inc., lands in the market of Mainland China to show its interests in China's booming market. Meanwhile, Tencent Holdings and Alibaba Group also both have announced their plans to expand their mobile payment service to regions/countries outside Mainland China. To show that cryptography plays a crucial role in online payment systems, please notice that *Authentication*, i.e., the **A.** in C.I.A. is a crucial step during the completion of online payment, and to guarantee

Fig. 7.3 WeChat Pay v.s. Alipay

Fig. 7.4 Apply Pay

authentication of each party involved, digital signature or other private-key plus public-key crypto-techniques must be applied. For instances, the MD5 invented and designed by Ronald Rivest [see Rivest (1992) and Wang and Yu (2005)], and the SHA-1 (Secure Hash Algorithm 1) designed by National Institute of Standards and Technology (NIST) and National Security Agency (NSA) (see Wikipedia https://en.wikipedia.org/wiki/SHA-1) are two classical crypto-techniques that have been adopted in digital signatures for many years. Crypto-currencies known as Bitcoin with block-chain technique embedded is also an interesting case in modern finance which adopts modern cryptography as one of the key parts in its realization. Besides these Fin-Tech cases, actually we would like to say that modern cryptography is everywhere in our daily life now (Figs. 7.3 and 7.4).

When we talk about *Fin-Tech* which is a field booming in the recent years, other than online payment systems, we would also like to mention Fin-Tech companies that focus on quantitatively managing capitals for their customers, with Betterment (www.betterment.com) and Wealthfront (www.wealthfront.com) be the benchmarking enterprises (see http://fintechinnovators.com/). As we know in practice, Black–Litterman model (see Black and Litterman 1992) is a classical model adopted in the basket of quantitative strategies of these enterprises; and in academia, theoretically, Black–Litterman model is an extension of Markowitz's mean-variance model (see Markowitz 1952) which could blend information collected from real market to mend the weights on each asset and thus to improve the performance of portfolios. We would like very much to draw your attention to the brilliant research works that Professor Duan Li and his collaborators have done in the field of portfolio selection theory, and for details please refer to their papers (Zhu et al. 2014; Gao et al. 2015), etc.

The concept "Quantum" is popular in the recent decade, and the possibilities of inventing Quantum Cryptosystems are also raised in the literature, which is a promising direction in the field of Modern Cryptosystem (please see Okamoto et al. (2000) and the literature therein).

The remainder of this book chapter is organized as follows. Section 7.2.1 briefly describes the Merkle–Hellman knapsack cryptosystem, and Shamir's attack in 1984, where a hands-on numerical example is given for illustration. Section 7.2.2 presents the hardest subset sum problem based cryptosystem, and shows a decryption method which adopts the lattice theory and the distinguished LLL algorithm (see Lenstra et al. 1982). By the end, Sect. 7.3 includes the conclusion and further discussion.

7.2 Two Classical Cryptosystems

In Sects. 7.2.1 and 7.2.2, Merkle–Hellman knapsack cryptosystem and hardest subset sum problem (SSP)-based cryptosystem will be introduced which have both been well studied in the literature. (Note that in the literature, the SSP-based cryptosystem is also sometimes called the knapsack cryptosystem.) The Merkle–Hellman knapsack cryptosystem is one of the classical public-key knapsack cryptosystems, and is invented by Merkle and Hellman in 1978 (see Merkle and Hellman 1978) which has been broken by Shamir in 1984 (see Shamir 1984). Meanwhile, since the subset sum problem belongs to NP-class in its nature theoretically (see Garey and Johnson 1979), and it has been proven that the subset sum problem with a density approximately equals 1 is hardest (see Lagarias and Odlyzko 1985), it could be adopted to construct a trapdoor cryptosystem. Whereas in order to break the trapdoor cryptosystem, a *hard* subset sum problem must be solved.

7.2.1 The Merkle–Hellman Knapsack Cryptosystem

In 1978, Merkle and Hellman published their seminal paper (Merkle and Hellman 1978) which discovered a public-key cryptosystem. Although compared with an RSA cryptosystem (see Rivest et al. 1978) which is two-way system and can be adopted for *Authentication* in cryptographic signing, Merkle–Hellman cryptosystem is one-way, i.e., the public key is used only for encryption and the private key is used only for decryption. But Merkle–Hellman cryptosystem is the first so-called *knapsack cryptosystem*. In their paper, Merkle and Hellman proposed a *singly iterated cryptosystem* together with a *multiply iterated cryptosystem*. Later in 1984, Shamir (1984) found a polynomial time algorithm to break the singly iterated cryptosystem.

In Sect. 7.2.1.1, we present a description of the basic singly iterated knapsack cryptosystem proposed by Merkle and Hellman. In Sect. 7.2.1.2, Shamir's attack on the singly iterated knapsack cryptosystem is studied in detail.

7.2.1.1 Singly Iterated Merkle–Hellman Knapsack Cryptosystem

Suppose that the sender Bob wants to send a secret message to the receiver Anna, the message is represented as a binary vector $x = (x_1, x_2, \ldots, x_n) \in \{0,1\}^n$ in the binary system. The question is: How could Bob send this message to Anna in a secure way? In Merkle–Hellman cryptosystem, a strategy is designed so that Bob can send this message to Anna against the potential eavesdropper. This strategy is described as follows:

1. Anna chooses a positive superincreasing integer sequence $a = (a_1, a_2, \ldots, a_n)^T$. Superincreasing is in the sense that

$$a_i > \sum_{j=1}^{i-1} a_j, \quad i = 2, 3, \ldots, n.$$

2. Anna chooses two relatively prime integers m and w, such that

$$m > \sum_{j=1}^{n} a_j, \quad \text{and} \quad \gcd(m, w) = 1.$$

3. Sequence $c = (c_1, c_2, \ldots, c_n)^T$ is calculated as follows:

$$c_i = a_i w \mod m.$$

4. The *public key* is sequence $c = (c_1, c_2, \ldots, c_n)^T$.
5. The *private key* consists of an integer pair (w, m).

Now, if Bob wants to send message x to Anna, he sends the number d instead of sending x directly, where $d = c^T x$. Anna receives d, and conducts the following calculation:

1. Calculates b, where $b = dw^{-1} \mod m$, and w^{-1} is the *modular multiplicative inverse* of w modulo m.
2. Solves the equation $a^T x = b$, where $x \in \{0,1\}^n$. Then the solution x is the message Bob sent. Actually, since a is superincreasing, the equation $a^T x = b$ can be solved in linear time.

While Anna could get the message easily, the eavesdropper needs to solve the equation $c^T x = d$ in order to get the message, which is much harder.

One thing should be noticed is that, actually, $a_i = c_i w^{-1} \mod m, i = 1, 2, \ldots, n$.

7.2.1.2 Analysis of Shamir's Attack on Singly Iterated Knapsack Cryptosystem

Basic Deductions

The public key $c = (c_1, \ldots, c_n)^T$ is known for everyone, what Shamir wanted to do is to find a positive and relatively prime integer pair $(\tilde{w}, \tilde{m})$, such that $a = c\tilde{w} \mod \tilde{m}$ is super-increasing. Actually (w^{-1}, m) is such a qualified pair, where w^{-1} is the *modular multiplicative inverse* of w modulo m and (w, m) is the private key. Notice that there may be qualified integer pairs other than (w^{-1}, m).

Let $\tilde{w} = w^{-1}$ and $\tilde{m} = m$, we do the following analysis. The super-increasing sequence a chosen by Anna is

$$a_i = c_i w^{-1} \mod m, \quad i = 1, 2, \ldots, n.$$

Divide both sides by m, an equivalent equation is obtained as follows:

$$\begin{aligned} \frac{a_i}{m} &= c_i \frac{w^{-1}}{m} \mod 1 \\ &= c_i \frac{w^{-1}}{m} - \left\lfloor c_i \frac{w^{-1}}{m} \right\rfloor, \quad i = 1, 2, \ldots, n. \end{aligned} \tag{7.1}$$

Since $a_i = c_i w^{-1} \mod m$, $i = 1, 2, \ldots, n$, there must exist positive integer q_i's such that

$$a_i = c_i w^{-1} - q_i m, \quad i = 1, 2, \ldots, n.$$

Divide both sides by m, we get the following equation,

$$\frac{a_i}{m} = c_i \frac{w^{-1}}{m} - q_i, \quad i = 1, 2, \ldots, n. \tag{7.2}$$

Relate Eqs. (7.1) and (7.2), we see that

$$q_i = \left\lfloor c_i \frac{w^{-1}}{m} \right\rfloor, \quad i = 1, 2, \ldots, n.$$

Moreover, $\frac{q_i}{c_i}$ is the closest minimum of the c_i-curve to the left of $\frac{w^{-1}}{m}$ (Fig. 7.5).

Observe the c_i-curve, or from Eq. (7.2), we see that the distance between $\frac{w^{-1}}{m}$ and $\frac{q_i}{c_i}$ is

$$\frac{w^{-1}}{m} - \frac{q_i}{c_i} = \frac{a_i}{mc_i}, \quad i = 1, 2, \ldots, n. \tag{7.3}$$

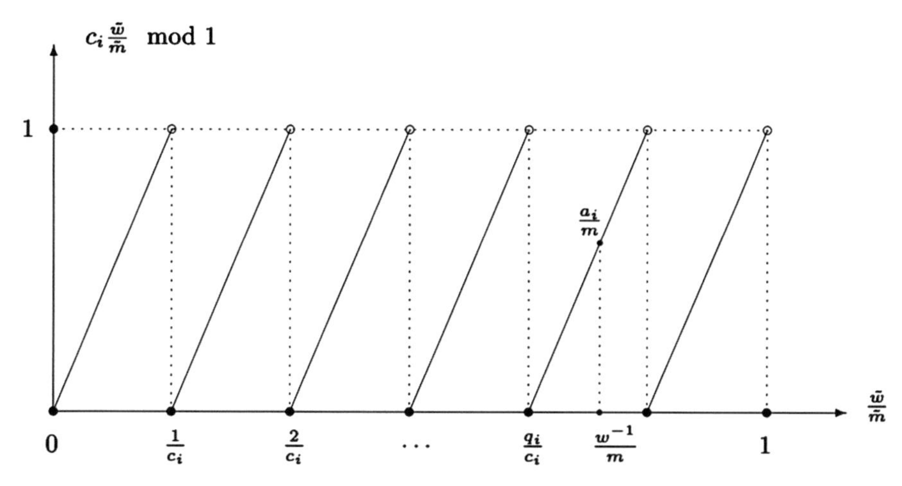

Fig. 7.5 c_i-curve: the relationship between $c_i \frac{\tilde{w}}{\tilde{m}}$ mod 1 and $\frac{\tilde{w}}{\tilde{m}}$

Then based on Eq. (7.3), we have

$$\frac{q_1}{c_1} - \frac{q_i}{c_i} = \frac{a_i}{mc_i} - \frac{a_1}{mc_1}, \quad i = 2, 3, \ldots, n, \tag{7.4}$$

and

$$q_1 c_i - q_i c_1 = \frac{a_i c_1}{m} - \frac{a_1 c_i}{m}, \quad i = 2, 3, \ldots, n. \tag{7.5}$$

How Many c_i-Curves Do We Need

According to Shamir's assumption, a_i is chosen to be a $dn - n + i - 1$ bit number, $i = 1, 2, \ldots, n$ and m is chosen to be a dn bit number. Here we just simply treat d as a parameter, and in Shamir's attack, $1 < d < 2$. (Actually d has much to do with the *density of a subset sum problem*, which will not be studied here. We just point out the relationship between them, which is: The lower of d the higher of the density of the subset sum problem, vice versa.)

Based on the assumption on the sizes of a_i and m, we choose a_i and m in the following way:

1. a_1 is a random integer number between 1 and 2^{dn-n}, with a uniform probability distribution.
2. a_i is a random integer number between $\sum_{j=1}^{i-1} a_j$ and $2^{dn-n+i-1}$, with a uniform probability distribution. Notice that there always has $\sum_{j=1}^{i-1} a_j < 2^{dn-n+i-1}$.
3. m is a random integer number between $\sum_{j=1}^{n} a_j$ and 2^{dn}, with a uniform probability distribution.

From Eq. (7.3), we have

$$\begin{aligned}\frac{w^{-1}}{m} - \frac{q_i}{c_i} = \frac{a_i}{mc_i} \\ &< \frac{2^{dn-n+i-1}}{mc_i} \\ &\approx \frac{2^{dn-n+i-1}}{2^{dn}c_i} \qquad (\because m \approx 2^{dn}) \\ &= \frac{2^{-n+i-1}}{c_i}.\end{aligned}$$

Hence,

$$\frac{q_i}{c_i} \in \left(\frac{w^{-1}}{m} - \frac{2^{-n+i-1}}{c_i}, \frac{w^{-1}}{m}\right), \quad i = 1, 2, \ldots, n.$$

For an arbitrary $\frac{\tilde{w}}{\tilde{m}}$, there must be a minimum of c_i-curve, such that the minimum belongs to the interval of,

$$\left(\frac{\tilde{w}}{\tilde{m}} - \frac{1}{c_i}, \frac{\tilde{w}}{\tilde{m}}\right).$$

Roughly, suppose that the minimum follows a uniform probability distribution in the above interval, then the probability that the minimum belongs to interval

$$\left(\frac{\tilde{w}}{\tilde{m}} - \frac{2^{-n+i-1}}{c_i}, \frac{\tilde{w}}{\tilde{m}}\right),$$

is

$$\frac{2^{-n+i-1}}{c_i} / \frac{1}{c_i} = 2^{-n+i-1}.$$

For an arbitrary c_1-curve's minimum $\frac{p}{c_1}$, choose $\frac{\tilde{w}}{\tilde{m}}$ and let it be in the following interval

$$\left(\frac{p}{c_1}, \frac{p}{c_1} + \frac{2^{-n}}{c_1}\right).$$

Suppose other $c_2, \ldots, c_l$-curves are chosen, then for the $\frac{\tilde{w}}{\tilde{m}}$, the probability that there exists one c_i-curve's minimum which belongs to the following interval

$$\left(\frac{\tilde{w}}{\tilde{m}} - \frac{2^{-n+i-1}}{c_i}, \frac{\tilde{w}}{\tilde{m}}\right),$$

at the same time for $i = 2, \ldots, l$ is

$$2^{-n+1} \cdot 2^{-n+2} \cdot \cdots \cdot 2^{-n+l-1} = 2^{l^2/2-nl-l/2+n}.$$

Let p run from 0 to $c_1 - 1$, then the expected number of $\frac{p}{c_1}$'s which satisfies the above condition is

$$\begin{aligned} c_1 \cdot 2^{l^2/2-nl-l/2+n} &= \alpha_1 \cdot m \cdot 2^{l^2/2-nl-l/2+n} \\ &\approx \alpha_1 \cdot 2^{dn+l^2/2-nl-l/2+n}, \end{aligned}$$

where $0 < \alpha_1 < 1$. When $2^{dn+l^2/2-nl-l/2+n} < 1$, we have $\alpha_1 \cdot 2^{dn+l^2/2-nl-l/2+n} < 1$. Simple mathematical deduction yields

$$2^{dn+l^2/2-nl-l/2+n} < 1,$$

which is equivalent to

$$l \in \left(n + \frac{1}{2} - \sqrt{n(n-1-2d) + \frac{1}{4}},\ n + \frac{1}{2} + \sqrt{n(n-1-2d) + \frac{1}{4}}\right).$$

Since $l \le n$ and $n + \frac{1}{2} + \sqrt{n(n-1-2d) + \frac{1}{4}} > n$, we have

$$l \in \left(n + \frac{1}{2} - \sqrt{n(n-1-2d) + \frac{1}{4}},\ n\right].$$

It can be checked that $n + \frac{1}{2} - \sqrt{n(n-1-2d) + \frac{1}{4}}$ is a convex and decreasing function with respect to n. When $n = 10$ and $d = 2$, $n + \frac{1}{2} - \sqrt{n(n-1-2d) + \frac{1}{4}} = 3.4113$. In this sense, 4 or 5 c_i-curves are enough for the analysis.

An Illustrative Example

Next we illustrate Shamir's attack based on the following concrete example.

Example 1 We generate a super-increasing sequence $a = (a_1, \ldots, a_n)$, $n = 10$,

$$a = (42, 64, 115, 263, 545, 1083, 2122, 4278, 8555, 17100)$$

where $a_i < 2^{dn-n+i-1}$, $i = 1, 2, \ldots, n$.

$m = 29193006$ is chosen such that $\sum_{i=1}^{n} a_i < m < 2^{dn}$.
$w = 11198095$ is randomly chosen such that $\gcd(w, m) = 1$.

c is calculated as follows,

$$\begin{aligned} c &= aw \mod m \\ &= (3231894, 16045936, 3288661, 25798385, 1623521, \\ &\quad 12439395, 28443712, 28920570, 17450039, 10498146). \end{aligned}$$

We have $w^{-1} = 1152457$, and

$$c/m = (0.1107, 0.5497, 0.1127, 0.8837, 0.0556, 0.4261, 0.9743, 0.9907, 0.5977, 0.3596).$$

□

7.2.2 Hardest Subset Sum Problem (SSP)-Based Cryptosystem

A subset sum problem is defined as follows:

$$ax^T = a_1x_1 + a_2x_2 + \cdots + a_nx_n = b, \tag{7.6}$$

with $a = (a_1, a_2, \ldots, a_n) \in \mathbb{R}_+^n$, $b \in \mathbb{R}_+^n$ be known, and $x = (x_1, x_2, \ldots, x_n) \in \{0, 1\}^n$ be unknown.

The concept *density* of a subset sum problem is defined as

$$density = \frac{n}{\max_{1 \le i \le n}(\log_2 a_i)}. \tag{7.7}$$

It has been revealed in the literature that subset sum problems in (7.6) with their density close to 1 constitute the hardest subclass of subset sum problems [see Lagarias and Odlyzko (1985), Coster et al. (1991) and Schnorr and Shevchenko (2012)]. Besides the *density* defined in (7.7), some other factors have also been proposed in the literature to describe the difficulty level of subset sum problems [see Jen et al. (2012b) and Jen et al. (2012a)].

Next we will review two decryption methods [see Lagarias and Odlyzko (1985), Coster et al. (1992), Schnorr and Euchner (1994)] which are designed based on the lattice theory and the distinguished LLL algorithm [see Lenstra et al. (1982), Nguyen and Vallée (2010)]. Lagarias and Odlyzko (1985) claimed that they could break "almost all" problems with a density < 0.645, and Coster et al. (1992) claimed that they could break "almost all" problems with a density < 0.941. It is worth mentioning that in Lu and Li's working paper (see Lu and Li 2016), and Lu's Ph.D. thesis (see Lu 2014), an algorithm that combines *disaggregation techniques* and *LLL algorithm* could break "almost all" problems with a density ≈ 1, compared with Lagarias and Odlyzko (1985) and Coster et al. (1992), for problems of the same dimension.

Here we spend a concise paragraph to elaborate the initial intuition of *disaggregation techniques* related work proposed in Lu's Ph.D. thesis (see Lu 2014) which aims to propose efficient algorithms equipped with *disaggregation techniques* together with *LLL algorithm*, for solving the following problem, i.e., a system of linear Diophantine equations:

$$Ax = b, \quad \text{with } x \text{ be unknown integer vectors and be bounded,} \tag{7.8}$$

which belongs to NP-class and where subset sum problems are special cases of Problem (7.8). The intuitions which stimulate us to conduct research work on *disaggregation techniques* are:

1. We are inspired by the time complexity of the cell enumeration method proposed by Prof. Duan Li and et al. in Li et al. (2011), which is bounded by $O((n \max\{u_1, \ldots, u_n\})^{n-m})$ and thus depends on the magnitude of $n - m$, where n is the number of unknown variables, m is the number of equations in the system $Ax = b$, and $(u_1, \ldots, u_n)$ are the upper bounds of the unknown variables. Obviously, reducing the magnitude of $n - m$ directly benefits us in the computing. Aiming to reduce the magnitude of $n - m$, we thus study possible solution schemes for disaggregation.
2. Glover and Woolsey formulated for the first time the inverse problem of aggregation, i.e., the *disaggregation problem*, in their paper (Glover and Woolsey 1972) in 1972. After presenting rich work on *aggregation* in Glover and Woolsey (1972), in their conclusion remarks, they strongly encouraged research on *disaggregation*: "The development of effective ways to do this (disaggregation) would be especially worthwhile." However, although Glover and Woolsey proposed this disaggregation problem, they actually didn't provide available and effective techniques to handle this problem, as evidenced by a sentence in their conclusion remarks in Glover and Woolsey (1972), "The theorems of this paper ..., but do not give an immediate clue about what multiples should be examined to effect the disaggregation." Though disaggregation problem is of importance, we discovered that the literature on proposing solutions to disaggregation problem is pretty limited. This fact encouraged us to study possible solution schemes for disaggregation.

For details of our research work on *disaggregation techniques*, please refer to Chap. 4 of Lu (2014).

Next we continue to spend our efforts to explain the two algorithms proposed by Lagarias and Odlyzko (1985) and Coster et al. (1992), respectively.

The $(n + 1) \times (n + 1)$ lattice proposed by Lagarias and Odlyzko (1985) is of the following form:

$$B_{LO} = \begin{pmatrix} I & \mathbf{0}^{n\times 1} \\ -a & b \end{pmatrix}. \tag{7.9}$$

We denote the column-wise LLL reduced matrix of B_{LO} by $\tilde{B}_{LO}$. The algorithm checks whether *any* column of $\tilde{B}_{LO}$ has the form of $\tilde{b}_{i,j} \in \{0, \lambda\}$, $i = 1, 2, \ldots, n$, for some fixed value λ and $\tilde{b}_{n+1,j} = 0$, where $1 \leq j \leq n+1$. If it fails, the algorithm repeats with b replaced by $\sum_{i=1}^{n} a_i - b$. If such a column appears, then we divide $\tilde{b}_{i,j} \in \{0, \lambda\}$, $i = 1, 2, \ldots, n$ by λ, and check whether the binary vector is a solution. We denote the method proposed in Lagarias and Odlyzko (1985) as **LO-Alg**. An analysis for **LO-Alg** method is presented in Frieze (1986) in 1986.

The $(n+1) \times (n+1)$ lattice proposed by Coster et al. (1992) is of the following form:

$$B_{CJOS} = \begin{pmatrix} I & \frac{1}{2} \times \mathbf{1}^{n \times 1} \\ aN & bN \end{pmatrix}, \tag{7.10}$$

with $N > \frac{1}{2}\sqrt{n}$. We denote the column-wise LLL reduced matrix of B_{CJOS} by $\tilde{B}_{CJOS}$. The algorithm checks whether *any* column of $\tilde{B}_{CJOS}$ has the form of $\tilde{b}_{i,j} \in \{-\frac{1}{2}, \frac{1}{2}\}$, $i = 1, 2, \ldots, n$, and $\tilde{b}_{n+1,j} = 0$, where $1 \leq j \leq n+1$. If yes, then we add back $\frac{1}{2}$ to $\tilde{b}_{i,j} \in \{-\frac{1}{2}, \frac{1}{2}\}$, $i = 1, 2, \ldots, n$, and check whether the binary vector is a solution. We denote the method proposed in Coster et al. (1992) as **CJOS-Alg**.

7.2.2.1 Review of the LLL Algorithm

In this part, we briefly introduce the mechanics behind LLL basis reduction algorithm and show how it works. We abbreviate LLL basis reduction algorithm to *LLL algorithm* and name the basis obtained by LLL algorithm as the *LLL-reduced basis*. Algebraically speaking, to obtain the LLL-reduced basis, a series of unimodular row operations need to be conducted on one ordered basis. Geometrically speaking, vectors in an LLL-reduced basis are relatively short and nearly orthogonal to one another. LLL algorithm has been proved to be very powerful as evidenced by its remarkable achievements in both theoretical advancement and successful applications, which is also an algorithm of polynomial time and arithmetic operation steps [see Sect. 4.3 of Bremner (2011)]. In theory, Lenstra (1983) proved that integer programming with a fixed dimension is polynomially solvable with the aid of the lattice basis reduction algorithm. In applications, many efficient algorithms have been developed in the last 30 years with LLL algorithm being their essential parts, including numerous cryptography-purpose algorithms for breaking knapsack public-key cryptosystems. By adopting LLL-based algorithms, e.g., the generalized LLL and the BKZ process (Lovász and Scarf 1992; Schnorr and Euchner 1994), efficient algorithms are designed in Brickell (1983), Lagarias and Odlyzko (1983), Lagarias and Odlyzko (1985), Coster et al. (1991), Coster et al. (1992), Schnorr and Euchner (1991) for breaking low density knapsack public-key cryptosystems. Among them, the algorithms in Lagarias and Odlyzko (1985) and Coster et al. (1992) represent two cornerstones of the development.

Definition 1 (Lattice) Given row vectors $b_1, b_2, \ldots, b_m \in \mathbb{R}^n$ with $m \leq n$. The set L defined as below

$$L = \mathbb{Z}b_1 + \mathbb{Z}b_2 + \cdots + \mathbb{Z}b_m = \left\{ \sum_{i=1}^{m} z_i b_i \mid z_i \in \mathbb{Z},\ i = 1, 2, \ldots, m \right\},$$

is called a lattice of dimension m. Moreover, $\{b_1, b_2, \ldots, b_m\}$ is called a basis for lattice L.

Theorem 1 *Given a lattice L, row vectors of B and row vectors of $\tilde{B}$ are two bases for L, if and only if there exists a unimodular matrix U, such that $B = U\tilde{B}$.*

Lemma 1 *If $\{b_1, b_2, \ldots, b_n\}$ is an α-reduced basis of the lattice $\Lambda \in \mathbb{R}^{\tilde{n}}$ with $\tilde{n} \geq n$, and $y_1, y_2, \ldots, y_t \in \Lambda$ are any t linearly independent lattice vectors, then for $1 \leq j \leq t$ we have*

$$||b_j||^2 \leq \beta^{n-1} \max\{||y_1||^2, ||y_2||^2, \ldots, ||y_t||^2\}.$$

The major steps of the LLL algorithm can be described as follows (Fig. 7.6).

- First, we conduct the Gram–Schmidt Orthogonalization (GSO) process on the input basis b_i, $i = 1, 2, \ldots, m$,

$$\begin{aligned}
b_1^* &= b_1, \\
b_2^* &= b_2 - \mu_{2,1} b_1^*, \quad \mu_{2,1} = \tfrac{b_2 \cdot b_1^*}{b_1^* \cdot b_1^*} \\
&\ldots \\
b_i^* &= b_i - \mu_{i,i-1} b_{i-1}^* - \mu_{i,i-2} b_{i-2}^* - \cdots - \mu_{i,1} b_1^*, \quad \mu_{i,j} = \tfrac{b_i \cdot b_j^*}{b_j^* \cdot b_j^*},\ 1 \leq j < i, \\
&\ldots \\
b_m^* &= b_m - \mu_{m,m-1} b_{m-1}^* - \mu_{m,m-2} b_{m-2}^* - \cdots - \mu_{m,1} b_1^*.
\end{aligned}$$

- Second, we conduct the following two main operations on basis vectors $b_1, \ldots, b_m$, which are called "Reduce" and "Exchange," respectively,
  - (Reduce) If $|\mu_{i,j}| > \frac{1}{2}$, then $b_i \leftarrow b_i - \lceil \mu_{i,j} \rfloor b_j$,
  - (Exchange) If $||b_i^* + \mu_{i,i-1} b_{i-1}^*||^2 < \alpha ||b_{i-1}^*||^2$, then exchange b_i and b_{i-1}, where $\frac{1}{4} < \alpha < 1$ is a parameter with the pre-given value.

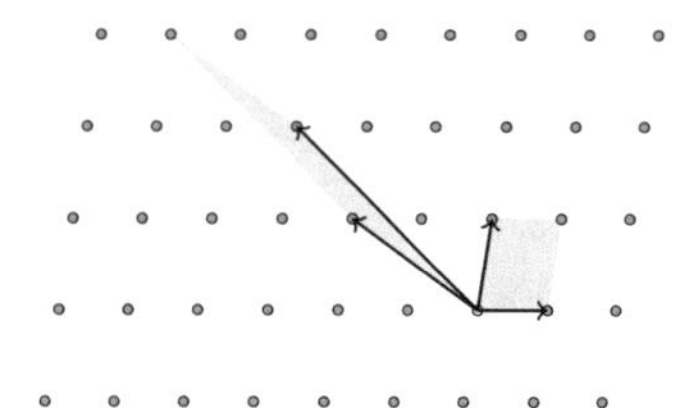

Fig. 7.6 Illustration of the LLL-reduced basis

As for the output, the LLL algorithm returns an α-reduced basis which satisfies the following conditions,

- $|\mu_{i,j}| \leq \frac{1}{2}$, $1 \leq j < i \leq m$,
- $||b_i^* + \mu_{i,i-1} b_{i-1}^*||^2 \geq \alpha ||b_{i-1}^*||^2$, $1 < i \leq m$.

The pseudocode for the LLL algorithm is presented in Algorithm 5 in Lu (2014). For a more detailed description of the LLL basis reduction algorithm, please refer to Chap. 4 of Bremner's book (Bremner 2011). As a remark, the book edited by Nguyen and Vallée (2010) is a more advanced introduction and survey for the theory and applications of the LLL basis reduction algorithm.

7.2.2.2 Illustrative Examples

Next we present a hands-on numerical example to illustrate how algorithms **LO-Alg** and **CJOS-Alg** work.

Example 2 Let's consider the following subset sum problem with $n = 3$,

$$3x_1 + 5x_2 + 7x_3 = 8,$$

where $x = (x_1, x_2, x_3) \in \{0, 1\}^3$, and density $= \frac{3}{\log_2 7} = 1.0686$.
The B_{LO} matrix defined in (7.9) is as follows:

$$B_{LO} = \begin{pmatrix} I & \mathbf{0}^{n \times 1} \\ -a & b \end{pmatrix} = \begin{pmatrix} 1 & 0 & 0 & 0 \\ 0 & 1 & 0 & 0 \\ 0 & 0 & 1 & 0 \\ -3 & -5 & -7 & 8 \end{pmatrix}.$$

Conducting GSO process yields the following decomposition,

$$\begin{pmatrix} d_1 \\ d_2 \\ d_3 \\ d_4 \end{pmatrix} = \begin{pmatrix} 1 & 0 & 0 & 0 \\ \mu_{2,1} & 1 & 0 & 0 \\ \mu_{3,1} & \mu_{3,2} & 1 & 0 \\ \mu_{4,1} & \mu_{4,2} & \mu_{4,3} & 1 \end{pmatrix} \begin{pmatrix} d_1^* \\ d_2^* \\ d_3^* \\ d_4^* \end{pmatrix}$$

$$= \begin{pmatrix} 1 & 0 & 0 & 0 \\ 1.5 & 1 & 0 & 0 \\ 2.1 & 1.9310 & 1 & 0 \\ -2.4 & -2.2069 & 1.6467 & 1 \end{pmatrix} \begin{pmatrix} d_1^* \\ d_2^* \\ d_3^* \\ d_4^* \end{pmatrix},$$

where d_i^T with $i = 1, 2, 3, 4$ is the ith column of matrix B_{LO}.
Setting $\alpha = 3/4$ in LLL algorithm yields the column-wise LLL reduced matrix as follows:

$$\tilde{B}_{LO} = \begin{pmatrix} 1 & 0 & -2 & -1 \\ 1 & 0 & 1 & 1 \\ 0 & 1 & 0 & 1 \\ 0 & 1 & 1 & -1 \end{pmatrix}.$$

where we could identify the following binary solution from the first column of matrix $\tilde{B}_{LO}$,

$$(x_1, x_2, x_3) = (1, 1, 0).$$

Remarks Recall that the **LO-Alg** algorithm identifies the binary solution in the following way,

"The algorithm checks whether any column of $\tilde{B}_{LO}$ has the form of $\tilde{b}_{i,j} \in \{0, \lambda\}$, $i = 1, 2, \ldots, n$, for some fixed value λ and $\tilde{b}_{n+1,j} = 0$, where $1 \leq j \leq n+1$. If it fails, the algorithm repeats with b replaced by $\sum_{i=1}^{n} a_i - b$. If such a column appears, then we divide $\tilde{b}_{i,j} \in \{0, \lambda\}$, $i = 1, 2, \ldots, n$ by λ, and check whether the binary vector is a solution. " □

Example 3 Let's reconsider the problem in Example 2. The B_{CJOS} matrix defined in (7.10) is as follows:

$$B_{CJOS} = \begin{pmatrix} I & \frac{1}{2} \times \mathbf{1}^{n\times 1} \\ aN & bN \end{pmatrix}.$$

We substitute the values of a and b into B_{CJOS}, choose $N = 10^2$, and set $\alpha = 3/4$ in LLL algorithm, then calculate the column-wise LLL reduced matrix as follows:

$$\tilde{B}_{CJOS} = \begin{pmatrix} -\frac{1}{2} & -\frac{1}{2} & \frac{1}{2} & 0 \\ \frac{1}{2} & -\frac{1}{2} & \frac{1}{2} & 0 \\ -\frac{1}{2} & -\frac{1}{2} & -\frac{1}{2} & 0 \\ 0 & 0 & 0 & -N \end{pmatrix}.$$

where we could identify the following binary solution from the third column of matrix $\tilde{B}_{CJOS}$,

$$(x_1, x_2, x_3) = (1, 1, 0).$$

Note Although the first and second column of $\tilde{B}_{CJOS}$ also satisfy the following condition,

" $\tilde{b}_{i,j} \in \{-\frac{1}{2}, \frac{1}{2}\}$, $i = 1, 2, \ldots, n$, and $\tilde{b}_{n+1,j} = 0$."
But after checking we discover that when we add back $\frac{1}{2}$, $(0, 1, 0)$ and $(0, 0, 0)$ are not binary solutions to the original problem.

Remarks Recall that the **CJOS-Alg** algorithm identifies the binary solution in the following way,

"The algorithm checks whether any column of $\tilde{B}_{CJOS}$ has the form of $\tilde{b}_{i,j} \in \{-\frac{1}{2}, \frac{1}{2}\}$, $i = 1, 2, \ldots, n$, and $\tilde{b}_{n+1,j} = 0$, where $1 \leq j \leq n + 1$. If yes, then we add back $\frac{1}{2}$ to $\tilde{b}_{i,j} \in \{-\frac{1}{2}, \frac{1}{2}\}$, $i = 1, 2, \ldots, n$, and check whether the binary vector is a solution."

□

7.3 Conclusion and Further Discussion

In this book chapter, we first go through, in introduction part, the development of Modern Cryptography from the era of World War II, to the prevailing Big Data Era now. The invention of "computer" empowers human computing ability, and together with wars between developed countries around 1930s, boost the development of theory and techniques of *Modern* Cryptography. Nowadays, applications of cryptography can be found everywhere in our daily life and multiple channels of industrial businesses, such as applications in Financial Technology (Fin-Tech), and Electric Power Industry, etc. We also use the "Authentication" step in online payment systems as an illustrative case to demonstrate the importance of cryptosystems in the newly emerging field *Fin-Tech*.

Later in Sect. 7.2, we review the classical knapsack cryptosystem designed by Merkle and Hellman in their seminal paper published in 1978 (Merkle and Hellman 1978) and also study the decryption technique proposed by Shamir (1984) in 1984. It's worth mentioning that Hellman is one of the winners of *Turing Award* in 2015 for his brilliant work together with Diffie in 1976 (Diffie and Hellman 1976). Besides, we also review and present the lattice theory based decryption technique proposed (Lagarias and Odlyzko 1985; Coster et al. 1992) to break the hardest subset sum problem (SSP)-based cryptosystem.

Cryptography existing as a science and art of secrecy communication has developed from Vintage Cryptography Era and Caesar's code adopted in *Gallic Wars* being one typical representative, to Enigma and Purple machines used in modern war, i.e., WW-II, to the MD5 and the SHA-1 techniques used in modern internet communication nowadays. We conjecture that one promising future development direction of cryptography theory and technique could be the quantum cryptosystems, equipped with the rapid development of quantum theory and quantum computer, which involves some notions from quantum mechanics explaining how objects behave at the microscopic level, and in the presence of a massive amount of "big data."

Last but not least, we would like to emphasize that this book chapter only serves as a modest spur to induce more valuable discussions, and is a starting point for readers to delve deeper into this promising field. We would like to thank all readers for their patience to go through this book chapter, and we would be more than happy to know that readers also find and believe modern cryptography is an interesting field with huge importance in the prevailing Big-Data Era.

Acknowledgements We would like to express our gratitude to Prof. Duan Li for sharing his comments and suggestions on this paper. We also would like to thank Dr. Junxian HUANG, the CEO of BeeCloud CO., Ltd. for sharing his knowledge on online payment systems, and Dr. Don HUANG for sharing his knowledge on Black–Litterman model.

References

F. Black, R. Litterman, Global portfolio optimization. Financ. Anal. J. **48**(5), 28–43 (1992)

M.R. Bremner, *Lattice Basis Reduction: An Introduction to the LLL Algorithm and Its Applications* (CRC, Boca Raton, FL, 2011)

E.F. Brickell, Solving low density knapsacks, in *Advances in Cryptology, Proceedings of CRYPTO '83, Santa Barbara, CA, August 21–24, 1983*, ed. by D. Chaum (Plenum, New York, 1983), pp. 25–37

V. Cerf, J.E. Hopcroft, R.E. Kahn, R.L. Rivest, A. Shamir, Information, data, security in a networked future, in *ACM-TURING'12 ACM Turing Centenary Celebration, San Francisco, CA, June 15–16*, no. 14 (2012)

M.J. Coster, B.A. LaMacchia, A.M. Odlyzko, C.P. Schnorr, An improved low-density subset sum algorithm, in *Advances in Cryptology: Proceedings of Eurocrypt '91* (1991), pp. 54–67

M.J. Coster, A. Joux, B.A. LaMacchia, A.M. Odlyzko, C.P. Schnorr, J. Stern, Improved low-density subset sum algorithms. Comput. Complex. **2**, 111–128 (1992)

W. Diffie, M.E. Hellman, New directions in cryptography. IEEE Trans. Inf. Theory **IT-22**, 644–654 (1976)

A.M. Frieze, On the Lagarias-Odlyzko algorithm for the subset sum problem. SIAM J. Comput. **15**, 536–539 (1986)

J. Gao, D. Li, X. Cui, S. Wang, Time cardinality constrained mean-variance dynamic portfolio selection and market timing: a stochastic control approach. Automatica **54**, 91–99 (2015)

M.R. Garey, D.S. Johnson, *Computers and Intractability: A Guide to the Theory of NP-Completeness* (W. H. Freeman, San Francisco, 1979)

Gartner, Gartner says solving 'big data' challenge involves more than just managing volumes of data. Gartner Special Report Examines How to Leverage Pattern-Based Strategy to Gain Value in Big Data (June 27, 2011)

F. Glover, R.E. Woolsey, Aggregating diophantine equations. Z. Oper. Res. **16**, 1–10 (1972)

M. Hillbert, Big data for development: a review of promises and challenges. Dev. Policy Rev. **34**, 1–41 (2015)

S.M. Jen, C.Y. Lu, T.L. Lai, J.F. Yang, Empirical exploration of lattice attacks for building secure knapsack cryptosystems, in *2012 International Conference on Anti-Counterfeiting, Security and Identification* (IEEE, New York, 2012a), pp. 1–5

S.M. Jen, T.L. Lai, C.Y. Lu, J.F. Yang, Knapsack cryptosystems and unreliable reliance on density, in *AINA*, ed. by L.T. Barolli, F. Enokido, Xhafa, M. Takizawa (IEEE, New York, 2012b), pp. 748–754

W. Jennifer, *Solving the Enigma: History of the Cryptanalytic Bombe* (Center for Cryptological History, National Security Agency, Fort George G. Meade, 2006)

J.C. Lagarias, A.M. Odlyzko, Solving low-density subset sum problems, in *24th Annual Symposium on Foundations of Computer Science, 7–9 November 1983, Tucson, AZ* (IEEE, New York, 1983), pp. 1–10

J.C. Lagarias, A.M. Odlyzko, Solving low-density subset sum problems. J. Assoc. Comput. Mach. **32**, 229–246 (1985)

H.W. Lenstra, Integer programming with a fixed number of variables. Math. Oper. Res. **8**, 538–548 (1983)

A.K. Lenstra, H.W. Lenstra Jr., L. Lovász, Factoring polynomials with rational coefficients. Math. Ann. **261**, 515–534 (1982)

D. Li, X. Sun, J. Gao, S. Gu, X. Zheng, Reachability determination in acyclic Petri nets by cell enumeration approach. Automatica **47**, 2094–2098 (2011)

L. Lovász, H.E. Scarf, The generalized basis reduction algorithm. Math. Oper. Res. **17**, 751–764 (1992)

B. Lu, Linear diophantine equations: integration of disaggregation with LLL algorithm. Ph.D. Thesis, The Chinese University of Hong Kong (2014)

B. Lu, D. Li, Tackling density one subset sum problems: integration of disaggregation technique with lattice attacks. Working paper (2016)

H. Markowitz, Portfolio selction. J. Financ. **7**(1), 77–91 (1952)

R.C. Merkle, M.E. Hellman, Hiding information and signatures in trapdoor knapsacks. IEEE Trans. Inf. Theory **IT-24**, 525–530 (1978)

P.Q. Nguyen, B. Vallée (eds.) *The LLL Algorithm: Survey and Applications*. Information Security and Cryptography. Springer, Berlin, Heidelberg (2010)

T. Okamoto, K. Tanaka, S. Uchiyama, Quantum public-key cryptosystems, in *Advances in Cryptology - CRYPTO 2000, 20th Annual International Cryptology Conference, Santa Barbara, CA, August 20–24, 2000, Proceedings*, ed. by M. Bellare. Lecture Notes in Computer Science, vol. 1880 (Springer, Berlin, 2000), pp. 147–165

R. Rivest, *The MD5 Message-Digest Algorithm* (MIT Laboratory for Computer Science and RSA Data Security, Cambridge, MA, 1992)

R.L. Rivest, A. Shamir, L.M. Adleman, A method for obtaining digital signatures and public-key cryptosystems. Commun. ACM **21**(2), 120–126 (1978)

C.P. Schnorr, M. Euchner, Lattice basis reduction: improved practical algorithms and solving subset sum problems, in *Fundamentals of Computation Theory, 8th International Symposium, FCT 91, Gosen, September 9–13, 1991, Proceedings*, ed. by L. Budach. Lecture Notes in Computer Science, vol. 529 (Springer, Berlin, 1991), pp. 68–85

C.P. Schnorr, M. Euchner, Lattice basis reduction: improved practical algorithms and solving subset sum problems. Math. Program. **66**, 181–191 (1994)

C.P. Schnorr, T. Shevchenko, Solving subset sum problems of density close to 1 by "randomized" BKZ-reduction. IACR Cryptol. ePrint Archive, 1–5 (2012)

A. Shamir, A polynomial-time algorithm for breaking the basic Merkle-Hellman cryptosystem. IEEE Trans. Inf. Theory **IT-30**, 699–704 (1984)

S. Singh, *The Code Book: The Science of Secrecy From Ancient Egypt to Quantum Cryptography* (Anchor, New York, 1999)

X. Wang, H. Yu, How to break MD5 and other hash functions, in *Advances in Cryptology - EUROCRYPT 2005*. Lecture Notes in Computer Science, vol. 3494 (Springer, Berlin, 2005), pp. 19–35

J. Watling, China's internet giants lead in online finance. The Financialist (Retrieved 15 February 2014), Credit Suisse

S. Zhu, M. Fan, D. Li, Portfolio management with robustness in both prediction and decision: a mixture model based learning approach. J. Econ. Dyn. Control **48**, 1–25 (2014)

Chapter 8
Modeling Supply Risk in the New Business Era: Supply Chain Competition and Cooperation

Xiang Li, Yongjian Li, and Linghua Zhao

Abstract In the current globalized supply chains, firms are more likely to suffer from supply risks caused by various sources, including internal production default and external disasters. This chapter focuses on the operational management problem related to the supply risk within the supply chain scope. We introduce a number of recent and important research developments, including problems in vertical supply chain interaction, horizontal supply chain competition, and supply chain network with both horizontal and vertical competitions. Analytical models are presented for each problem, and the main results are elucidated. Moreover, further research directions along with big data trends are emphasized as well.

Keywords Supply risk • Supply chain model • Competition and cooperation • Reliability improvement

8.1 Introduction

Owing to the rapid information technology development and increasingly intense global competition, the traditional perspective on firm operation management has given way to a new paradigm of supply chain management in consideration of the close multi-firm relations and interactions in the modern market place. Along with this trend, the world has become increasingly variant with inherent and exogenous uncertainties. Among them, supply uncertainty has become a major concern in global supply chain management. In traditional manufacturing processes, stochastic capacity, random yield, and uncertain transportation delay are the main causes of supply uncertainty. Unexpected disruption is another type of uncertainty that

---

X. Li • L. Zhao
College of Economic and Social Development, Nankai University, Tianjin 300071, People's Republic of China
e-mail: xiangli@nankai.edu.cn; sadallday@sina.com

Y. Li (✉)
Business School, Nankai University, Tianjin 300071, People's Republic of China
e-mail: liyongjian@nankai.edu.cn

T.-M. Choi et al. (eds.), *Optimization and Control for Systems in the Big-Data Era*, International Series in Operations Research & Management Science 252,
DOI 10.1007/978-3-319-53518-0_8

commonly leads to a total supply default, which severely damages several supply chain operations. A well-known industrial example is Ericsson losing 400 million Euros after a fire on the semiconductor plant of their supplier in 2000, as well as Apple losing numerous customer orders during a supply shortage of DRAM chips after an earthquake hit Taiwan in 1999. A more recent incident occurred during the March 2011 earthquake in Japan triggering a massive 23-foot tsunami and a nuclear crisis, further leading to a global supply disruption. An industrial survey conducted by Protiviti and the American Production and Inventory Control Society (APICS) showed that 66% of the respondents considered supply uncertainty as one of their most significant concern among all supply chain-related risks (O'Keeffe 2006).

Conversely, the generated supply chain field data contains highly rich information caused by technological changes. Statistics and forecasts have long been recognized as useful tools in supply chain risk analysis, as well as in the corresponding decision making (Choi et al. 2003, 2006, 2008a); whereas other powerful methodologies are being developed in this area as technologies, such as data mining, machine learning, and cloud computing, are updated (Lee et al. 2014; Fan et al. 2015). Supply chain risk management can particularly, largely benefit from new data technologies and analytic methods for collecting, analyzing, and monitoring both supply chain internal and environmental data. The increasing complexity calls for increasing the attention paid to data processing and analysis, as well as in the development of new optimization models to analyze supply chain competition and cooperation and to especially enhance the robustness of the supply chain in the presence of supply risk. Such task can be realized by sufficiently using the data derived from these advanced information programs and systems.

To achieve this goal, many researchers have developed optimization models that aim at mitigating the respective supply uncertainty and the associated risk. This chapter focuses on the related supply chain problems in the complex business environment of firm competition and cooperation to provide recent research models and results. The selected papers are not comprehensive; however, they are typical and representative to instigate contemplation and illuminate future study directions. The following studies are incorporated:

- We consider three streams of research from the supply chain structure perspective.
  - The first stream is on the vertical supply chain interaction modeled by a Stackelberg leader–follower game. Different from the traditional supply chain channel game in which the upstream supplier is the Stackelberg leader, the buyer is commonly the leader under an unreliable supply, and the supplier is the follower, as shown in Keren (2009), Li et al. (2012, 2013), and Tang et al. (2014), which will be discussed in our chapter.
  - The second stream is on the horizontal supply chain competition modeled from a Nash game in which multiple firms simultaneously act under supply uncertainty. The Nash equilibrium solution derivation and analysis is the main point of this type of problems, as shown in Qin et al. (2014), Tang and Kouvelis (2011), Chen and Guo (2013), Huang and Xie (2015), and Lee and Lu (2015).

  - The third stream is of a more complex structure, with a semblance to supply chain networks. Babich et al. (2007) and Qi et al. (2015) investigated multiple suppliers competing with one another while simultaneously interacting with a downstream buyer, whereas Fang and Shou (2015) explored a chain-to-chain competition with two suppliers and two buyers. In these models, vertical channel game is explored combined with horizontal competition game, thereby creating a generally and relatively tedious solving procedure.

- We consider three major types of supply risks widely adopted in literature from the supply risk model perspective.

  - The first type, random yield, refers to an uncertain production loss that actually delivers only part of the planned production size. Hence, for input quantity q, the output quantity $S(q)$ conforms to one of the two following forms: $S(e) = e \cdot \xi$ or $S(e) = e + \xi$, where ξ is a random variable with a known distribution. The former form is called proportional random yield, which depicts the situation of finally delivering a random fraction of input; the latter is called additive random yield, which demonstrates the situation of a random disturbance fluctuating around the input quantity. This random yield model is adopted in Keren (2009), Li et al. (2012, 2013), Tang and Kouvelis (2011), and Fang and Shou (2015).
  - The second type, random capacity, implies that an uncertain upper bound on actual delivery is independent of the planned production size. Hence, for input quantity q, the output quantity becomes $S(q) = \min[q, K]$, in which K is the random capacity. This random capacity model is adopted by Qin et al. (2014) and Chen and Guo (2013). Furthermore, Wang et al. (2010) considered both random yield and random capacity risks.
  - The third type is the "all-or-nothing" random disruption, wherein the supply process is of an "on" or "off" state that includes some probabilities, with a 100% output of a planned production under the "on" state but with nothing delivered under the "off" state. Specifically, for input quantity q, the output quantity becomes $S(q) = \begin{cases} q & \textit{probability}\, a \\ 0 & \textit{probability}\, 1 - a. \end{cases}$

    Mathematically, this is a special case of random yield or capacity, with the actual delivery limited by an all-or-nothing Bernoulli trial. This supply disruption model is adopted by Tang et al. (2014), Babich et al. (2007), and Qi et al. (2015). Furthermore, Lee and Lu (2015) considered a generalized random yield model while also making the all-or-nothing random disruption a special and important case.

- We should also note that the supply risk can be divided into two categories from the maneuverability perspective.

  - The supply risk is traditionally regarded as an exogenous factor that can only be statistically counted but not controlled. For example, the supply process is interrupted by irresistible forces, such as natural disasters, labor strikes, terrorist attacks, and government regulation changes. Thus, the random factor assumes a given probability distribution.

– Both industrial and academic fields recently regard the internal supply process to be possibly enhanced, such that, by exerting production/technology/labor efforts, the supply reliability can be improved (of course at a certain expense). Correspondingly, the probability distribution of random supply disturbance is then affected by the reliability effort, which poses new research questions. The endogenous supply effort model first used by Wang et al. (2010) is then employed by a number of recent studies, such as those of Tang et al. (2014), Huang and Xie (2015), Lee and Lu (2015), and Qi et al. (2015).

Table 8.1 presents the model features of the main papers discussed in our chapter.

The vast literature on the optimization models and techniques for the centralized production system with random supply is notable. However, few research has considered supply chain models, and much of this has been recently conducted. This chapter aims to classify and to describe the research to date regarding supply chain models under supply risk. Hence, we exclude the stream of research focusing on the centralized operations management model for facility location design, production planning, and inventory optimization. Most studies included in this paper incorporate the multi-firm interactions, with the exception of Wang et al. (2010). Nevertheless, Wang et al. (2010) considered a problem under the supply chain environment; furthermore, they are the first to explore the endogenous supply reliability improvement effort, hence, their study is incorporated into our chapter. In addition, apart from the uncertainty in supply quantity, other uncertain factors are notably observed within the area of supply risk. These factors include procurement cost (Babich 2006; Alexandrov 2015) and lead time risks (Lin 2016), as well as procurement product quality uncertainty (Cai et al. 2010). However, these studies are excluded given the required limit in scope on the quantity risk of the supply side in this chapter.

8.2 Vertical Supply Chain Interaction

8.2.1 Exogenous Supply Risk

This section investigates how the supply risk affects the downside order and the profits of firms within a vertical supply chain channel. We first consider the scenario in which the supply risk is of a random yield type and the market side is of a deterministic demand. The supply chain specifically consists of a buyer facing the known demand d and a supplier with the random production yield model introduced in the previous section. The supplier's output quantity $S(e)$ particularly conforms to either $S(e) = e + \xi$ or $S(e) = e \cdot \xi$, for the planned production quantity e.

Table 8.1 Model features of the main papers discussed in this chapter

Paper	Supply chain structure	Supply model	Demand model	Main focus
Keren (2009), Li et al. (2012)	Supplier–buyer vertical channel	Random yield (exogenous)	Deterministic demand	Relation between order and production quantities
Li et al. (2013)	Supplier–buyer vertical channel	Random yield (exogenous)	Deterministic and random demand	Supply chain coordination contract design
Wang et al. (2010)	Supplier–buyer vertical channel	Proportional random yield, random capacity (endogenous)	Random demand	Strategic choice of dual sourcing or reliability improvement
Tang et al. (2014)	Supplier–buyer vertical channel	All-or-nothing disruption (endogenous)	Random demand	Strategic choice of dual sourcing or reliability improvement
Qin et al. (2014)	Competing suppliers	Random capacity: (exogenous)	Price-setting deterministic demand	Impact of random capacity on supply chain game with competing suppliers
Tang and Kouvelis (2011)	Competing buyers	Proportional random yield (exogenous)	Deterministic demand (Cournot game)	Supplier diversification strategy with buyer competition
Chen and Guo (2013)	Competing buyers	Random capacity (exogenous)	Deterministic demand (hotelling model)	Incentives of adopting dual sourcing with buyer competition
Huang and Xie (2015)	Horizontal competition	Proportional random yield (endogenous)	Deterministic demand (Cournot game)	Price and cost reduction effects due to competition
Lee and Lu (2015)	Horizontal competition	Proportional random yield, all-or-nothing disruption (endogenous)	Random demand (newsvendor game)	Game properties and effect of reliability on order decisions and profits
Babich et al. (2007)	N competing suppliers +1 buyer	All-or-nothing disruption: (exogenous)	Deterministic and random demand	Supplier competition effect and diversification effects
Qi et al. (2015)	2 competing suppliers +1 buyer	All-or-nothing disruption (endogenous)	Random demand	Combined competition strategy of both price and reliability
Fang and Shou (2015)	Chain-to-chain competition: 2 suppliers +2 buyers	Proportional random yield: (exogenous)	Deterministic demand; quantity competition	The strategic choice of centralization or decentralization

The sequence of events is described as follows: The buyer first submits an order quantity q to the supplier at an exogenous wholesale price w; afterwards, the producer responsively determines the planned production quantity e. After realizing the random yield, the producer delivers the minimum output production quantity and the order, with the wholesale price w paid for each delivery. Other parameters include selling price p for the distributor and production cost c for the producer (incurred by each planned unit even when not converted to the final yield), as well as leftover holding costs h_1 and h_2 for the distributor and the producer (these can be negative if the leftover earns a salvage value).

The problem can be handled in accordance with the classical procedure employed for the Stackelberg game, in which the supplier's maximizing problem is initially solved as follows:

$$\pi_p(e) = E\left\{w \min\left[q, S(e)\right] - h_2[S(e) - q]^+ - ce\right\},$$

and then optimizing the buyer's objective as

$$\pi_d(q) = E\{p \min\left[d, q, S\left(e*(q)\right)\right] - w \min\left[q, S\left(e*(q)\right)\right] \\ - h_1[\min\left[q, S\left(e*(q)\right)\right] - d]^+\},$$

where $e_*(q)$ is the optimal response from the supplier for order quantity q.

Keren (2009) analyzed this problem and derived analytical solutions to the Buyer's ordering decision, assuming that the supply random yield follows the uniform distribution. The numerical examples provided showed that under the uniform distribution assumption, the optimal ordering quantity is shown as possibly beyond the known demand. However, Keren (2009) failed to address the questions whether ordering more is consistently optimal for the distributor or when to order more. Furthermore, the scenarios of other distributions for the random yield are neglected.

Li et al. (2012) revisited this problem and further examined supply chain decisions and profits under the generalized distribution of yield randomness. They derived analytical solutions to the optimal decisions for supply chain members and provided explicit conditions under which the buyer should order beyond the demand. These conditions are found relevant in different means to the yield distribution of additive and multiplicative risks, which indicate the importance of recognizing the production yield risk type. Furthermore, analytical solutions of the profit losses caused by the random production yield are derived for the supply chain members. The performances of the buyer and the entire supply chain are shown to be constantly worse off. However, the supplier can benefit from this random yield under certain conditions, which indicates the importance of deriving a more effective risk-sharing mechanism rather than a simple wholesale price scheme.

Hence, the next question is how to design such a coordination mechanism under the random yield of the supplier's side. Note that demand is deterministic in the above model. Under this situation, Li et al. (2013) showed that a shortage penalty contract enables the supply chain coordination and the arbitrary profit allocation between buyer and supplier. In this contract, the supplier is paid with the wholesale price for each delivered unit within the deterministic demand, as well as charged a penalty for each order shortage for the demand. However, under the random demand situation, Li et al. (2013) found that an "accept-all" type of contract is required to coordinate the supply chain, which is a much more complicated situation than that with the deterministic demand. The coordination contract specifically requires the buyer to accept all yielded units from the supplier in response to the random disturbance on the demand side. The derived coordination contracts are notably applicable to extremely generalized settings, such as the nonlinear production cost $C(e)$ rather than that in the above model, ce. Hence, they can be adopted in some other specific industrial cases, such as random yield, uncertain capacity, and stochastic used product collection. Moreover, they can easily be extended into a multiple-supplier scenario, such as decentralized assembly systems with suppliers subject to random component yields.

8.2.2 Endogenous Supply Effort

The previous subsection discusses the vertical supply chain with a primary focus on using coordination mechanism to cope with supply risk. The underlying assumption under this scenario is that the supply risk is exogenous and inherent within the production system. Conversely, the supply risk can be affected by endogenous effort in some real practical scenarios. Consequently, the buyer has an incentive to invest in improving the suppliers' processes to lessen costs, enhance quality, and improve reliability. For example, companies in the automotive industry, such as Honda, Toyota, BMW, and Hyundai commonly, work with their suppliers to improve performance (Handfield et al. 2000; Krause et al. 2007).

Wang et al. (2010) explored a model in which a buyer can source from two suppliers and/or exert effort to improve supplier reliability. For both random capacity and random yield types of supply uncertainty, a modeling framework of process improvement is established in which improvement efforts (if successful) increase supplier reliability by demonstrating that the delivered quantity (for any given order quantity) is stochastically large after the improvement. The specific model is presented as follows:

A buyer faces a newsvendor style random demand X for a product over a single selling season. Let r, v, and p denote the product's per unit revenue, salvage value, and penalty cost (for unfilled demand), respectively. The firm can source from two suppliers, $i = 1, 2$. Suppliers are unreliable in that the quantity y_i delivered by supplier i is less than or equal to the quantity q_i ordered by the buyer. The incurred procurement cost is $(\eta_i q_i + (1 - \eta_i) y_i) c_i$, where c_i is the supplier i's unit cost, and $0 \leq \eta_i \leq 1$ is the supplier i's committed cost. The supply risk is the random capacity model introduced in Sect. 8.1. Thus, for a given order quantity q_i, supplier i's delivery quantity is then given by $y_i = \min\{q_i, (K_i - \xi_i)^+\}$, where K_i is the supplier i's design capacity, and ξ_i is the supplier i's random capacity loss.

The model also incorporates a reliability index a_i with supplier i. A higher a_i implies a lower ξ_i, which increases reliability relative to the stochastic order. Let supplier i's initial reliability index be given by a_i^0. A feature of this problem lies in the buyer's capacity to exert effort to increase supplier i's reliability index. However, improvement efforts can and do fail. If the firm exerts an effort level $z_i \geq 0$, then supplier i's capability improves to $a_i(z_i) \geq a_i^0$ with probability θ_i and remains at a_i^0 with probability 1 - θ_i. The reliability improvement cost is linear in its effort and denoted as $m_i z_i$ for improving supplier i. The core problem for the buyer is deciding its improvement efforts $z = (z_1, z_2)$ first followed by determining the order quantities,$q = (q_1, q_2)$ after observing the success or failure of these efforts. Therefore, the process can be formulated as the following two-stage stochastic programming:

$$\begin{aligned}\prod\nolimits_1(a) = \sum\nolimits_{i=1}^2 -m_i z_i(a_i) + \theta_1\theta_2 \prod\nolimits_2^*(a_1, a_2) + \theta_1(1-\theta_2)\prod\nolimits_2^*\left(a_1, a_2^0\right) \\ + (1-\theta_1)\theta_2 \prod\nolimits_2^*\left(a_1^0, a_2\right) + (1-\theta_1)(1-\theta_2)\prod\nolimits_2^*\left(a_1^0, a_2^0\right),\end{aligned}$$

where $\prod_2^*(a^r) = \max_{q \geq 0}\left\{\mathrm{E}_{\xi(a^r),X}\left[\pi(q)\right]\right\}$ and

$$\begin{aligned}\pi(q) = -\sum\nolimits_i(\eta_i q_i + (1-\eta_i)y_i)c_i + r\min\left\{x, \sum\nolimits_i y_i\right\} \\ + v\left(\sum\nolimits_i y_i - x\right)^+ - p\left(x - \sum\nolimits_i y_i\right)^+.\end{aligned}$$

The above modeling framework facilitates the examination of two typical supply risk mitigation strategies, namely single sourcing with process improvement or dual sourcing without improvement. A number of typical supplier attributes, such as cost and reliability, are considered as factors influencing the strategy preference of the buyer. The benefits of both strategies are more pronounced with the growth of the heterogeneity of the cost or of the reliability between the two suppliers. However, comparison results indicate that improvement is increasingly favored over dual sourcing as the supplier cost heterogeneity increases; however, dual sourcing is favored over improvement if the supplier reliability heterogeneity is high. Furthermore, if both improvement and dual-sourcing strategies can be jointly used, then its value is more significant if the suppliers are extremely unreliable or if they have low capacities relative to demand.

A similar model can also be proposed to analyze the random yield model situation, which is consistent with the modeling approach discussed in Sect. 8.1. The result is quite interesting. In the random yield model, increasing cost heterogeneity can reduce the attractiveness of improvement. Furthermore, improvement can be favored over dual sourcing if the reliability heterogeneity is high, which sharply contrasts with the situation of random capacity.

The above model guides when the dual-sourcing approach is favored relative to the process improvement approach. This comparison assumes that the buyer has developed a close relationship with the supplier, thereby enabling the adoption of a particular production process in the production facility of the supplier. In reality, such close partnership between supply chain members may not constantly be easily achieved. In some cases, each member has autonomy over its operational decisions, such as process and technology choices, as well as production and order quantities. Tang et al. (2014) investigated such a problem in which the buyer may provide incentives to influence supply reliability; however, the supplier firm makes process/technology choices and production decisions. The study by Tang et al. (2014) differed from that of Wang et al. (2010) in the adoption of the random disruption model instead of random capacity or random yield of the former along with the assumption of a deterministic demand in the base model. The sequence of event is as follows: the supplier first proposes an incentive contract consisting of the order quantity and the sharing fraction of reliability improvement cost incurred by the supplier; then, the supplier exerts the reliability improvement effort accordingly.

For the all-or-nothing disruption model, the buyer is shown to prefer using the subsidy option only, which removes the need to inflate order quantity. However, both incentives, namely subsidy and order inflations, may be simultaneously used in the partial disruption model. Another central issue is the comparison between the effectiveness of process improvement and dual-sourcing strategies, which is also the core research question in Wang et al. (2010). However, in this case, the improvement effort is undertaken by the supplier and can only be indirectly induced by the buyer, such that, it is exerted anyway even under the dual-sourcing strategy. Hence, the basic tradeoff for the buyer is different with that in Wang et al. (2010). If the buyer places the entire order in a single supplier and possibly offers subsidy to reduce supply risk, the buyer ensures great supplier effort, high reliability, and a good chance of meeting the demand. In contrast, if the buyer diversifies, it lowers supply risk because both suppliers have no tendency to experience disruption simultaneously. However, a potential downside of supply diversification exists in endogenous reliability choice; this implies that a lower order allocation to each supplier may reduce the incentive of the supplier to invest in the reliability-improving effort. The results indicate that despite the benefit of a large order in the single-sourcing mode, dual sourcing may lead to higher expected profit for the buyer under the same wholesale price. This phenomenon can be accounted to the benefit of risk diversification together with the savings from the lower overage cost that can outweigh the loss resulting from less supplier reliability in some cases. Conversely, cases in which dual sourcing is attractive only if wholesale price is low are observed when sourcing from two suppliers. The above insights are also verified to be valid in the newsvendor type random demand situation. In conclusion, although single sourcing provides great indirect incentive to the selected supplier because order splitting is avoided, the buyer may prefer the diversification strategy under certain circumstances.

8.3 Horizontal Supply Chain Competition

8.3.1 Exogenous Supply Risk

8.3.1.1 Supplier Competition

In this section, we turn our attention to the horizontal competition within the supply chains. A supplier competition issue is investigated by Qin et al. (2014) using a model with the following features: first, suppliers are competing on the wholesale price w in the supply chain, which sharply contrasts with the previous models in which the wholesale price(s) is assumed exogenous; second, the supply risk is a random capacity type, that is, supplier i has a stochastic delivery capacity K_i with a known distribution; third, the market price is endogenous and determined by the buyer, which influences market demand. Specifically, the price-dependent market demand is assumed as a linear function of price p, i.e., $D(p) = \alpha - \beta p$.

The sequence of events is as follows: first, supplier i sets the unit wholesale price w_i; second, the buyer sets the order quantity q_i; third, the supplier i plans to produce quantity q_i. Supply capacity k_i is realized at value k_i, and the supplier produces and ships $z_i = \min(q_i, k_i)$ to the buyer; finally, the buyer receives shipments and sets retail price p, with demand materialized and all revenues and costs incurred.

A basic model of single supplier and single buyer can be first analyzed as a benchmark for the supplier competition problem, which should be solved in a chronologically reverse order. Thus, the first optimization problem determines price p to maximize the expected revenue of the buyer as follows:

$$\underset{p}{Max}\mathrm{Re} = E(ps), \quad s.t. \;\; s = \min(D(p), z).$$

The second problem is deciding the order quantity q to maximize the expected profit of the buyer as follows:

$$\underset{q}{Max}\prod_{B} \equiv E\left(\mathrm{Re}^* - wz\right), \quad s.t. \;\; z = \min(k, q).$$

Finally, the (single) supplier's problem is determining an optimal wholesale price to maximize the profit as follows:

$$\underset{w}{Max}\prod_{S} \equiv E\left[(w - c)\,z\right], \quad s.t. \;\; z = \min\left(k, q^*\right), w > c.$$

Solving the above problems yields the result that the introduction of risk to a decentralized supply chain does not alter the relationship between the buyer's order size and wholesale price; instead, it leads to the supplier charging a high wholesale price, sequentially decreasing the order quantity of the buyer. Consequently, both the supplier and the buyer suffer from low profits under the supply capacity risk. Consumer surplus and welfare are also low because of the increased retail price.

Consistent with the above modeling framework, the dual-sourcing case can be analyzed under supplier competition. Two cases of dual sourcing can be considered. One case suggests that one supplier is perfectly reliable, whereas the other is unreliable. The other case indicates that both suppliers are subject to random capacities. In the dual-sourcing case, random capacity risk clearly affects wholesale pricing differently than in the single sourcing because of the suppliers' competition for the buyer's order. Reducing capacity uncertainty may not constantly benefit a supplier competing for a monopolistic buyer's orders; the benefit of the reduction fundamentally depends on the cost heterogeneity between the suppliers.

Moreover, a supplier-duopoly case, in which both suppliers directly sell to the market without the monopolistic buyer, is explored. In this case, the unreliable supplier is proven to constantly benefit from reduced capacity variability, which deviates from the result under the two suppliers selling through a buyer. These findings highlight the role of the buyer's diversification strategy in distorting a supplier's incentive for reducing capacity uncertainty under supplier price competition.

8.3.1.2 Buyer Competition

The above work investigates unreliable suppliers competing in wholesale prices. Another issue on horizontal competition is the competition between downstream buyers given an uncertain supply. The strategic sourcing decision of a firm can initiate the chain effect to the demand-side competitor under supply risk. Consequently, the effect of supply uncertainty on firm profitability should be evaluated in the context of the vertical buyer–supplier relationship and the horizontal buyer market competition.

Tang and Kouvelis (2011) investigated this issue by adopting the supply risk model as random yield type. Thus, for an order of size q received by supplier i, the actual quantity delivered is $Y_i * q$, where Y_i is a random variable with support on [0, 1]. The supply chain structure forms a two-echelon configuration, where competing buyers order a critical component from outside suppliers and use it to produce substitutable products for the end market. A buyer's procurement cost for an order of size q includes a fixed ordering cost f and a variable cost proportional to the quantity of the item being ordered at an agreed wholesale price w. The assumption that the buyer pays for the ordered item is slightly different from the previously introduced model; however, it is plausible and possibly observed in actual practices, such as agricultural industries. The buyer also incurs unit production cost c to produce one unit final product to satisfy demand.

The market demand is price sensitive. For a monopolist buyer, the inverse demand function is given by $P(Q) = a - bQ$, where P is the market price determined by the total available-to-sell quantity Q. In the duopoly model, the competition between buyers is modeled as the Cournot quantity competition. The inverse demand function faced by firm i is assumed to be $P_i(Q_i, Q_j) = a - b(Q_i + Q_j)$, where Q_i and Q_j are the available-to-sell quantities by buyers i and j, respectively. This downside demand competition model particularly fits a limited end-market situation, where the market prices for buyers are highly influenced by their output.

Industrial examples include the electronic chip manufacturers of Xilinx and Alter, who use different sourcing strategies and the personal computer firms HP and Dell, who utilize various sourcing channels.

A benchmark is the monopoly model, in which a single-sourcing buyer determines the order q to maximize the expected profit as follows:

$$\pi_{ms} = E_Y\left[(a - bq\gamma) - cq\gamma - wq\right] - f.$$

If the buyer adopts dual sourcing, the quantities q_1 and q_2 from Suppliers 1 and 2 should be determined, respectively, to maximize the expected profit as follows:

$$\begin{aligned}\pi_{md}(q_1, q_2) = E_{\gamma_1,\gamma_2}\big\{&\left[a - b\left(q_1\gamma_1 + q_2\gamma_2\right)\right]\left(q_1\gamma_1 + q_2\gamma_2\right)\\ &- c\left(q_1\gamma_1 + q_2\gamma_2\right) - w\left(q_1 + q_2\right)\big\} - 2f.\end{aligned}$$

Solving the above two problems and comparing their results indicate that dual sourcing can bring value to the monopolist buyer by reducing the variability in market output, thereby diminishing the market output inefficiency caused by the random yield. This benefit is defined as the diversification effect. Furthermore, a more diverse supply base leads to a larger diversification benefit.

Under the duopoly model of buyer competition, the buyers simultaneously choose the order quantity to be placed with their supplier(s). The end-market price is determined by the total quantity delivered by suppliers after yield realization. Three cases are under consideration, namely, *Case 1*: both buyers with a sole source; *Case 2*: both buyers with dual sources; *Case 3*: one buyer with a sole source and the other with dual sources. The Nash equilibrium solutions of order quantities can be derived for the competing buyers. Dual sourcing is proven to improve the expected profit over sole sourcing when the fixed ordering cost and the supplier correlation are relatively low. Therefore, buyer competition does not change the logic of choice between sole versus dual sourcing. However, the variability reduction in market output is an inconsistent desirable target in terms of supplier selection and order allocation caused by its occasional failure to increase expected buyer profit, which differs from the monopoly case. For example, the buyer equally splits the order between two identical suppliers regardless of their supply process correlation in the monopoly model, which is not the optimal response for a buyer competing with a sole-sourcing opponent using a common supplier.

The above work mainly focuses on the benefits of supplier diversification in the context of dual-sourcing duopolies, as well as the related effects of supplier correlation. Chen and Guo (2013) studied competing buyers under supply risk from another angle, i.e., considering the incentives of firms in choosing a dual-sourcing strategy from both risk mitigation and strategic-sourcing perspectives. They examined how different sourcing strategies affect firm performance given both supply uncertainty and retail competition. Their model assumed that the yield uncertainty interdependently affects the order fulfillment of competing firms, which is also different from the findings by Tang and Kouvelis (2011).

We specifically consider a supply chain model consisting of a common supplier selling an essential input at unit wholesale price w to two buyers, labeled as Firms A and B; these firms transform the essential input into differentiated retail products and sell them at unit retail price p_i, for $i = A\,, B$, at the end of the consumer market. The two firms differ in their sourcing options. Although Firm A relies solely on the common supplier for the essential input, Firm B has an alternative supplier that can provide unlimited supply at unit price S. Thus, Firm A adopts a single-sourcing strategy, whereas Firm B uses a dual-sourcing strategy. This representative supply chain structure captures a class of real-world scenarios in which competing firms adopt distinct sourcing strategies (in relation to a common supplier), similar to the case of Nokia and Ericsson in the famous fire event that occurred in early 2000 at the Philips Electronics plant, a major microchip supplier for the two cell phone producers.

This model has two issues that require clarification. First is on the demand side. The two buyer's competition is supposed to be a Hotelling's horizontal product differentiation model, which yields simple linear demand functions with a pricing competition for both firms. On the supply side, the common supplier is subject to a random yield, which causes uncertain supply to the two firms. More specifically, the supplier has high (infinite) capacity with the probability $\alpha\,, \alpha \in (0, 1)$, as well as a realized finite capacity Q with the probability $1 - \alpha$. In the latter, the common supplier adopts a uniform allocation rule because of its desirable properties, such as fair and strategy-proof.

The sequence of events is as follows: (1) Given the price pair $(w; s)$, both firms decide on their retail prices $(p_A; p_B)$ and place orders $(q_A; q_B)$ to the common supplier. (2) The common supplier fully fulfills the orders of both firms under the situation of high capacity, whereas the supplier rations the orders from the two firms in accordance with the uniform allocation rule under the situation when capacity Q is realized, and firm B can temporarily acquire additional supply from its alternative source. (3) The market clears based on the realized delivery of products from the two firms. The firms are risk neutral, and the supply chain structure is common knowledge. Each firm optimally chooses its retail price and order quantity, anticipating the action of its rival. A Nash game is consequently induced, with the objective functions of the two firms as

$$\max_{p_A, q_A} E\pi_A = [\alpha q_A + (1 - \alpha)\, g_A]\,(p_A - w)\,,$$

and

$$\max_{p_B, q_B} E\pi_B = [\alpha q_B + (1 - \alpha)\, g_B]\,(p_B - w) + [\alpha\,(D_B - q_B) + (1 - \alpha)\,(1 - g_A - g_B)]\,(p_B - s)\,.$$

Chen and Guo (2013) solved the above model by considering two scenarios. One scenario is $w \le s$, i.e., the wholesale price is lower than the alternative supply price for Firm B. In this scenario, the price of Firm B is shown to be higher than that of Firm A, which in turn is priced higher when both firms adopt a single-sourcing strategy. This finding is accounted for by the following: with the option of dual sourcing, Firm B obtains a "monopoly" of power over the residual demand and induces it to raise its retail price. Consequently, this price increase by Firm B reduces the pressure on Firm A's pricing. Firm A then raises its retail price as well, but not to the extent that Firm B does because of Firm B's competitive advantage over the residual demand. Furthermore, by comparing the firm's expected profits with the single-sourcing benchmark, Firm B's dual-sourcing strategy is shown to probably benefit itself, as well as Firm A. This result is expected for Firm B, given that an alternative supply secures more for its order fulfillment. However, such result is relatively interesting for Firm A because under Firm B's dual-sourcing strategy, Firm A charges a relatively lower price than Firm B, which yields higher demands and expected sales, compared with the single-source benchmark case. The increased price and sales lead to a higher expected profit for Firm A. Thus, the alternative sourcing of one firm creates a positive externality for its rival.

Another scenario under consideration is $w \leq s$, i.e., when the wholesale price is lower than the alternative supply price for Firm B. In this scenario, as long as the wholesale price is within a certain interval, Firm B has an incentive to order from the common supplier even at a relatively higher cost compared with the alternative supply. Accordingly, Firm B limits its rival's supply to the market in the event of a supply shortage, the benefit of which can outweigh the extra cost paid to the common supplier. This finding indicates a strategic sourcing incentive for other effective retail completion. Under this scenario, both firms charge higher prices and earn higher expected profits in the dual-sourcing environment than in the single-sourcing benchmark, and Firm B still charges a higher price than Firm A does. These insights are similar to that in the former scenario.

8.3.2 Endogenous Supply Effort

8.3.2.1 Cournot Quantity Competition

The previous subsection discusses vertical competition under supply uncertainty; however, the random factors in the supply side are exogenous. We currently investigate the problems through which the supply reliability can be improved with endogenous effort. In this aspect, Huang and Xie (2015) considered two unreliable firms who endogenously exert effort to improve their reliability through a Cournot quantity game competition.

Consider two symmetric firms, i and j, who produce identical products in a market characterized by Cournot competition. The production process is unreliable in terms of the quantity of qualified output for either firm $i(j)$. Suppose the input quantity is $q_i(q_j)$ for manufacture $i(j)$; then, the output quantity is $q_i y_i(q_j y_j)$, where y_i and y_j are random yield rates independent and identically distributed over support $[0, 1]$. Dropping the subscripts because of symmetry, the yield rate for each firm is assumed to be a uniform distributed random variable y $\mathrm{U}(0, a(e))$. Here, $a(e)$

is a concave function that increases in e with $a(0)=a^0$ and $\lim_{e\to\infty}a(e)=1$. a measures the reliability after improvement over (0, 1), and a^0 is the initial reliability without improvement. Furthermore, the disutility of effort e is denoted by an increasing convex function $z(e)$. On the demand side, the inverse demand function is $p=d-bQ(b>0)$, where Q is the total quantity supplied to the market, d is the market potential, and b is the sensitivity parameter. The total production cost is given by $c_i=(1-(1-\eta_i)(1-y_i))q_i w$, where w is the unit production cost, and $\eta\in(0,1]$ measures the loss associated with the defective product.

The sequence of the events is as follows: (1) The two firms simultaneously determine the reliability improvement efforts; (2) The firms decide the input quantities after observing the realized reliabilities; (3) The firms engage in quantity competition on the market with output quantities. Suppose that the firm is unaware of the opponent's realized yield when making input quantity decision, hence, a two-stage dynamic game is established.

The game can be solved using a backward approach, such that, the second-stage game should be considered first. For firm i, given q_j, the second-stage profit can be maximized by inputting q_i as follows:

$$\prod\nolimits_{d2}\left(q_i;q_j,a_i^r,a_j^r\right)$$
$$=E_{y_i\left(a_i^r\right),y_j\left(a_j^r\right)}\left[\left(d-b\left(q_iy_i+q_jy_j\right)\right)q_iy_i-\left(1-\left(1-\eta_i\right)\left(1-y_i\right)\right)q_iw\right].$$

The problem can be solved with analytical solutions of Nash equilibriums for the firms' input quantities under the following four possible scenarios after firms exert efforts: both firms succeed, both firms fail, firm i succeeds, but j fails, and firm i fails whereas j succeeds. The comparison results of firm input quantities under two scenarios (firm success versus failure) are closely related to market potential. When the market potential is low, the successful firm inputs additional quantities than the failed firm; however, when the market potential is high, the successful firm inputs less quantities. On the relationship between optimal input and realized reliability, the firm's optimal input quantity decreases in the competitor's realized reliability. Furthermore, the firm's optimal input quantity increases in its own realized reliability when the market potential is low, although its realized reliability decreases when the market potential is high. This phenomenon is explained by the possible two contradictory effects when the realized reliability of the firm increased, namely the price reduction (negative effect) and cost reductions (positive effect). Under low market potential, the firm prefers to exploit the cost reduction effect and inputs additional quantity expecting to lower average cost. In contrast, under large market potential, the firm inputs less quantity to diminish the price reduction effect and to maintain high margins on products sold.

For the first-stage problem, the problem of choosing an optimal effort is converted into that of choosing an optimal reliability. Thus, the firm determines reliability a to maximize the first-stage profit function as follows:

$$\prod\nolimits_{d1}(a) = \theta^2\prod\nolimits_{d2}^{*}(a,a) + \theta(1-\theta)\prod\nolimits_{d2}^{*}(a,a^0) + (1-\theta)\theta\prod\nolimits_{d2}^{*}(a^0,a) + (1-\theta)^2\prod\nolimits_{d2}^{*}(a^0,a^0) - z(e_d(a))$$

Two aspects of results can be obtained by analyzing the Nash equilibrium solution for this problem. First, on the effect of quantity competition on reliability improvement, the optimal effort the firm exerts in the duopoly case is less than that in the monopoly case, and the difference between the optimal efforts under the two cases increases with the probability of improvement success. Second, on the effect of reliability improvement on quantity competition, the endogenous behavior of reliability improvement intensifies competition by making firms increase inputs under the low market potential in terms of expectation, while weakening competition under the high market potential. This insight is similar to the relationship presented in the second stage as follows: when the market potential is small, firms tend to use the cost reduction effect from reliability improvement by increasing the input quantity; when the market potential is large, firms depend more on the price reduction effect than saving costs, and thus input a smaller quantity.

8.3.2.2 Newsvendor Inventory Competition

Inventory competition, also commonly referred to as newsvendor game, is a commonly observed phenomenon in a competitive market initially studied by Parlar (1988). Lee and Lu (2015) investigated this horizontal inventory competition under yield uncertainty, in which two firms with random yields compete for a substitutable demand as follows: If one firm suffers a stock-out, which can be caused by yield failure, its unsatisfied customers may switch to its competitor. On the supply side, each firm is subject to a random yield, with the modeling similar to that in the Cournot competition problem. The stochastic yield rate y_i of firm $i = (1, 2)$ is related to the yield reliability a_i, which can be endogenously enhanced by the firm. Let q_i denote the input ordering quantity of firm i; then, the output stocking quantity is $q_i y_i$. On the demand side, let D_i denote the initial demand share of firm i. If firm i suffers a stock-out, that is, $q_i y_i$ turns out to be less than D_i, then a fixed fraction of the excess demand will switch to its competitor, firm j ($j \neq i$). Let D_i^s denote the effective demand of firm i; and it can be expressed as $D_i^s = D_i + \gamma_{ji}(D_j - q_j)^+$, where γ_{ij} ($0 \leq \gamma_{ij} \leq 1$) is the switching rate of the unsatisfied customers of firm i going to purchase from firm j.

The sequence of event is also similar to the Cournot competition as follows: first, the firms select reliability levels (a_1, a_2) to improve and to incur the improvement costs. Afterwards, these reliability levels are observed, and the firms decide the initial order quantities (q_1, q_2). The actual output is then realized, and unsatisfied customers switch to the other firm. A two-stage game is hence established and can be solved in a reverse order.

Given a fixed pair of reliability index, $a = (a_1, a_2)$, the expected profit of firm i in the quantity game can be written as

$$\pi_i^q\left(q_i|q_j, a\right) = E\left[p_i \min\left(D_i^s, q_i y_i\right) + s_i y_i - D_i^s\right]^+ - c_i q_i + \delta_i c_i q_i\left(1 - y_i\right)\Big].$$

This stage of game is proven to be a submodular game, which means that a firm will reduce its order quantity if its competitor increases the order. Random supply yield noticeably gives rise to multiple equilibria, which differs from the traditional result of unique equilibrium without yield uncertainty (shown in Parlar 1988). Nevertheless, a unique equilibrium does exist if the random yield follows a Bernoulli distribution. Quantity and yield reliability also serve as complementary instruments for the competing firms. The firm can increase its expected profit with a higher reliability level, through which its competitor's profit is simultaneously reduced.

Let $\left(q_1^*(a), q_2^*(a)\right)$ be the equilibrium quantities in the second stage, then, firm $i$maximizes the first-stage profit by choosing a reliability level a_i. The first-stage optimization problem of firm i can be written as

$$\max_{a_i \geq a_i^0} \pi_i^r\left(a_i|a_j\right) = \pi_i^q\left(q_i^*(a)|q_j^*(a), a\right) - z_i\left(a_i\right),$$

where $z_i(a_i)$ is an increasing convex cost function of exerting effort to raise the reliability level to a_i. This first-stage reliability game can be analyzed if the firm's initial demand is deterministic and if the random yield follows a Bernoulli distribution. Under this situation, this reliability game is also submodular. Furthermore, competing firms are found to be possibly reluctant to pursue a high-reliability level as a monopoly does. This result indicates that the competition weakens the incentive to improve yield reliability. This finding is explained by the fact that the potential market share of a competitive firm is smaller than that of a monopoly; thus, the marginal gain from improving reliability is relatively small for the competitive firm. Furthermore, the equilibrium reliability levels are also sensitive to the customer-switching rate. The firm would exert a higher reliability level if more customers can switch to this firm from its competitor and vice versa. Hence, raising the reliability level is preferred if more of its competitor's customers regard itself as a backup vendor.

8.4 Supply Chain Networks

8.4.1 Supplier Competition + Buyer Diversification (N Suppliers + One Buyer)

The previous sections have provided preliminary models on one buyer dealing with multiple-competing suppliers, who may fail to deliver order quantities because of supply disruptions. However, those models exclusively focus on horizontal supplier competition. In this subsection, we incorporate both horizontal supplier competition and vertical channel competition between suppliers and their downstream buyer. Consider a simple supply chain model with one buyer and N suppliers perfectly producing substitutable products. The suppliers are unreliable because they are subject to random defaults modeled as "all-or-nothing" disruptions. Let δ_i be a binary random variable denoting the disruption of supplier i with a joint distribution of $\delta_1, \ldots, \delta_N$ determined by the probabilities $p_{d_1 d_2 \ldots d_N} = P[\delta_1 = d_1, \ldots, \delta_N = d_N]$, $d_i \in \{0, 1\}, i = 1, \ldots, N$. This modeling approach is adopted because it highlights the correlation among the disruptions of these risky suppliers.

Demand D can be deterministic or random, with unit retail sales price s as the predetermined parameter. The event sequence is similar to the typical supplier–buyer interaction within a supply chain channel as follows: The suppliers first determine their wholesale prices w_i, and then, the buyer responds by choosing order quantities q_i. Thus, the suppliers compete with one another for the buyer's business, and collectively, they serve as the Stackelberg leaders in a game where the buyer is the Stackelberg follower. The per unit production cost for supplier i is c_i.

The optimization problem of the buyer placing orders with N suppliers is

$$\max_{q_1 \geq 0, q_2 \geq 0, \ldots q_N \geq 0} \left(sE\left\{ \min\left[D, \sum_{i=1}^{N} (1 - \delta_i) q_i \right] \right\} - \sum_{i=1}^{N} c_i q_i \right),$$

whereas the suppliers compete with one another for the buyer's business and solve the following optimization problems:

$$\sup_{w_i \geq 0} (w_i - c_i) z_i (q_i, \ldots, q_N), \qquad i = 1, 2, \ldots, N.$$

Babich et al. (2007) analyzed the above model by considering the codependence among the suppliers' random disruptions. For the two-supplier problem with deterministic demand ($N = 2$ and D is deterministic), the buyer is shown to prefer suppliers with highly positive correlated disruptions. This result contradicts the intuition that negative correlation generates a diversification advantage to the buyer. With competition, the positive correlation between supplier disruptions leads to lower wholesale prices, thereby compensating the buyer for losing diversification benefits. Conversely, all things being equal, each supplier prefers a highly negative correlation between their own default processes and those of their competitors, leading to less competition and more profits extracted from the buyer. Alternately, simultaneously obtaining diversification benefits and low wholesale prices with over two suppliers ($N \geq 3$) is possible for the buyer. For example, if two competing suppliers are highly correlated and the third supplier being negatively correlated with the others, the buyer can benefit from the low wholesale price induced by the competition between the two highly codependent suppliers and use the third supplier to hedge against disruption risk.

The analysis increases in difficulty when considering models of random demand (D is a random variable); however, the overall direction of the results remains unchanged. Thus, contrary to the initial intuition regarding the advantages of diversification, positive default correlation can benefit the buyer, which outweighs the losses from a weak diversification. Simultaneously, a negative disruption correlation benefits the suppliers and the channel in general. Therefore, the preferences of the buyer and the channel for default correlation are misaligned.

The above model assumes that the supplier competition is on wholesale pricing under exogenous supply disruption risks. Qi et al. (2015) considered the situation in which the suppliers' reliabilities are endogenous and can be enhanced at some expenses. Thus, a buyer procures a product from two suppliers competing not only through pricing strategy but also through reliability improvement efforts. The framework is approximately similar to Babich et al. (2007), with some differences on supply and demand modeling. For example, the demand is assumed to be a newsvendor random one, D. The reliability of supplier i is assumed to be q_i when the market is on and $a_i\, q_i$ when the market is off, in which q_i is the reliability decision of supplier i, and the market state is shared by both suppliers with either "on" or "off," with given respective probabilities. The sequence of events is as follows: (1) The suppliers simultaneously decide on their reliabilities; (2) The suppliers observe the reliability decisions made by their competitors, respectively, and then determine the wholesale prices; (3) Based on the suppliers' wholesale prices and reliabilities, the buying firm places orders to the suppliers; (4) All uncertainties are resolved, and the transactions are completed.

Studies have shown that the reliability of suppliers, as an endogenous decision variable, frequently plays a more important role than the wholesale price in supplier competition. In fact, maintaining the reliability and wholesale price both high is the ideal strategy for suppliers with multiple options. Noticeably, when the demand uncertainty is relatively high or when the supply reliability is low, the competition among suppliers on both price and reliability may render the sole-sourcing strategy optimal in some cases, depending on the format of suppliers' cost functions. This phenomenon is a counterintuitive result opposed to the conventional wisdom that low supply reliability and high demand uncertainty motivate dual sourcing. Moreover, a supplier's profit and that of the buyer may unnecessarily decrease under supplier competition as the cost or vulnerability of this supplier increases.

8.4.2 Chain-to-Chain Competition (Two Suppliers + Two Buyers)

Chain-to-chain competition is regarded as the current business conception replacing the traditional model of firm-to-firm competition. Combined with supply uncertainty, this problem may require a more complex analysis. Fang and Shou (2015) systematically examined how to design and operate supply chains to deal with supply uncertainty effectively by considering the interaction between two competing supply chains. Each chain consists of one buyer and an exclusive supplier. Both chains are subject to supply uncertainty, which is modeled by a random yield a_i between 0 and 1. On the demand side, the market demand of chain i is determined by $p_i(Q_i, Q_j) = A - a_iQ_i - \gamma a_jQ_j$, where A is the market base, and $\gamma \in (0, 1)$ is the competition intensity, whereas Q_i and Q_j are the buyers' order quantities in chains i and j, where $i, j \in \{1, 2\}, i \neq j$.

Three types of competition games are explored, namely centralized, hybrid, and decentralized games. In the centralized game, central planners for both supply chains simultaneously determine the order quantities Q_i and Q_j to maximize their own expected profits. In the decentralized game, each supplier announces its contract term consisting of a wholesale price per unit of successful delivery and a penalty paid to the buyer per unit of unfilled order, and then, the respective buyer accordingly chooses the production quantity. The hybrid game is a mix of the centralized and decentralized chains, with the supplier in the decentralized chain making contract term before the quantity competition commences.

The obtained equilibrium solutions for the above three games provide the following observations: first, the expected order quantity and profit of a supply chain increase if its competing supply chain becomes less reliable or if its own supply becomes more reliable. Thus, a supply chain with a reliable supply can significantly maximize the high supply risk of its competing chain. Second, higher competition intensity results in lower equilibrium order quantities and expected profits for both supply chains. Third, order quantities are upper-bounded by those in the standard monopoly game without uncertainty.

Another question of interest on the strategic level is whether supply chain centralization provides a competitive advantage when dealing with competition and supply uncertainty. The answer is not necessarily. In fact, a supply chain is consistently better off by choosing to centralize, which implies that centralization is a dominant strategy. However, if the supply risk is low and the chain competition is intensive, centralization can actually decrease the supply chain profit compared with the case of the decentralized game. This phenomenon leads to a prisoner's dilemma. Alternatively, if the supply risk is high and/or the competition level is low, centralization constantly increases the supply chain profit. Hence, the desirability of supply chain centralization is enhanced by high supply uncertainty or low chain competition.

8.5 Potential Research Directions

Supply risk management has grown in importance because of the need for designing, coordinating, and operating extended supply chains. The risk can be the consequence of a host of random factors; it can also severely damage the supply chain firms. This chapter discusses supply chain models under supply risks, followed by these three classes of problems:

- First, a wholesale price contracts the risk allocation imbalance among supply chain members; thus, the channel coordination contract design under supply uncertainty is an important yet complicated problem. The strategic choice of dual-sourcing or reliability improvement is also vital for the firms within a supply chain, given that the supplier's process reliability can be endogenously improved.
- Second, the effect of supply uncertainty on firm profitability should be evaluated in the context of the horizontal market competition. With supplier competition, supply uncertainty affects the retailer's diversification strategy for replenishment and changes the suppliers' wholesale price competition and the incentive to reduce capacity uncertainty. With buyer competition, the strategic choice of single or dual sourcing is crucial for both the buyer under consideration and its competitor. The effect of reliability competition and its relation with pricing competition is also a hot topic when the supply effort is endogenous.
- Third, under a more complex system of N suppliers plus one buyer, the diversification and the price competition effects should be carefully weighed as they are closely related to the number of the supplier and the correlations among their disruptions. For a chain-to-chain network system, channel centralization inconsistently offers a competitive advantage. Thus, the choice of channel centralization also depends on system parameter.

A number of other issues require further exploration for future research directions:

- Information asymmetry: A common assumption in the above research is the existence of information symmetry within the supply chain system, i.e., both supplier and buyer share common knowledge. However, this finding may not apply in reality. For example, the suppliers may be vaguely aware of the market state, whereas the buyers may have incomplete information of the suppliers' attributes, such as costs and reliabilities. Hence, incentive theory, including adverse section and moral hazard, can be adopted to establish and analyze such models. Some studies such as those of Yang et al. (2012) and Huang et al. (2016) have looked into this research domain, which suggests a promising future direction.
- Firms' behavior: Behavior operations management has recently been in the spotlight. Hence, incorporating the features of supply chain firm behavior is another interesting topic. A major subject concerns the risk attitude of firms toward supply uncertainty. In this aspect, possible modeling tools include expected utility theory, mean-variance theory, VaR and CVaR, and prospect theory (Choi et al. 2008b; Choi and Ruszczynski 2011; Choi and Chiu 2012; Liu et al. 2013). For example, Li and Li (2016) studied a lot-sizing problem in the presence of random yield supply under loss aversion, whereas Madadi et al. (2014) investigated a centralized supply network design problem with an unreliable supply under both risk neutrality and aversion. On the supply chain interaction, other behavior characteristics can be adopted. For example, Chen et al. (2015) studied a supply chain-contracting problem with yield uncertainty and horizontal fairness concerns. We believe the study of supply chain model is potentially great by considering firms' behavior toward supply uncertainty.

- Channel power and cooperation: The above research fails to investigate specifically the issue of channel power. In fact, the effect of channel member power on the supply chain decisions and profits are interesting problems worth investigating. For example, Hwang et al. (2016) showed that the simple wholesale price contract leads to different performances under different channel power structures. Another future research issue, the supply chain cooperation and profit allocation in the presence of supply risk, is linked to the channel power problem.
- Supply risk assessment in the big data era: In our present supply chain modeling papers, the probability information of supply risk should be provided. In the real industry, such information comes from the risk assessment process, which integrates all identified knowledge of experts' opinion, historical data, and supply chain structure. Thus, measuring and quantifying supply chain risk has proven to be an enormous challenge in both the industry and the academe. According to a literature survey by Tang and Musa (2011), of the 138 papers they identified within this research domain, less than a quarter are empirical or quantitative. This finding corresponds with the comment by Wagner and Neshat (2012), "ways of measuring and quantifying supply chain risk are just beginning to emerge." Along with today's big data trend, the current process of maximizing more transparent information and revolutionary big data approach to more accurately identify and evaluate the likelihood of supply risk becomes a problem of substantial significance and interest. Innovative supply risk modeling frameworks using big data analytics are regarded extremely valuable, considering that integrating big data in operations and supply chains aids firms in improving intra- and inter-firm efficiency and effectively manages risks as well (Sanders and Ganeshan 2015).

Acknowledgment This work is supported by National Natural Science Foundation of China (NSFC) Nos. 71372002 and 71372100.

References

A. Alexandrov, When should firms expose themselves to risk? Manag. Sci. **61**(2), 3001–3008 (2015)

V. Babich, Vulnerable options in supply chains: effects of supplier competition. Nav. Res. Logist. **53**(7), 656–673 (2006)

V. Babich, A.N. Burnetas, P.H. Ritchken, Competition and diversification effects in supply chains with supplier default risk. Manuf. Serv. Oper. Manage. **9**(2), 46–123 (2007)

X.Q. Cai, J. Chen, Y.B. Xiao, X.L. Xu, Optimization and coordination of fresh product supply chains with freshness keeping efforts. Prod. Oper. Manag. **19**(3), 78–261 (2010)

J. Chen, Z. Guo, Strategic sourcing in the presence of uncertain supply and retail competition. Prod. Oper. Manag. **23**(10), 1748–1760 (2013)

J. Chen, X. Zhao, Z. Shen, Risk mitigation benefit from backup suppliers in the presence of the horizontal fairness concern. Decis. Sci. **46**(4), 663–696 (2015)

T.M. Choi, C.H. Chiu, Mean-downside-risk and mean-variance newsvendor models: Implications for sustainable fashion retailing. Int. J. Prod. Econ. **135**(2), 552–560 (2012)

S. Choi, A. Ruszczynski, A multi-product risk-averse newsvendor with exponential utility function. Eur. J. Oper. Res. **214**(1), 78–84 (2011)
T.M. Choi, D. Li, H.M. Yan, Optimal two-stage ordering policy with Bayesian information updating. J. Oper. Res. Soc. **54**(8), 846–859 (2003)
T.M. Choi, D. Li, H.M. Yan, Quick response policy with Bayesian information updates. Eur. J. Oper. Res. **170**(3), 788–808 (2006)
T.M. Choi, D. Li, H.M. Yan, Mean–variance analysis of a single supplier and retailer supply chain under a returns policy. Eur. J. Oper. Res. **184**(1), 356–376 (2008a)
T.M. Choi, D. Li, H.M. Yan, Channel coordination in supply chains with agents having mean-variance objectives. Omega Int. J. Manage. Sci. **36**(4), 565–576 (2008b)
Y. Fan, L. Heilig, S. Voss, Supply chain risk management in the era of big data design, user experience, and usability: design discourse. Lect. Notes Comput. Sci. **9186**, 283–294 (2015)
Y. Fang, B. Shou, Managing supply uncertainty under supply chain Cournot competition. Eur. J. Oper. Res. **243**(1), 156–176 (2015)
R.B. Handfield, D.R. Krause, T.V. Scannell, R.M. Monczka, Avoid the pitfalls in supplier development. Sloan Manage. Rev. **41**(2), 37–49 (2000)
H. Huang, T. Xie, Reliability improvement and production decision under Cournot competition. Int. J. Prod. Res. **53**(15), 4754–4768 (2015)
H. Huang, X.Y. Shen, H.Y. Xu, Procurement contracts in the presence of endogenous disruption risk. Decis. Sci. **47**(3), 437–472 (2016)
W. Hwang, N. Bakshi, V. DeMiguel, *Simple Contracts for Reliable Supply: Capacity Versus Yield Uncertainty Reliability*. Working paper (London Business School, London, 2016)
B. Keren, The single-period inventory problem: extension to random yield from the perspective of the supply chain. Omega Int. J. Manage. Sci. **37**(4), 801–810 (2009)
D.R. Krause, R.B. Handfield, B.B. Tyler, The relationships between supplier development, commitment, social capital accumulation and performance improvement. J. Oper. Manag. **25**(2), 528–545 (2007)
H. Lee, S.G. Kimm, H. Park, P. Kang, Pre-launch new product demand forecasting using the bass model: a statistical and machine learning-based approach. Technol. Forecast. Soc. Chang. **86**(1), 49–64 (2014)
C.Y. Lee, T. Lu, Inventory competition with yield reliability improvement. Nav. Res. Logist. **62**(2), 107–126 (2015)
X. Li, Y.J. Li, On lot-sizing problem in a random yield production system under loss aversion. Ann. Oper. Res. **240**(2), 415–434 (2016)
X. Li, Y.J. Li, X.Q. Cai, A note on the random yield from the perspective of the supply chain. Omega Int. J. Manage. Sci. **40**(5), 601–610 (2012)
X. Li, Y.J. Li, X.Q. Cai, Double marginalization in the supply chain with uncertain supply and coordination contract design. Eur. J. Oper. Res. **226**(2), 228–236 (2013)
H.J. Lin, Investing in lead-time variability reduction in a collaborative vendor–buyer supply chain model with stochastic lead time. Comput. Oper. Res. **72**, 43–49 (2016)
W. Liu, S.J. Song, C. Wu, Impact of loss aversion on the newsvendor game with product substitution. Int. J. Prod. Econ. **141**(1), 352–359 (2013)
A.R. Madadi, M.E. Kurz, K.M. Taaffe, J.L. Sharp, S.J. Mason, Supply network design: risk- averse or risk-neutral? Comput. Ind. Eng. **78**, 55–65 (2014)
P. O'Keeffe, *Understanding Supply Chain Risk Areas, Solutions, and Plans: A Five-Paper Series* (APICS, The Association of Operations Management, Alexandria, VA, 2006)
M. Parlar, Game theoretic analysis of the substitutable product inventory problem with random demands. Nav. Res. Logist. **35**(3), 397–409 (1988)
L. Qi, J.J. Shi, X. Xu, Supplier competition and its impact on firm's sourcing strategy. Omega Int. J. Manage. Sci. **55**, 91–110 (2015)
F. Qin, U.S. Rao, H. Gurnani, B. Ramesh, Role of random capacity risk and the retailer in decentralized supply chains with competing suppliers. Decis. Sci. **45**(2), 255–279 (2014)
N.R. Sanders, R. Ganeshan, Special issue of production and operations management on "big data in supply chain management". Prod. Oper. Manag. **24**(5), 852–853 (2015)

S. Tang, P. Kouvelis, Supplier diversification strategies in the presence of yield uncertainty and buyer competition. Manuf. Serv. Oper. Manage. **13**(4), 439–451 (2011)

O. Tang, S.N. Musa, Identifying risk issues and research advancements in supply chain risk management. Int. J. Prod. Econ. **133**(1), 25–34 (2011)

S. Tang, H. Gurnani, D. Gupta, Managing disruptions in decentralized supply chains with endogenous supply process reliability. Prod. Oper. Manag. **23**(7), 1198–1211 (2014)

S.M. Wagner, N. Neshat, A comparison of supply chain vulnerability indices for different categories of firms. Int. J. Prod. Res. **50**(11), 2877–2891 (2012)

Y. Wang, W. Gilland, B. Tomlin, Mitigating supply risk: dual sourcing or process improvement? Manuf. Serv. Oper. Manage. **12**(3), 489–510 (2010)

Z. Yang, G. Aydin, V. Babich, D. Beil, Using a dual-sourcing option in the presence of asymmetric information about supplier reliability: competition vs. diversification. Manuf. Serv. Oper. Manage. **14**(2), 202–217 (2012)

Part III
Financial Optimization Analysis

Chapter 9
A Parameterized Method for Optimal Multi-Period Mean-Variance Portfolio Selection with Liability

Xun Li, Zhongfei Li, Xianping Wu, and Haixiang Yao

Abstract Big data is being generated by everything around us at all times. The massive amount and corresponding data of assets in the financial market naturally form a big data set. In this paper, we tackle the multi-period mean-variance portfolio of asset-liability management using the parameterized method addressed in Li et al. (SIAM J. Control Optim. 40:1540–1555, 2002) and the state variable transformation technique. By this simple yet efficient method, we derive the analytical optimal strategies and efficient frontiers accurately. A numerical example is presented to shed light on the results established in this work.

Keywords Multi-period portfolio • Mean-variance formulation • Asset-liability management

9.1 Introduction

Portfolio selection is concerned with finding the most desirable group of funds to hold. The mean-variance model proposed by Markowitz (1952) aims to seek a balance between the gain and the risk, which are expressed by expectation and

X. Li (✉)
Department of Applied Mathematics, The Hong Kong Polytechnic University, Hong Kong, China
e-mail: malixun@polyu.edu.hk

Z. Li
Department of Finance and Investment, Sun Yat-Sen Business School, Sun Yat-Sen University, Guangzhou, China
e-mail: lnslzf@mail.sysu.edu.cn

X. Wu
School of Mathematical Sciences, South China Normal University, Guangzhou, China
e-mail: pphappe@sina.com

H. Yao
School of Finance, Guangdong University of Foreign Studies, Guangzhou, China
e-mail: yaohaixiang@gdufs.edu.cn

T.-M. Choi et al. (eds.), *Optimization and Control for Systems in the Big-Data Era*, International Series in Operations Research & Management Science 252,
DOI 10.1007/978-3-319-53518-0_9

variance of the investment return, respectively. In order to trace out the efficient frontier for this bi-objective optimization problem, one typically puts weights on the two criteria and transforms the problem into a single-objective optimization problem.

After Markowitz's vanward work in a single-period setting, the mean-variance portfolio selection framework was extended to multi-period setting by Li and Ng (2000) using an embedding technique. Zhou and Li (2000) considered a continuous-time mean-variance problem while Li et al. (2002) investigated the problem with no short setting. As any nonlinear term of expectation operator, the term $(\mathbb{E}[x_T])^2$ in the mean-variance case, induces nonseparability, the spirit of both the embedding scheme proposed by Li and Ng (2000) and Zhou and Li (2000) and the parameterized method developed by Li et al. (2002) is to embed $(\mathbb{E}[x_T])^2$ into an auxiliary function or to replace $\mathbb{E}[x_T]$ by an auxiliary variable in mean-variance models to deal with mean-variance problems in dynamic programming. Besides the above, Cui et al. (2014) presented another powerful tool named mean-field formulation to tackle the nonseparability of multi-period mean-variance portfolio selection problem and derived analytical optimal strategies and efficient frontiers. Yi et al. (2014) developed the mean-field formulation method to solve the multi-period mean-variance portfolio selection problem with an uncertain exit horizon.

Big data is being generated by everything around us at all times. The number of assets in the financial market and the corresponding data constitute a typical big data. Big data is also changing the way people investing. Insights from big data and extracting meaningful value from big data can enable all investors to make better profit. It is well known that the stability of financial institutions depends crucially on the matching of assets, and liabilities. Liability is being brought more and more into the limelight when investors establish their portfolios. The mean-variance framework of asset-liability management was first investigated by Sharp and Tint (1990) in a single-period setting. For the multi-period setting and by the embedding technique, Leippold et al. (2004) derived the closed form optimal policies and mean-variance frontiers under exogenous and endogenous liabilities using a geometric approach; Chiu and Li (2006) employed the stochastic optimal control theory to analytically solve the asset-liability management in a continuous time setting; Yi et al. (2008) considered the situation of uncertain investment horizon; Chen and Yang (2011) studied the case with regime switching; Zeng and Li (2011) investigated the model under benchmark and mean-variance criteria in a jump diffusion market; Li and Li (2012) took the risk control over bankruptcy into account; Yao et al. (2013) re-considered the uncertain time-horizon model of Yi et al. (2008) by adding an uncontrolled cash flow.

Most of the papers for multi-period mean-variance portfolio selection of asset-liability management mentioned above are based on the embedding technique. The embedding scheme is indeed an efficient way to deal with problems having the nonseparable property. However, it is prone to involve inefficient and complicated calculation during the derivation of the optimal strategies and efficient frontiers by embedding. Therefore, research is naturally required on developing a simple yet accurate method. In this paper, we study asset-liability management under a multi-period mean-variance portfolio selection framework using the parameterized method addressed in Li et al. (2002). We first deduce the case when the returns of assets and liability are correlated. Then we reduce it to the uncorrelated setting. One prominent feature of the dynamic mean-variance formulations is that the optimal portfolio policy is always linear with respect to the current wealth and liability. According to this feature, we derive the analytical optimal policies and efficient frontiers. The analytical form of the Lagrange multiplier is also given in expression of the expectation of the final surplus.

The rest of the paper is organized as follows. In Sect. 9.2, we present the mean-variance formulation of the multi-period portfolio selection model for asset-liability management. The optimal strategies and efficient frontiers are derived in Sect. 9.3. Section 9.4 provides some numerical examples to illustrate the results developed in this paper. Section 9.5 concludes this paper.

9.2 Mean-Variance Formulation

Assume that an investor joining the market at the beginning of period 0 with an initial wealth x_0 and initial liability l_0, plans to invest his/her wealth within a time horizon T. He/she can reallocate his/her portfolio at the beginning of each of the following $T-1$ consecutive periods. The capital market consists of one risk-free asset, n risky assets and one liability. At time period t, the given deterministic return of the risk-free asset, the random returns of the n risky assets, and the random return of the liability are denoted by s_t (> 1), vector $\mathbf{e}_t = [e_t^1, \cdots, e_t^n]'$ and q_t, respectively. The random vector $\mathbf{e}_t = [e_t^1, \cdots, e_t^n]'$ and the random variable q_t are defined over the probability space $(\Omega, \mathcal{F}, P)$ and are supposed to be statistically independent at different time periods.

Suppose that M and N are symmetric matrices with the same order. We denote $M \succ N$ ($M \succcurlyeq N$) if and only if $M - N$ is positive definite (semidefinite). We assume that the only information known about $\mathbf{e}_t$ and q_t are their first two unconditional moments, $\mathbb{E}[\mathbf{e}_t] = \left(\mathbb{E}[e_t^1], \cdots, \mathbb{E}[e_t^n]\right)'$, $\mathbb{E}[q_t]$ and $(n+1) \times (n+1)$ positive definite covariance

$$\mathrm{Cov}\left(\begin{pmatrix} \mathbf{e}_t \\ q_t \end{pmatrix}\right) = \mathbb{E}\left[\begin{pmatrix} \mathbf{e}_t \\ q_t \end{pmatrix} (\mathbf{e}_t' \; q_t)\right] - \mathbb{E}\left[\begin{pmatrix} \mathbf{e}_t \\ q_t \end{pmatrix}\right] \mathbb{E}\left[(\mathbf{e}_t' \; q_t)\right].$$

From the above assumptions, we have

$$\begin{pmatrix} s_t^2 & s_t\mathbb{E}[\mathbf{e}_t'] & s_t\mathbb{E}[q_t] \\ s_t\mathbb{E}[\mathbf{e}_t] & \mathbb{E}[\mathbf{e}_t\mathbf{e}_t'] & \mathbb{E}[\mathbf{e}_t q_t] \\ s_t\mathbb{E}[q_t] & \mathbb{E}[q_t\mathbf{e}_t'] & \mathbb{E}[q_t^2] \end{pmatrix} \succ 0.$$

We further define the excess return vector of risky assets $\mathbf{P}_t = (P_t^1, \cdots, P_t^n)'$ as $(e_t^1 - s_t, \cdots, e_t^n - s_t)'$. The following is then true for $t = 0, 1, \cdots, T-1$:

$$\begin{pmatrix} s_t^2 & s_t\mathbb{E}[\mathbf{P}_t'] & s_t\mathbb{E}[q_t] \\ s_t\mathbb{E}[\mathbf{P}_t] & \mathbb{E}[\mathbf{P}_t\mathbf{P}_t'] & \mathbb{E}[\mathbf{P}_t q_t] \\ s_t\mathbb{E}[q_t] & \mathbb{E}[q_t\mathbf{P}_t'] & \mathbb{E}[q_t^2] \end{pmatrix} = \begin{pmatrix} 1 & \mathbf{0}' & 0 \\ -\mathbf{1} & I & \mathbf{0} \\ 0 & \mathbf{0}' & 1 \end{pmatrix} \begin{pmatrix} s_t^2 & s_t\mathbb{E}[\mathbf{e}_t'] & s_t\mathbb{E}[q_t] \\ s_t\mathbb{E}[\mathbf{e}_t] & \mathbb{E}[\mathbf{e}_t\mathbf{e}_t'] & \mathbb{E}[\mathbf{e}_t q_t] \\ s_t\mathbb{E}[q_t] & \mathbb{E}[q_t\mathbf{e}_t'] & \mathbb{E}[q_t^2] \end{pmatrix} \begin{pmatrix} 1 & -\mathbf{1}' & 0 \\ \mathbf{0} & I & \mathbf{0} \\ 0 & \mathbf{0}' & 1 \end{pmatrix} \succ 0,$$

where $\mathbf{1}$ and $\mathbf{0}$ are the n-dimensional all-one and all-zero vectors, respectively, and I is the $n \times n$ identity matrix, which further implies, for $t = 0, 1, \cdots, T-1$,

$$\begin{pmatrix} \mathbb{E}[\mathbf{P}_t\mathbf{P}_t'] & \mathbb{E}[\mathbf{P}_t q_t] \\ \mathbb{E}[q_t\mathbf{P}_t'] & \mathbb{E}[q_t^2] \end{pmatrix} \succ 0,$$

and $s_t^2(1 - B_t) > 0$, where $B_t \triangleq \mathbb{E}[\mathbf{P}_t']\mathbb{E}^{-1}[\mathbf{P}_t\mathbf{P}_t']\mathbb{E}[\mathbf{P}_t]$. This implies that $0 < B_t < 1$ for $t = 0, 1, \cdots, T-1$.

Let x_t and l_t be the wealth and liability of the investor at the beginning of period t, respectively, then $x_t - l_t$ is the net wealth. At period t, if π_t^i, $i = 1, 2, \cdots, n$ is the amount invested in the ith risky asset, then $x_t - \sum_{i=1}^n \pi_t^i$ is the amount invested in the risk-free asset. We assume in this paper that the liability is exogenous, which means it is uncontrollable and cannot be affected by the investor's strategies. Denote the information set at the beginning of period t, $t = 1, 2, \cdots, T-1$, as $\mathcal{F}_t = \sigma(\mathbf{P}_0, \mathbf{P}_1, \cdots, \mathbf{P}_{t-1}, q_0, q_1, \cdots, q_{t-1})$ and the trivial σ-algebra over Ω as $\mathcal{F}_0$. Therefore, $\mathbb{E}[\cdot|\mathcal{F}_0]$ is just the unconditional expectation $\mathbb{E}[\cdot]$. We confine all admissible investment strategies to be the $\mathcal{F}_t$-adapted Markov controls, i.e., $\pi_t = (\pi_t^1, \ \pi_t^2, \ \cdots, \ \pi_t^n)' \in \mathcal{F}_t$. Then, $\mathbf{P}_t$ and π_t are independent, $\{x_t, l_t\}$ is an adapted Markovian process and $\mathcal{F}_t = \sigma(x_t, l_t)$.

The multi-period mean-variance model of asset-liability management is to seek the best strategy, $\pi_t^* = [(\pi_t^1)^*, (\pi_t^2)^*, \cdots, (\pi_t^n)^*]'$, $t = 0, 1, \cdots, T-1$, which is the solution of the following dynamic stochastic optimization problem,

$$\begin{cases} \min \mathrm{Var}(x_T - l_T) \equiv \mathbb{E}[(x_T - l_T - d)^2], \\ \text{s.t. } \mathbb{E}[x_T - l_T] = d, \\ \qquad x_{t+1} = s_t\left(x_t - \sum_{i=1}^n \pi_t^i\right) + \sum_{i=1}^n e_t^i \pi_t^i \\ \qquad\qquad = s_t x_t + \mathbf{P}_t'\pi_t, \\ \qquad l_{t+1} = q_t l_t, \qquad\qquad t = 0, 1, \cdots, T-1. \end{cases} \tag{9.1}$$

Introducing a Lagrange multiplier $2\omega > 0$ yields

$$\begin{cases} \min \mathbb{E}[(x_T - l_T - d)^2] - 2\omega(\mathbb{E}[x_T - l_T] - d), \\ \quad \text{s.t. } \{x_t, l_t, \pi_t\} \text{ satisfies the dynamic system of problem (9.1)}, \end{cases} \tag{9.2}$$

which is equivalent to the following problem,

$$\begin{cases} \min \mathbb{E}[(x_T - l_T - d - \omega)^2], \\ \text{s.t. } \{x_t, l_t, \pi_t\} \text{ satisfies the dynamic system of problem (9.1)}, \end{cases} \tag{9.3}$$

in the sense that the two problems have the same optimal strategy. It can be rewritten as

$$\begin{cases} \min \mathbb{E}[(x_T - \gamma - l_T)^2], \\ \text{s.t. } \{x_t, l_t, \pi_t\} \text{ satisfies the dynamic system of problem (9.1)}, \end{cases} \tag{9.4}$$

where $\gamma = d + \omega$. Set

$$y_t := x_t - \gamma \prod_{k=t}^{T-1} s_k^{-1}, \tag{9.5}$$

and denote $\prod_{k=T}^{T-1} s_k^{-1} := 1$. Then the dynamic system of problem (9.1) turns to

$$\begin{cases} y_{t+1} = s_t y_t + \mathbf{P}_t' \pi_t, \\ l_{t+1} = q_t l_t, \qquad t = 0, 1, \cdots, T-1, \end{cases} \tag{9.6}$$

where $y_0 = x_0 - \gamma \prod_{k=0}^{T-1} s_k^{-1}$. Problem (9.4) can be reformulated as

$$\begin{cases} \min \mathbb{E}[(y_T - l_T)^2], \\ \text{s.t. } \{y_t, l_t, \pi_t\} \text{ satisfies Eq. (9.6)}, \end{cases} \tag{9.7}$$

and it is the 'same' as the following problem:

$$\begin{cases} \min \mathbb{E}[y_T^2 - 2l_T y_T], \\ \text{s.t. } \{y_t, l_t, \pi_t\} \text{ satisfies Eq. (9.6)}, \end{cases} \tag{9.8}$$

The 'same' here means that they have the same optimal strategy. By studying problem (9.8), we can obtain the optimal strategy of the original problem (9.1).

9.3 The Optimal Strategies

9.3.1 The Optimal Strategy with Correlation of Assets and Liability

In this subsection, assume that the returns of assets and liability are correlated at every period, i.e., $\mathbf{P}_t$ and q_t are dependent on each other at period $t = 0, 1, \cdots, T-1$. Before we derive the optimal strategy, we denote

$$\widehat{B}_t \triangleq \mathbb{E}[q_t\mathbf{P}_t']\mathbb{E}^{-1}[\mathbf{P}_t\mathbf{P}_t']\mathbb{E}[\mathbf{P}_t],$$

$$\widetilde{B}_t \triangleq \mathbb{E}[q_t\mathbf{P}_t']\mathbb{E}^{-1}[\mathbf{P}_t\mathbf{P}_t']\mathbb{E}[q_t\mathbf{P}_t],$$

for $t = 0, 1, 2, \cdots, T-1$.

Theorem 1 *Assume that the returns of assets and liability are correlated at every period. Then the optimal strategy of problem (9.1) is given by*

$$\pi_t^* = -\mathbb{E}^{-1}[\mathbf{P}_t\mathbf{P}_t']\mathbb{E}[\mathbf{P}_t]s_t\left(x_t - \gamma^* \prod_{k=t}^{T-1} s_k^{-1}\right) + \left(\prod_{k=t+1}^{T-1} \frac{\mathbb{E}[q_k] - \widehat{B}_k}{(1-B_k)s_k}\right)\mathbb{E}^{-1}[\mathbf{P}_t\mathbf{P}_t']\mathbb{E}[q_t\mathbf{P}_t]l_t, \tag{9.9}$$

where

$$\gamma^* = \frac{x_0 \prod_{k=0}^{T-1}(1-B_k)s_k - d - l_0 \prod_{k=0}^{T-1}\left(\mathbb{E}[q_k] - \widehat{B}_k\right)}{\prod_{k=0}^{T-1}(1-B_k) - 1}. \tag{9.10}$$

Proof We prove it by making use of the dynamic programming approach. For the information set $\mathcal{F}_t$, the cost-to-go functional of problem (9.8) at period t is

$$J_t(y_t, l_t) = \min_{\pi_t} \mathbb{E}\left[J_{t+1}(y_{t+1}, l_{t+1}) \middle| \mathcal{F}_t\right],$$

where the terminal condition $J_T(y_T, l_T) = y_T^2 - 2l_T y_T$. □

We start from the last stage $T-1$. While $t = T-1$, we have

$$\begin{aligned}
&\mathbb{E}\left[J_T(y_T, l_T) \middle| \mathcal{F}_{T-1}\right] \\
&= \mathbb{E}\left[y_T^2 - 2l_T y_T \middle| \mathcal{F}_{T-1}\right] \\
&= s_{T-1}^2 y_{T-1}^2 + 2s_{T-1}y_{T-1}\mathbb{E}[\mathbf{P}_{T-1}']\pi_{T-1} + \pi_{T-1}'\mathbb{E}[\mathbf{P}_{T-1}\mathbf{P}_{T-1}']\pi_{T-1} \\
&\quad - 2\mathbb{E}[q_{T-1}]s_{T-1}l_{T-1}y_{T-1} - 2\mathbb{E}[q_{T-1}\mathbf{P}_{T-1}']l_{T-1}\pi_{T-1}.
\end{aligned}$$

Minimizing it with respect to π_{T-1} yields the optimal decision at period $T-1$ as follows:

$$\pi^*_{T-1} = -\mathbb{E}^{-1}[\mathbf{P}_{T-1}\mathbf{P}'_{T-1}]\mathbb{E}[\mathbf{P}_{T-1}]s_{T-1}y_{T-1} + \mathbb{E}^{-1}[\mathbf{P}_{T-1}\mathbf{P}'_{T-1}]\mathbb{E}[q_{T-1}\mathbf{P}_{T-1}]l_{T-1}.$$

Substituting π^*_{T-1} to $\mathbb{E}\big[J_T(y_T, l_T)\big|\mathcal{F}_{T-1}\big]$, we obtain

$$\begin{aligned} J_{T-1}(y_{T-1}, l_{T-1}) &= \min_{\pi_{T-1}} \mathbb{E}\big[J_T(y_T, l_T)\big|\mathcal{F}_{T-1}\big] \\ &= (1-B_{T-1})s^2_{T-1}y^2_{T-1} - 2\big(\mathbb{E}[q_{T-1}] - \widehat{B}_{T-1}\big)s_{T-1}l_{T-1}y_{T-1} - \widetilde{B}_{T-1}l^2_{T-1}. \end{aligned}$$

In order to derive the cost-to-go functional and the optimal decision at period t clearly, we patiently repeat the procedure at time $T-2$. While $t = T-2$, we have

$$\begin{aligned} &\mathbb{E}\big[J_{T-1}(y_{T-1}, l_{T-1})\big|\mathcal{F}_{T-2}\big] \\ &= \mathbb{E}\big[(1-B_{T-1})s^2_{T-1}y^2_{T-1} - 2\big(\mathbb{E}[q_{T-1}] - \widehat{B}_{T-1}\big)s_{T-1}l_{T-1}y_{T-1} - \widetilde{B}_{T-1}l^2_{T-1}\big|\mathcal{F}_{T-2}\big] \\ &= (1-B_{T-1})s^2_{T-1}\Big(s^2_{T-2}y^2_{T-2} + 2s_{T-2}y_{T-2}\mathbb{E}[\mathbf{P}'_{T-2}]\pi_{T-2} + \pi'_{T-2}\mathbb{E}[\mathbf{P}_{T-2}\mathbf{P}'_{T-2}]\pi_{T-2}\Big) \\ &\quad - 2\big(\mathbb{E}[q_{T-1}] - \widehat{B}_{T-1}\big)\mathbb{E}\big[q_{T-2}]s_{T-1}s_{T-2}l_{T-2}y_{T-2} \\ &\quad - 2\big(\mathbb{E}[q_{T-1}] - \widehat{B}_{T-1}\big)\mathbb{E}[q_{T-2}\mathbf{P}'_{T-2}]s_{T-1}l_{T-2}\pi_{T-2} \\ &\quad - \widetilde{B}_{T-1}\mathbb{E}[q^2_{T-2}]l^2_{T-2}. \end{aligned}$$

We derive the following optimal decision at period $T-2$ by minimizing the above functional with respect to π_{T-2}

$$\begin{aligned} \pi^*_{T-2} = &-\mathbb{E}^{-1}[\mathbf{P}_{T-2}\mathbf{P}'_{T-2}]\mathbb{E}[\mathbf{P}_{T-2}]s_{T-2}y_{T-2} \\ &+ \frac{\mathbb{E}[q_{T-1}] - \widehat{B}_{T-1}}{(1-B_{T-1})s_{T-1}}\mathbb{E}^{-1}[\mathbf{P}_{T-2}\mathbf{P}'_{T-2}]\mathbb{E}[q_{T-2}\mathbf{P}_{T-2}]l_{T-2}. \end{aligned}$$

Then the cost-to-go functional at period $T-2$ is

$$\begin{aligned} J_{T-2}(y_{T-2}, l_{T-2}) &= \min_{\pi_{T-2}} \mathbb{E}\big[J_{T-1}(y_{T-1}, l_{T-1})\big|\mathcal{F}_{T-2}\big] \\ &= (1-B_{T-1})(1-B_{T-2})s^2_{T-1}s^2_{T-2}y^2_{T-2} \\ &\quad - 2\big(\mathbb{E}[q_{T-1}] - \widehat{B}_{T-1}\big)\big(\mathbb{E}\big[q_{T-2}] - \widehat{B}_{T-2}\big)s_{T-1}s_{T-2}l_{T-2}y_{T-2} \\ &\quad - \left(\frac{\big(\mathbb{E}[q_{T-1}] - \widehat{B}_{T-1}\big)^2}{1-B_{T-1}}\widetilde{B}_{T-2} + \widetilde{B}_{T-1}\mathbb{E}[q^2_{T-2}]\right)l^2_{T-2}. \end{aligned}$$

While $t = T-3$, we can similarly get

$$
\begin{aligned}
&\mathbb{E}\big[J_{T-2}(y_{T-2}, l_{T-2})\big|\mathcal{F}_{T-3}\big] \\
&= \mathbb{E}\Big[(1-B_{T-1})(1-B_{T-2})s_{T-1}^2 s_{T-2}^2 y_{T-2}^2 \\
&\quad - 2\big(\mathbb{E}[q_{T-1}] - \widehat{B}_{T-1}\big)\big(\mathbb{E}[q_{T-2}] - \widehat{B}_{T-2}\big)s_{T-1}s_{T-2}l_{T-2}y_{T-2} \\
&\quad - \left(\frac{\big(\mathbb{E}[q_{T-1}] - \widehat{B}_{T-1}\big)^2}{1-B_{T-1}}\widetilde{B}_{T-2} + \widetilde{B}_{T-1}\mathbb{E}[q_{T-2}^2]\right) l_{T-2}^2 \Big| \mathcal{F}_{T-3}\Big] \\
&= (1-B_{T-1})(1-B_{T-2})s_{T-1}^2 s_{T-2}^2 \\
&\quad \times \left(s_{T-3}^2 y_{T-3}^2 + 2s_{T-3}y_{T-3}\mathbb{E}[\mathbf{P}'_{T-3}]\pi_{T-3} + \pi'_{T-3}\mathbb{E}[\mathbf{P}_{T-3}\mathbf{P}'_{T-3}]\pi_{T-3}\right) \\
&\quad - 2\big(\mathbb{E}[q_{T-1}] - \widehat{B}_{T-1}\big)\big(\mathbb{E}[q_{T-2}] - \widehat{B}_{T-2}\big)\mathbb{E}[q_{T-3}]s_{T-1}s_{T-2}s_{T-3}l_{T-3}y_{T-3} \\
&\quad - 2\big(\mathbb{E}[q_{T-1}] - \widehat{B}_{T-1}\big)\big(\mathbb{E}[q_{T-2}] - \widehat{B}_{T-2}\big)\mathbb{E}[q_{T-3}\mathbf{P}'_{T-3}]s_{T-1}s_{T-2}l_{T-3}\pi_{T-3} \\
&\quad - \left(\frac{\big(\mathbb{E}[q_{T-1}] - \widehat{B}_{T-1}\big)^2}{1-B_{T-1}}\widetilde{B}_{T-2} + \widetilde{B}_{T-1}\mathbb{E}[q_{T-2}^2]\right)\mathbb{E}[q_{T-3}^2]l_{T-3}^2.
\end{aligned}
$$

Thus the optimal decision at period $T-3$ is

$$
\begin{aligned}
\pi_{T-3}^* =& -\mathbb{E}^{-1}[\mathbf{P}_{T-3}\mathbf{P}'_{T-3}]\mathbb{E}[\mathbf{P}_{T-3}]s_{T-3}y_{T-3} \\
&+ \frac{\mathbb{E}[q_{T-1}] - \widehat{B}_{T-1}}{(1-B_{T-1})s_{T-1}}\frac{\mathbb{E}[q_{T-2}] - \widehat{B}_{T-2}}{(1-B_{T-2})s_{T-2}}\mathbb{E}^{-1}[\mathbf{P}_{T-3}\mathbf{P}'_{T-3}]\mathbb{E}[q_{T-3}\mathbf{P}_{T-3}]l_{T-3},
\end{aligned}
$$

and the cost-to-go functional at period $T-3$ is

$$
\begin{aligned}
J_{T-3}&(y_{T-3}, l_{T-3}) = \min_{\pi_{T-3}} \mathbb{E}\big[J_{T-2}(y_{T-2}, l_{T-2})\big|\mathcal{F}_{T-3}\big] \\
&= (1-B_{T-1})(1-B_{T-2})(1-B_{T-3})s_{T-1}^2 s_{T-2}^2 s_{T-3}^2 y_{T-3}^2 \\
&\quad - 2\big(\mathbb{E}[q_{T-1}] - \widehat{B}_{T-1}\big)\big(\mathbb{E}[q_{T-2}] - \widehat{B}_{T-2}\big)\big(\mathbb{E}[q_{T-3}] - \widehat{B}_{T-3}\big)s_{T-1}s_{T-2}s_{T-3}l_{T-3}y_{T-3} \\
&\quad - \Bigg[\frac{\big(\mathbb{E}[q_{T-1}] - \widehat{B}_{T-1}\big)^2}{1-B_{T-1}}\frac{\big(\mathbb{E}[q_{T-2}] - \widehat{B}_{T-2}\big)^2}{1-B_{T-2}}\widetilde{B}_{T-3} \\
&\quad + \left(\frac{\big(\mathbb{E}[q_{T-1}] - \widehat{B}_{T-1}\big)^2}{1-B_{T-1}}\widetilde{B}_{T-2} + \widetilde{B}_{T-1}\mathbb{E}[q_{T-2}^2]\right)\mathbb{E}[q_{T-3}^2]\Bigg] l_{T-3}^2.
\end{aligned}
$$

Inspired by the above three stages, we conjecture that the cost-to-go functional at period t can be expressed in the following form:

$$J_t(y_t,l_t) = \Big(\prod_{k=t}^{T-1}(1-B_k)s_k^2\Big)y_t^2 - 2\Big(\prod_{k=t}^{T-1}(\mathbb{E}[q_k]-\widehat{B}_k)s_k\Big)l_t y_t \\ -\sum_{j=t}^{T-1}\Big(\prod_{k=j+1}^{T-1}\frac{(\mathbb{E}[q_k]-\widehat{B}_k)^2}{1-B_k}\Big)\widetilde{B}_j\Big(\prod_{m=t}^{j-1}\mathbb{E}[q_m^2]\Big)l_t^2. \tag{9.11}$$

Next, we prove it in mathematical induction. Assume that the cost-to-go functional (9.11) holds at period $t+1$. Then we shall prove that it still holds at time t. For the given information set $\mathcal{F}_t$, we have

$$\begin{aligned}
&\mathbb{E}\big[J_{t+1}(y_{t+1},l_{t+1})\big|\mathcal{F}_t\big]\\
&=\mathbb{E}\Big[\Big(\prod_{k=t+1}^{T-1}(1-B_k)s_k^2\Big)y_{t+1}^2 - 2\Big(\prod_{k=t+1}^{T-1}(\mathbb{E}[q_k]-\widehat{B}_k)s_k\Big)l_{t+1}y_{t+1}\\
&\quad -\sum_{j=t+1}^{T-1}\Big(\prod_{k=j+1}^{T-1}\frac{\big(\mathbb{E}[q_k]-\widehat{B}_k\big)^2}{1-B_k}\Big)\widetilde{B}_j\Big(\prod_{m=t+1}^{j-1}\mathbb{E}[q_m^2]\Big)l_{t+1}^2\Big|\mathcal{F}_t\Big]\\
&=\Big(\prod_{k=t+1}^{T-1}(1-B_k)s_k^2\Big)\big(s_t^2y_t^2+2s_ty_t\mathbb{E}[\mathbf{P}_t']\pi_t+\pi_t'\mathbb{E}[\mathbf{P}_t\mathbf{P}_t']\pi_t\big)\\
&\quad -2\Big(\prod_{k=t+1}^{T-1}(\mathbb{E}[q_k]-\widehat{B}_k)s_k\Big)\big(\mathbb{E}[q_t]s_tl_ty_t+\mathbb{E}[q_t\mathbf{P}_t']l_t\pi_t\big)\\
&\quad -\sum_{j=t+1}^{T-1}\Big(\prod_{k=j+1}^{T-1}\frac{\big(\mathbb{E}[q_k]-\widehat{B}_k\big)^2}{1-B_k}\Big)\widetilde{B}_j\Big(\prod_{m=t+1}^{j-1}\mathbb{E}[q_m^2]\Big)\mathbb{E}[q_t^2]l_t^2.
\end{aligned}$$

Minimizing the above functional with respect to π_t, we get the optimal strategy decision at time t as follows:

$$\pi_t^* = -\mathbb{E}^{-1}[\mathbf{P}_t\mathbf{P}_t']\mathbb{E}[\mathbf{P}_t]s_ty_t + \Big(\prod_{k=t+1}^{T-1}\frac{\mathbb{E}[q_k]-\widehat{B}_k}{(1-B_k)s_k}\Big)\mathbb{E}^{-1}[\mathbf{P}_t\mathbf{P}_t']\mathbb{E}[q_t\mathbf{P}_t]l_t.$$

Substituting it to $\mathbb{E}\big[J_{t+1}(y_{t+1},l_{t+1})\big|\mathcal{F}_t\big]$ yields

$$\begin{aligned}
J_t(y_t,l_t) &= \min_{\pi_t}\mathbb{E}\big[J_{t+1}(y_{t+1},l_{t+1})\big|\mathcal{F}_t\big]\\
&=\Big(\prod_{k=t+1}^{T-1}(1-B_k)s_k^2\Big)s_t^2y_t^2 - 2\Big(\prod_{k=t+1}^{T-1}(\mathbb{E}[q_k]-\widehat{B}_k)s_k\Big)\mathbb{E}[q_t]s_tl_ty_t
\end{aligned}$$

$$
\begin{aligned}
&-\left(\prod_{k=t+1}^{T-1}(1-B_k)s_k^2\right)\mathbb{E}[\mathbf{P}_t']\mathbb{E}^{-1}[\mathbf{P}_t\mathbf{P}_t']\mathbb{E}[\mathbf{P}_t]s_t^2y_t^2\\
&+2\left(\prod_{k=t+1}^{T-1}(\mathbb{E}[q_k]-\widehat{B}_k)s_k\right)\mathbb{E}[q_t\mathbf{P}_t']\mathbb{E}^{-1}[\mathbf{P}_t\mathbf{P}_t']\mathbb{E}[\mathbf{P}_t]s_tl_ty_t\\
&-\left(\prod_{k=t+1}^{T-1}\frac{\left(\mathbb{E}[q_k]-\widehat{B}_k\right)^2}{1-B_k}\right)\mathbb{E}[q_t\mathbf{P}_t']\mathbb{E}^{-1}[\mathbf{P}_t\mathbf{P}_t']\mathbb{E}[q_t\mathbf{P}_t]l_t^2\\
&-\sum_{j=t+1}^{T-1}\left(\prod_{k=j+1}^{T-1}\frac{\left(\mathbb{E}[q_k]-\widehat{B}_k\right)^2}{1-B_k}\right)\widetilde{B}_j\left(\prod_{m=t+1}^{j-1}\mathbb{E}[q_m^2]\right)\mathbb{E}[q_t^2]l_t^2\\
&\quad=\left(\prod_{k=t}^{T-1}(1-B_k)s_k^2\right)y_t^2-2\left(\prod_{k=t}^{T-1}(\mathbb{E}[q_k]-\widehat{B}_k)s_k\right)l_ty_t\\
&\quad-\sum_{j=t}^{T-1}\left(\prod_{k=j+1}^{T-1}\frac{\left(\mathbb{E}[q_k]-\widehat{B}_k\right)^2}{1-B_k}\right)\widetilde{B}_j\left(\prod_{m=t}^{j-1}\mathbb{E}[q_m^2]\right)l_t^2,
\end{aligned}
$$

which proves (9.11).

To derive the expression (9.10) of γ, we first consider the value of the optimal objective function in (9.8). In fact,

$$
\begin{aligned}
&\mathbb{E}\left[y_T^2-2l_Ty_T\right]=\mathbb{E}\left[y_T^2-2l_Ty_T\middle|\mathcal{F}_0\right]=J_0(y_0,l_0)\\
&=y_0^2\prod_{k=0}^{T-1}(1-B_k)s_k^2-2l_0y_0\prod_{k=0}^{T-1}(\mathbb{E}[q_k]-\widehat{B}_k)s_k\\
&\quad-l_0^2\sum_{j=0}^{T-1}\left(\prod_{k=j+1}^{T-1}\frac{\left(\mathbb{E}[q_k]-\widehat{B}_k\right)^2}{1-B_k}\right)\widetilde{B}_j\left(\prod_{m=0}^{j-1}\mathbb{E}[q_m^2]\right).
\end{aligned}
$$

Then

$$
\begin{aligned}
\mathrm{Var}(x_T-l_T)&=\mathbb{E}[(x_T-l_T-d)^2]\\
&=\mathbb{E}[(x_T-l_T-d)^2]-2\omega(\mathbb{E}[x_T-l_T]-d)+\omega^2-\omega^2\\
&=\mathbb{E}[(x_T-l_T-d)^2-2\omega(x_T-l_T-d)+\omega^2]-\omega^2\\
&=\mathbb{E}[(x_T-l_T-d-\omega)^2]-\omega^2\\
&=\mathbb{E}[(y_T-l_T)^2]-\omega^2\\
&=\mathbb{E}[y_T^2-2l_Ty_T]+\mathbb{E}[l_T^2]-\omega^2
\end{aligned}
$$

$$= y_0^2 \prod_{k=0}^{T-1} (1-B_k) s_k^2 - 2 l_0 y_0 \prod_{k=0}^{T-1} \big(\mathbb{E}[q_k] - \widehat{B}_k\big) s_k$$
$$- l_0^2 \sum_{j=0}^{T-1} \Bigg(\prod_{k=j+1}^{T-1} \frac{\big(\mathbb{E}[q_k] - \widehat{B}_k\big)^2}{1-B_k} \Bigg) \widetilde{B}_j \Bigg(\prod_{m=0}^{j-1} \mathbb{E}[q_m^2] \Bigg)$$
$$+ l_0^2 \prod_{k=0}^{T-1} \mathbb{E}[q_k^2] - \omega^2.$$

Since

$$y_0 = x_0 - \gamma \prod_{k=0}^{T-1} s_k^{-1} = x_0 - (d+\omega) \prod_{k=0}^{T-1} s_k^{-1},$$

we have

$$y_0^2 \prod_{k=0}^{T-1} (1-B_k) s_k^2 = \Bigg(x_0 - (d+\omega) \prod_{k=0}^{T-1} s_k^{-1} \Bigg)^2 \prod_{k=0}^{T-1} (1-B_k) s_k^2$$
$$= \Bigg(x_0 \prod_{k=0}^{T-1} s_k - (d+\omega) \Bigg)^2 \prod_{k=0}^{T-1} (1-B_k)$$

and

$$y_0 \prod_{k=0}^{T-1} (\mathbb{E}[q_k] - \widehat{B}_k) s_k = \Bigg(x_0 - (d+\omega) \prod_{k=0}^{T-1} s_k^{-1} \Bigg) \prod_{k=0}^{T-1} \big(\mathbb{E}[q_k] - \widehat{B}_k\big) s_k$$
$$= \Bigg(x_0 \prod_{k=0}^{T-1} s_k - (d+\omega) \Bigg) \prod_{k=0}^{T-1} \big(\mathbb{E}[q_k] - \widehat{B}_k\big).$$

Hence,

$$\mathrm{Var}(x_T - l_T)$$
$$= \Bigg(x_0 \prod_{k=0}^{T-1} s_k - (d+\omega) \Bigg)^2 \prod_{k=0}^{T-1} (1-B_k) - 2 l_0 \Bigg(x_0 \prod_{k=0}^{T-1} s_k - (d+\omega) \Bigg) \prod_{k=0}^{T-1} \big(\mathbb{E}[q_k] - \widehat{B}_k\big)$$
$$- l_0^2 \sum_{j=0}^{T-1} \Bigg(\prod_{k=j+1}^{T-1} \frac{\big(\mathbb{E}[q_k] - \widehat{B}_k\big)^2}{1-B_k} \Bigg) \widetilde{B}_j \Bigg(\prod_{m=0}^{j-1} \mathbb{E}[q_m^2] \Bigg) + l_0^2 \prod_{k=0}^{T-1} \mathbb{E}[q_k^2] - \omega^2$$
$$= \Bigg[\prod_{k=0}^{T-1} (1-B_k) - 1 \Bigg] \Bigg(\omega - \frac{\big(x_0 \prod_{k=0}^{T-1} s_k - d\big) \prod_{k=0}^{T-1} (1-B_k) - l_0 \prod_{k=0}^{T-1} \big(\mathbb{E}[q_k] - \widehat{B}_k\big)}{\prod_{k=0}^{T-1} (1-B_k) - 1} \Bigg)^2$$
$$+ \frac{\prod_{k=0}^{T-1} (1-B_k)}{1 - \prod_{k=0}^{T-1} (1-B_k)} \Bigg(d - x_0 \prod_{k=0}^{T-1} s_k + l_0 \prod_{k=0}^{T-1} \frac{\mathbb{E}[q_k] - \widehat{B}_k}{1-B_k} \Bigg)^2 + l_0^2 C_0, \tag{9.12}$$

where

$$C_0 = -\prod_{k=0}^{T-1} \frac{\left(\mathbb{E}[q_k] - \widehat{B}_k\right)^2}{1 - B_k} - \sum_{j=0}^{T-1} \left(\prod_{k=j+1}^{T-1} \frac{(\mathbb{E}[q_k] - \widehat{B}_k)^2}{1 - B_k} \right) \widetilde{B}_j \left(\prod_{m=0}^{j-1} \mathbb{E}[q_m^2] \right) + \prod_{k=0}^{T-1} \mathbb{E}[q_k^2]. \tag{9.13}$$

Since $0 < B_t < 1$ for $t = 0, 1, \cdots, T-1$,

$$0 < \prod_{k=0}^{T-1} (1 - B_k) < 1.$$

This implies that the variance term $\mathrm{Var}(x_T - l_T)$ in (9.12) is concave in ω. To obtain the minimum variance $\mathrm{Var}(x_T - l_T)$ and the optimal strategy for the original portfolio selection problem (9.1), one needs to maximize the value in (9.12) over $\omega \in \mathbb{R}$ according to the Lagrange duality theorem in Luenberger (1968). Taking the first order derivative for (9.12) with respect to ω yields

$$\omega^* = \frac{\left(x_0 \prod_{k=0}^{T-1} s_k - d \right) \prod_{k=0}^{T-1} (1 - B_k) - l_0 \prod_{k=0}^{T-1} \left(\mathbb{E}[q_k] - \widehat{B}_k\right)}{\prod_{k=0}^{T-1} (1 - B_k) - 1}.$$

A simple calculation of $\gamma^* = d + \omega^*$ implies the desired result (9.10). □

9.3.2 *Efficient Frontier*

For any matrix M, we denote by M^+ the Moore-Penrose pseudoinverse of M satisfying

$$MM^+M = M, M^+MM^+ = M^+, (MM^+)' = MM^+, (M^+M)' = M^+M.$$

It can be proved that M^+ is unique for any matrix M and if the inverse M^{-1} of M exists, then $M^+ = M^{-1}$.

Let M be a square matrix partitioned as

$$M = \begin{pmatrix} M_{11} & M_{12} \\ M_{21} & M_{22} \end{pmatrix}. \tag{9.14}$$

Then we have

Lemma 1 *If M_{22} is invertible, then* $|M| = |M_{22}|\left|M_{11} - M_{12}M_{22}^{-1}M_{21}\right|$.

Suppose that the square matrix M is symmetrical and partitioned as (9.14), where M_{11} and M_{22} are also symmetrical square matrices, then the following two lemmas hold.

Lemma 2 *The matrix $M \succcurlyeq 0$ is equivalent to $M_{22} \succcurlyeq 0, M_{22}M_{22}^{+}M_{21} = M_{21}$ and $M_{11} - M_{12}M_{22}^{+}M_{21} \succcurlyeq 0$, where $M_{21} = M_{12}'$.*

Lemma 3 *If $M \succcurlyeq N \succcurlyeq 0$, then $|M| \geq |N|$.*

The proof of Lemmas 1 and 3 can be found in Zhang (2011). And the proof of Lemma 2 can be found in Albert (1969).

Before we analyze the efficient frontier, we prove the following important result.

Lemma 4 *If $\mathbb{E}\left[\begin{pmatrix}\mathbf{P}_k \\ q_k\end{pmatrix}(\mathbf{P}_k' \; q_k)\right]$ is positive definite for $k = 0, 1, \cdots, T-1$, then*

$$C_0 \geq 0. \tag{9.15}$$

Proof Let $L_k = \begin{pmatrix}\mathbf{P}_k \\ 1\end{pmatrix}$ and $Q_k = \begin{pmatrix}\mathbf{P}_k \\ q_k\end{pmatrix}$, then

$$\begin{pmatrix}\mathbb{E}[\mathbf{P}_k\mathbf{P}_k'] & \mathbb{E}[\mathbf{P}_k] \\ \mathbb{E}[\mathbf{P}_k'] & 1\end{pmatrix} = \mathbb{E}\left[\begin{pmatrix}\mathbf{P}_k \\ 1\end{pmatrix}(\mathbf{P}_k' \; 1)\right] = \mathbb{E}[L_kL_k'], \tag{9.16}$$

$$\begin{pmatrix}\mathbb{E}[\mathbf{P}_k\mathbf{P}_k'] & \mathbb{E}[q_k\mathbf{P}_k] \\ \mathbb{E}[q_k\mathbf{P}_k'] & \mathbb{E}[q_k^2]\end{pmatrix} = \mathbb{E}\left[\begin{pmatrix}\mathbf{P}_k \\ q_k\end{pmatrix}(\mathbf{P}_k' \; q_k)\right] = \mathbb{E}[Q_kQ_k'], \tag{9.17}$$

$$\begin{pmatrix}\mathbb{E}[\mathbf{P}_k\mathbf{P}_k'] & \mathbb{E}[\mathbf{P}_k] \\ \mathbb{E}[q_k\mathbf{P}_k'] & \mathbb{E}[q_k]\end{pmatrix} = \mathbb{E}\left[\begin{pmatrix}\mathbf{P}_k \\ q_k\end{pmatrix}(\mathbf{P}_k' \; 1)\right] = \mathbb{E}[Q_kL_k']. \tag{9.18}$$

Taking determinant on both sides for (9.16)–(9.18) and according to Lemma 1, we get

$$\begin{vmatrix}\mathbb{E}[\mathbf{P}_k\mathbf{P}_k'] & \mathbb{E}[\mathbf{P}_k] \\ \mathbb{E}[\mathbf{P}_k'] & 1\end{vmatrix} = \left(1 - \mathbb{E}[\mathbf{P}_k']\mathbb{E}^{-1}[\mathbf{P}_k\mathbf{P}_k']\mathbb{E}[\mathbf{P}_k]\right)\left|\mathbb{E}[\mathbf{P}_k\mathbf{P}_k']\right| = \left|\mathbb{E}[L_kL_k']\right|, \tag{9.19}$$

$$\begin{vmatrix}\mathbb{E}[\mathbf{P}_k\mathbf{P}_k'] & \mathbb{E}[q_k\mathbf{P}_k] \\ \mathbb{E}[q_k\mathbf{P}_k'] & \mathbb{E}[q_k^2]\end{vmatrix} = \left(\mathbb{E}[q_k^2] - \mathbb{E}[q_k\mathbf{P}_k']\mathbb{E}^{-1}[\mathbf{P}_k\mathbf{P}_k']\mathbb{E}[q_k\mathbf{P}_k]\right)\left|\mathbb{E}[\mathbf{P}_k\mathbf{P}_k']\right| = \left|\mathbb{E}[Q_kQ_k']\right|, \tag{9.20}$$

$$\begin{vmatrix}\mathbb{E}[\mathbf{P}_k\mathbf{P}_k'] & \mathbb{E}[\mathbf{P}_k] \\ \mathbb{E}[q_k\mathbf{P}_k'] & \mathbb{E}[q_k]\end{vmatrix} = \left(\mathbb{E}[q_k] - \mathbb{E}[q_k\mathbf{P}_k']\mathbb{E}^{-1}[\mathbf{P}_k\mathbf{P}_k']\mathbb{E}[\mathbf{P}_k]\right)\left|\mathbb{E}[\mathbf{P}_k\mathbf{P}_k']\right| = \left|\mathbb{E}[Q_kL_k']\right|. \tag{9.21}$$

By the assumption of $\mathbb{E}[Q_kQ_k'] \succ 0$, the inverse $\mathbb{E}^{-1}[Q_kQ_k']$ of $\mathbb{E}[Q_kQ_k']$ exists. Then $\mathbb{E}^{+}[Q_kQ_k'] = \mathbb{E}^{-1}[Q_kQ_k']$. Since

$$\mathbb{E}\left[\begin{pmatrix} L_k \\ Q_k \end{pmatrix} \begin{pmatrix} L_k' & Q_k' \end{pmatrix}\right] = \begin{pmatrix} \mathbb{E}[L_kL_k'] & \mathbb{E}[L_kQ_k'] \\ \mathbb{E}[Q_kL_k'] & \mathbb{E}[Q_kQ_k'] \end{pmatrix} \succcurlyeq 0, \tag{9.22}$$

it follows from Lemma 2 that

$$\mathbb{E}[L_kL_k'] - \mathbb{E}[L_kQ_k']\mathbb{E}^{-1}[Q_kQ_k']\mathbb{E}[Q_kL_k'] \succcurlyeq 0.$$

Obviously,

$$\mathbb{E}[L_kQ_k']\mathbb{E}[Q_kQ_k']^{-1}\mathbb{E}[Q_kL_k'] = \mathbb{E}[L_kQ_k']\mathbb{E}^{-1}[Q_kQ_k']\left(\mathbb{E}[L_kQ_k']\right)' \succcurlyeq 0.$$

Consequently,

$$\mathbb{E}[L_kL_k'] \succcurlyeq \mathbb{E}[L_kQ_k']\mathbb{E}^{-1}[Q_kQ_k']\mathbb{E}[Q_kL_k']. \tag{9.23}$$

Then according to (9.23) and Lemma 3, it follows that

$$\left|\mathbb{E}[L_kL_k']\right| \geq \left|\mathbb{E}[L_kQ_k']\mathbb{E}^{-1}[Q_kQ_k']\mathbb{E}[Q_kL_k']\right| = \left|\mathbb{E}[L_kQ_k']\right|\left|\mathbb{E}^{-1}[Q_kQ_k']\right|\left|\mathbb{E}[Q_kL_k']\right|. \tag{9.24}$$

Notice that $\left|\mathbb{E}[Q_kL_k']\right| = \left|\mathbb{E}[L_kQ_k']\right|$ and $\left|\mathbb{E}^{-1}[Q_kQ_k']\right| = \left|\mathbb{E}[Q_kQ_k']\right|^{-1}$, then (9.24) implies

$$\left|\mathbb{E}[Q_kL_k']\right|^2 \leq \left|\mathbb{E}[Q_kQ_k']\right|\left|\mathbb{E}[L_kL_k']\right|. \tag{9.25}$$

By (9.19)–(9.21) and (9.25), we obtain

$$\begin{aligned}&\left(1 - \mathbb{E}[\mathbf{P}_k']\mathbb{E}^{-1}[\mathbf{P}_k\mathbf{P}_k']\mathbb{E}[\mathbf{P}_k]\right)\left(\mathbb{E}[q_k^2] - \mathbb{E}[q_k\mathbf{P}_k']\mathbb{E}^{-1}[\mathbf{P}_k\mathbf{P}_k']\mathbb{E}[q_k\mathbf{P}_k]\right)\\ &\geq \left(\mathbb{E}[q_k] - \mathbb{E}[q_k\mathbf{P}_k']\mathbb{E}^{-1}[\mathbf{P}_k\mathbf{P}_k']\mathbb{E}[\mathbf{P}_k]\right)^2.\end{aligned}$$

Namely,

$$\left(\mathbb{E}[q_k] - \widehat{B}_k\right)^2 \leq \left(\mathbb{E}[q_k^2] - \widetilde{B}_k\right)(1 - B_k).$$

Then

$$\widetilde{B}_k \leq \mathbb{E}[q_k^2] - \frac{\left(\mathbb{E}[q_k] - \widehat{B}_k\right)^2}{1 - B_k}.$$

Therefore,

$$\begin{aligned}
&\sum_{j=0}^{T-1}\Big(\prod_{k=j+1}^{T-1}\frac{(\mathbb{E}[q_k]-\widehat{B}_k)^2}{1-B_k}\Big)\widetilde{B}_j\Big(\prod_{m=0}^{j-1}\mathbb{E}[q_m^2]\Big)\\
&\leq\sum_{j=0}^{T-1}\Big(\prod_{k=j+1}^{T-1}\frac{(\mathbb{E}[q_k]-\widehat{B}_k)^2}{1-B_k}\Big)\Big(\mathbb{E}[q_j^2]-\frac{(\mathbb{E}[q_j]-\widehat{B}_j)^2}{1-B_j}\Big)\Big(\prod_{m=0}^{j-1}\mathbb{E}[q_m^2]\Big)\\
&=\sum_{j=0}^{T-1}\Big(\prod_{k=j+1}^{T-1}\frac{(\mathbb{E}[q_k]-\widehat{B}_k)^2}{1-B_k}\Big)\mathbb{E}[q_j^2]\Big(\prod_{m=0}^{j-1}\mathbb{E}[q_m^2]\Big)\\
&\quad-\sum_{j=0}^{T-1}\Big(\prod_{k=j+1}^{T-1}\frac{(\mathbb{E}[q_k]-\widehat{B}_k)^2}{1-B_k}\Big)\frac{(\mathbb{E}[q_j]-\widehat{B}_j)^2}{1-B_j}\Big(\prod_{m=0}^{j-1}\mathbb{E}[q_m^2]\Big)\\
&=\sum_{j=0}^{T-1}\Big(\prod_{k=j+1}^{T-1}\frac{(\mathbb{E}[q_k]-\widehat{B}_k)^2}{1-B_k}\Big)\Big(\prod_{m=0}^{j}\mathbb{E}[q_m^2]\Big)-\sum_{j=0}^{T-1}\Big(\prod_{k=j}^{T-1}\frac{(\mathbb{E}[q_k]-\widehat{B}_k)^2}{1-B_k}\Big)\Big(\prod_{m=0}^{j-1}\mathbb{E}[q_m^2]\Big)\\
&=\Big(\prod_{k=T}^{T-1}\frac{(\mathbb{E}[q_k]-\widehat{B}_k)^2}{1-B_k}\Big)\Big(\prod_{m=0}^{T-1}\mathbb{E}[q_m^2]\Big)-\Big(\prod_{k=0}^{T-1}\frac{(\mathbb{E}[q_k]-\widehat{B}_k)^2}{1-B_k}\Big)\Big(\prod_{m=0}^{-1}\mathbb{E}[q_m^2]\Big)\\
&=\Big(\prod_{m=0}^{T-1}\mathbb{E}[q_m^2]\Big)-\Big(\prod_{k=0}^{T-1}\frac{(\mathbb{E}[q_k]-\widehat{B}_k)^2}{1-B_k}\Big)\\
&=\Big(\prod_{k=0}^{T-1}\mathbb{E}[q_k^2]\Big)-\Big(\prod_{k=0}^{T-1}\frac{(\mathbb{E}[q_k]-\widehat{B}_k)^2}{1-B_k}\Big).
\end{aligned}$$

As a result, it follows from the above inequality that

$$\begin{aligned}
C_0=&-\prod_{k=0}^{T-1}\frac{(\mathbb{E}[q_k]-\widehat{B}_k)^2}{1-B_k}-\sum_{j=0}^{T-1}\Big(\prod_{k=j+1}^{T-1}\frac{(\mathbb{E}[q_k]-\widehat{B}_k)^2}{1-B_k}\Big)\widetilde{B}_j\Big(\prod_{m=0}^{j-1}\mathbb{E}[q_m^2]\Big)\\
&+\prod_{k=0}^{T-1}\mathbb{E}[q_k^2]\geq 0.
\end{aligned}$$

This completes the proof of Lemma 4. □

It follows from Eq. (9.12) with ω^* that we have the following minimum variance theorem.

Theorem 2 *Assume that the returns of assets and liability are correlated at every period. Then the efficient frontier is given by*

$$Var(x_T-l_T)=\frac{\prod_{k=0}^{T-1}(1-B_k)}{1-\prod_{k=0}^{T-1}(1-B_k)}\Big(d-x_0\prod_{k=0}^{T-1}s_k+l_0\prod_{k=0}^{T-1}\frac{\mathbb{E}[q_k]-\widehat{B}_k}{1-B_k}\Big)^2+l_0^2C_0.$$

Setting the expected terminal surplus $d = x_0 \prod_{k=0}^{T-1} s_k - l_0 \prod_{k=0}^{T-1} \mathbb{E}[q_k]$, we obtain the global minimum variance as

$$\mathrm{Var}_{\min}(x_T - l_T) := C_0 {l_0}^2. \tag{9.26}$$

By Lemma 4, it follows that the global minimum variance $\mathrm{Var}_{\min}(x_T - l_T) \geq 0$.

9.3.3 *The Optimal Strategy with Uncorrelation of Assets and Liability*

Assume that the returns of asset and liability are uncorrelated at every period. Then

$$\widehat{B}_t = \mathbb{E}[q_t]B_t \quad \text{and} \quad \widetilde{B}_t = (\mathbb{E}[q_t])^2 B_t.$$

Hence, we have the following results:

$$\prod_{k=t}^{T-1} \frac{\mathbb{E}[q_k] - \widehat{B}_k}{(1-B_k)s_k} = \prod_{k=t}^{T-1} \mathbb{E}[q_k] s_k^{-1},$$

$$\prod_{k=t}^{T-1} \left(\mathbb{E}[q_k] - \widehat{B}_k\right) = \prod_{k=t}^{T-1} \mathbb{E}[q_k](1-B_k),$$

$$\prod_{k=t}^{T-1} \frac{\mathbb{E}[q_k] - \widehat{B}_k}{1-B_k} = \prod_{k=t}^{T-1} \mathbb{E}[q_k],$$

$$\prod_{k=t}^{T-1} \frac{\left(\mathbb{E}[q_k] - \widehat{B}_k\right)^2}{1-B_k} = \prod_{k=t}^{T-1} \left(\mathbb{E}[q_k]\right)^2 (1-B_k)$$

and

$$C_0 = -\prod_{k=0}^{T-1} \left(\mathbb{E}[q_k]\right)^2 (1-B_k) - l_0^2 \sum_{j=0}^{T-1} \left(\prod_{k=j+1}^{T-1} \left(\mathbb{E}[q_k]\right)^2 (1-B_k) \right) \left(\mathbb{E}[q_j]\right)^2$$
$$\times B_j \left(\prod_{m=0}^{j-1} \mathbb{E}[q_m^2] \right) + \prod_{k=0}^{T-1} \mathbb{E}[q_k^2].$$

Therefore, we have the following two theorems.

Theorem 3 *Assume that the returns of assets and liability are uncorrelated at every period. Then the optimal strategy of problem (9.1) is given by*

$$\pi_t^* = -\mathbb{E}^{-1}[\mathbf{P}_t\mathbf{P}_t']\mathbb{E}[\mathbf{P}_t]s_t\left(x_t - \gamma^*\prod_{k=t}^{T-1}s_k^{-1} - l_t\prod_{k=t}^{T-1}\mathbb{E}[q_k]s_k^{-1}\right), \tag{9.27}$$

where

$$\gamma^* = \frac{x_0\prod_{k=0}^{T-1}(1-B_k)s_k - d - l_0\prod_{k=0}^{T-1}\mathbb{E}[q_k](1-B_k)}{\prod_{k=0}^{T-1}(1-B_k) - 1}. \tag{9.28}$$

Theorem 4 *Assume that the returns of assets and liability are uncorrelated at every period. Then the efficient frontier is given by*

$$Var(x_T - l_T) = \frac{\prod_{k=0}^{T-1}(1-B_k)}{1-\prod_{k=0}^{T-1}(1-B_k)}\left(d - x_0\prod_{k=0}^{T-1}s_k + l_0\prod_{k=0}^{T-1}\mathbb{E}[q_k]\right)^2 + l_0^2C_0.$$

9.4 Numerical Examples

We consider an example of constructing a pension fund consisting of S&P 500 (SP), the index of Emerging Market (EM), Small Stock (MS) of the US market, and a bank account. Based on the data provided in Elton et al. (2007), Table 9.1 presents the expected values, variances, and correlation coefficients of the annual return rates of these indices.

Table 9.1 Data for the asset allocation example

	SP	EM (%)	MS (%)	Liability (%)
Expected return	14	16	17	10
Standard deviation	18.5	30	24	20
Correlation coefficient				
SP	1	0.64	0.79	ρ_1
EM	0.64	1	0.75	ρ_2
MS	0.79	0.75	1	ρ_3
Liability	ρ_1	ρ_2	ρ_3	1

Thus, for any time t, we have

$$\mathbb{E}[\mathbf{P}_t] = \begin{pmatrix} 0.09 \\ 0.11 \\ 0.12 \end{pmatrix}, \quad \mathrm{Cov}(\mathbf{P}_t) = \begin{pmatrix} 0.0342 & 0.0355 & 0.0351 \\ 0.0355 & 0.0900 & 0.0540 \\ 0.0351 & 0.0540 & 0.0576 \end{pmatrix},$$

$$\mathbb{E}[\mathbf{P}_t\mathbf{P}_t'] = \begin{pmatrix} 0.0423 & 0.0454 & 0.0459 \\ 0.0454 & 0.1021 & 0.0672 \\ 0.0459 & 0.0672 & 0.0720 \end{pmatrix}.$$

We consider five time periods and an annual risk free rate 5% ($s_t = 1.05$). Assume that the investor has an initial wealth $x_0 = 3$ and an initial liability $l_0 = 1$. Furthermore, for $t = 0, 1, 2, 3, 4$, the correlation of assets and the liability is $\boldsymbol{\rho} = (\rho_1, \rho_2, \rho_3)$, where

$$\rho_i = \frac{\mathrm{Cov}(q_t, P_t^i)}{\sqrt{\mathrm{Var}(q_t)}\sqrt{\mathrm{Var}(P_t^i)}}$$

is the correlation coefficient of the ith asset and the liability. This means

$$\mathbb{E}[q_t P_t^i] = \mathbb{E}[q_t]\mathbb{E}[P_t^i] + \rho_i \sqrt{\mathrm{Var}(q_t)}\sqrt{\mathrm{Var}(P_t^i)}.$$

9.4.1 Correlation

In this subsection, assume that the returns of the assets and liability are correlated with $\boldsymbol{\rho} = (\rho_1, \rho_2, \rho_3) = (-0.25, 0.5, 0.25)$. Hence,

$$\mathrm{Cov}\left(\begin{pmatrix} \mathbf{P}_t \\ q_t \end{pmatrix}\right) = \begin{pmatrix} \mathrm{Cov}(\mathbf{P}_t) & \mathrm{Cov}(q_t, \mathbf{P}_t) \\ \mathrm{Cov}(q_t, \mathbf{P}_t') & \mathrm{Var}(q_t) \end{pmatrix} = \begin{pmatrix} 0.0342 & 0.0355 & 0.0351 & -0.0092 \\ 0.0355 & 0.0900 & 0.0540 & 0.0300 \\ 0.0351 & 0.0540 & 0.0576 & 0.0120 \\ -0.0092 & 0.0300 & 0.0120 & 0.0400 \end{pmatrix} \succ 0.$$

Using the above formula of $\mathbb{E}[q_t P_t^i]$, we have $\mathbb{E}[q_t \mathbf{P}_t] = (0.0898, 0.1510, 0.1440)'$. We seek for the expected terminal target with $d = 3.5$. According to Theorem 1, we can derive $\gamma^* = 4.0470$ and the optimal strategy of problem (9.1) is specified as follows:

$$\begin{aligned}
\pi_0^* &= -1.05(x_0 - 3.1710)\mathbf{K}_1 + 1.2053\mathbf{K}_2 l_0, \\
\pi_1^* &= -1.05(x_1 - 3.3295)\mathbf{K}_1 + 1.1503\mathbf{K}_2 l_1, \\
\pi_2^* &= -1.05(x_2 - 3.4960)\mathbf{K}_1 + 1.0979\mathbf{K}_2 l_2, \\
\pi_3^* &= -1.05(x_3 - 3.6708)\mathbf{K}_1 + 1.0478\mathbf{K}_2 l_3, \\
\pi_4^* &= -1.05(x_4 - 3.8543)\mathbf{K}_1 + 1.0000\mathbf{K}_2 l_4,
\end{aligned}$$

where

$$\mathbf{K}_1 = \mathbb{E}^{-1}[\mathbf{P}_t\mathbf{P}_t']\mathbb{E}[\mathbf{P}_t] = \begin{bmatrix} 1.0580 \\ -0.1207 \\ 1.1052 \end{bmatrix}, \quad \mathbf{K}_2 = \mathbb{E}^{-1}[\mathbf{P}_t\mathbf{P}_t']\mathbb{E}[q_t\mathbf{P}_t] = \begin{bmatrix} -0.2398 \\ 0.4374 \\ 1.7446 \end{bmatrix}.$$

The variance of the final optimal surplus is $\mathrm{Var}(x_5 - l_5) = 0.7289$.

9.4.2 *Uncorrelation*

In this subsection, assume that the returns of the assets and liability are uncorrelated. Hence,

$$\mathrm{Cov}\left(\begin{pmatrix} \mathbf{P}_t \\ q_t \end{pmatrix}\right) = \begin{pmatrix} \mathrm{Cov}(\mathbf{P}_t) & \mathrm{Cov}(q_t, \mathbf{P}_t) \\ \mathrm{Cov}(q_t, \mathbf{P}_t') & \mathrm{Var}(q_t) \end{pmatrix} = \begin{pmatrix} 0.0342 & 0.0355 & 0.0351 & 0 \\ 0.0355 & 0.0900 & 0.0540 & 0 \\ 0.0351 & 0.0540 & 0.0576 & 0 \\ 0 & 0 & 0 & 0.04 \end{pmatrix} \succ 0.$$

We still seek to attain the same expected terminal target with $d = 3.5$. According to Theorem 3, we can derive $\gamma^* = 4.0464$ and the optimal strategy of problem (9.1) is specified as follows:

$$\begin{aligned}
\pi_0^* &= -1.05(x_0 - 3.1705 + 1.1472 l_0)\mathbf{K}_1, \\
\pi_1^* &= -1.05(x_1 - 3.3290 + 1.0950 l_1)\mathbf{K}_1, \\
\pi_2^* &= -1.05(x_2 - 3.4955 + 1.0452 l_2)\mathbf{K}_1, \\
\pi_3^* &= -1.05(x_3 - 3.6702 + 0.9977 l_3)\mathbf{K}_1, \\
\pi_4^* &= -1.05(x_4 - 3.8538 + 0.9524 l_4)\mathbf{K}_1,
\end{aligned}$$

where $\mathbf{K}_1$ is the same as in Sect. 9.4.1, and the variance of the final optimal surplus is $\mathrm{Var}(x_5 - l_5) = 1.0043$.

9.5 Conclusion

Using the parameterized method, the state variable transformation technique, and the dynamic programming approach, we obtain in this paper the closed-form expressions for the optimal investment strategy and the efficient frontier of our multi-period mean-variance asset-liability management problem. Compared with previous studies in the literature, our method is simpler yet more efficient, and the result is more concise and powerful since we do not need to solve an auxiliary problem and investigating the relationship of the auxiliary problem and the original one. Our method is hence especially useful in the big data era. In the future, we will try to use the parameterized method to solve the portfolio selection problem when the returns are correlated in every period, with probability constraint, with uncertain exit time and with Markov jumps.

Acknowledgements This work was partially supported by Research Grants Council of Hong Kong under grants 519913, 15209614 and 15224215, by National Natural Science Foundation of China (Nos. 71231008, 71471045), by China Postdoctoral Science Foundation (No. 2014M560658 and No. 2016M592505), and by Characteristic and Innovation Foundation of Guangdong Colleges and Universities (Humanity and Social Science Type).

References

A. Albert, Conditions for positive and nonnegative definiteness in terms of pseudoinverses. SIAM J. Appl. Math. **17**, 434–440 (1969)

P. Chen, H.L. Yang, Markowitz's mean-variance asset-liability management with regime switching: a multi-period model. Appl. Math. Financ. **18**, 29–50 (2011)

M.C. Chiu, D. Li, Asset and liability management under a continuous-time mean-variance optimization framework. Insur. Math. Econ. **39**, 330–355 (2006)

X.Y. Cui, X. Li, D. Li, Unified framework of mean-field formulations for optimal multi-period mean-variance portfolio selection. IEEE Trans. Autom. Control **59**, 1833–1844 (2014)

E.J. Elton, M.J. Gruber, S.J. Brown, W.N. Goetzmann, *Modern Portfolio Thoery and Investment Analysis* (Wiley, New York, 2007)

M. Leippold, F. Trojani, P. Vanini, A geometric approach to multi-period mean-variance optimization of assets and liabilities. J. Econ. Dyn. Control. **28**, 1079–1113 (2004)

C.J. Li, Z.F. Li, Multi-period portfolio optimization for asset-liability management with bankrupt control. Appl. Math. Comput. **218**, 11196–11208 (2012)

D. Li, W.L. Ng, Optimal dynamic portfolio selection: multi-period mean-variance formulation. Math. Financ. **10**, 387–406 (2000)

X. Li, X.Y. Zhou, A.E.B. Lim, Dynamic mean-variance portfolio selection with no-shorting constraints. SIAM J. Control. Optim. **40**, 1540–1555 (2002)

D.G. Luenberger, *Optimization by Vector Space Methods* (Wiley, New York, 1968)

H.M. Markowitz, Portfolio selection. J. Financ. **7**, 77–91 (1952)

W.F. Sharp, L.G. Tint, Liabilities–a new approach. J. Portf. Manag. **16**, 5–10 (1990)

H.Y. Yao, Y. Zeng, S. Chen, Multi-period mean-variance asset-liability management with uncontrolled cash flow and uncertain time-horizon. Econ. Model. **30**, 492–500 (2013)

L. Yi, D. Li, Z.F. Li, Multi-period portfolio selection for asset-liability management with uncertain investment horizon. J. Ind. Manag. Optim. **4**, 535–552 (2008)

L. Yi, X.P. Wu, X. Li, X.Y. Cui, Mean-field formulation for optimal multi-period mean-variance portfolio selection with uncertain exit time. Oper. Res. Lett. **42**(8), 489–94 (2014)

Y. Zeng, Z.F. Li, Asset-liability management under benchmark and mean-variance criteria in a jump diffusion market. J. Syst. Sci. Complex. **24**, 317–327 (2011)

F.Z. Zhang, *Matrix Theory: Basic Results and Techniques*, 2nd edn. (Springer, New York, 2011)

X.Y. Zhou, D. Li, Continuous-time mean-variance portfolio selection: a stochastic LQ framework. Appl. Math. Optim. **42**, 19–33 (2000)

Chapter 10
Sparse and Multiple Risk Measures Approach for Data Driven Mean-CVaR Portfolio Optimization Model

Jianjun Gao and Weiping Wu

Abstract This paper studies the out-of-sample performance of the data driven Mean-CVaR portfolio optimization(DDMC) model, in which the historical data of the stock returns are regarded as the realized returns and used directly in the mean-CVaR portfolio optimization formulation. However, in practical portfolio management, due to a limited number of monthly or weekly based historical data, the out-of-sample performance of the DDMC model is quite unstable. To overcome such a difficulty, we propose to add the penalty on the sparsity of the portfolio weight and combine the variance term in the DDMC formulation. Our experiments demonstrate that the proposed method mitigates the fragility of out-of-sample performance of the DDMC model significantly.

Keywords Conditional value-at-risk • Portfolio optimization • Multiple risk measures • Sparse portfolio • Out-of-sample stability

10.1 Introduction

The mean-variance(MV) portfolio selection model proposed by Markowitz (1952) laid the foundation of the modern investment theory. It suggests to balance the profit and the risk in portfolio decision. Following the spirit of Markowtiz's MV model, the framework of mean-risk portfolio analysis has been extended in various directions, e.g., see Li et al. (2006), Kolm et al. (2014), Gao and Li (2013) and the references therein. However, using variance as the risk measure has some drawbacks, i.e., it

J. Gao (✉)
School of Information Management and Engineering, Shanghai University of Finance and Economics, Shanghai, People's Republic of China
e-mail: gao.jianjun@shufe.edu.cn

W. Wu
Department of Automation, Shanghai Jiao Tong University, Shanghai, People's Republic of China
e-mail: godream@sjtu.edu.cn

T.-M. Choi et al. (eds.), *Optimization and Control for Systems in the Big-Data Era*, International Series in Operations Research & Management Science 252,
DOI 10.1007/978-3-319-53518-0_10

penalizes both profit and loss of the random return symmetrically. Realizing the variance is not a perfect term for risk measure, a large amount of new risk measures have been proposed since the development of the MV portfolio selection model. Among these risk measures, the *Value-at-Risk* (VaR), defined as the quantile of a specified exceeding probability of the loss, becomes popular in the financial industry since the mid-90s. However, the VaR fails to satisfy the axiomatic system of coherent risk measures proposed by Artzner et al. (1999), and it suffers from the non-convexity property in the corresponding portfolio optimization problems. On the other hand, the conditional Value-at-Risk (CVaR), defined as the expected value of the loss exceeding the VaR (Rockafellar and Uryasev 2000, 2002), possesses several good properties, such as convexity, monotonicity and homogeneity, which also proved to be in the class of coherent risk measures (Pflug 2000; Artzner et al. 1999). Rockafellar and Uryasev (2000, 2002) developed an equivalent formulation to compute the CVaR which leads to a convex optimization problem. Due to these nice properties, CVaR has been widely applied in various applications of portfolio selection and risk management, e.g., derivative portfolio (Alexander et al. 2006), credit risk optimization (Andersson et al. 2001), and robust portfolio management (Zhu and Fukushima 2009).

Although the mean-risk portfolio optimization model has been studied extensively in the academic society, translating these models as some useful tools in the real world financial practice is not a trivial task. Even for the classical MV portfolio selection model, it is well known that estimating the expected return and covariance matrix are not an easy task, especially when the size of the portfolio is large (e.g., see Merton 1980; Demiguel et al. 2009a,b). Highly related to the estimation problem of the stock return statistics, the stableness of the out-of-sample performance of the portfolio optimization model is another issue. Demiguel et al. (2009b) checked several portfolio construction methods rooted from the MV portfolio selection formulation. However, these models cannot significantly or consistently outperform the naive portfolio strategy which allocates wealth evenly in all assets. As for the mean-CVaR portfolio optimization model, since CVaR measures just a small portion of the whole distribution, a large number of samples is needed to guarantee the statistical stability. Takeda and Kanamori (2009) and Kondor et al. (2007) showed that the mean-CVaR portfolio optimization model has more serious problems of instability regarding the out-of-sample performance than the MV model. Recently, Lim et al. (2011) reported the similar results that the correspondent portfolio of the mean-CVaR portfolio decision model is extremely unreliable due to the estimation errors. Furthermore, Lim et al. (2011) showed that this problem is even worse when the distribution of the return has a heavy tail. To deal with unstable out-of-sample performance of the mean-CVaR portfolio optimization model, several methods have been proposed. Gotoh and Takeda (2011) introduced the norm-regularity in the mean-risk portfolio decision model to reduce the sparsity of the portfolio decision. Gotoh et al. (2013) further adopted the robust mean-CVaR portfolio optimization technique to overcome such an instability problem.

Motivated by the above research (Lim et al. 2011; Gotoh and Takeda 2011; Gotoh et al. 2013), we propose to use the sparse portfolio and multiple risk measures to mitigate the fragility of the CVaR based data driven portfolio selection model. More specifically, we add the l_1-norm penalty of the portfolio decision vector and the variance of the portfolio return in the mean-CVaR portfolio selection model. To enhance the sparsity of the solution, we also adopt the reweighted-l_1 norm method by computing the weights iteratively. Our numerical experiments show that the resulted out-of-sample performance is significantly enhanced comparing with the traditional DDMC portfolio optimization model.

This paper is organized as follows. The alternative formulations of the DDMC portfolio optimization problems are proposed in Sect. 10.2. The out-of-sample performance of these different models is evaluated by using the simulation approach in Sect. 10.3. The paper is concluded in Sect. 10.4.

10.2 The Data Driven Mean-Risk Portfolio Optimization

We consider a portfolio constructed by n candidate risky assets, whose random returns are denoted as $\mathbf{R} \in \mathbb{R}^n$. Let $\mathbf{x} = (x_1, \cdots, x_n)' \in \mathbb{R}^n$ be the portfolio decision vector, which represents the weight of the allocation of the wealth in each securities. Let $f(\mathbf{x}, \mathbf{R})$ be the portfolio loss associated with $\mathbf{x}$ and $\mathbf{R}$, e.g., we can simply set $f(\mathbf{x}, \mathbf{R}) = b - \mathbf{R}'\mathbf{x}$, where b is the benchmark return. To define the CVaR of the loss $f(\mathbf{x}, \mathbf{R})$ for a given confidence level β(i.e., $\beta = 95\%$), we need the cumulative distribution function of $f(\mathbf{x}, \mathbf{R})$,

$$\Psi(y) = \mathbb{P}(f(\mathbf{x}, \mathbf{R}) \leq y),$$

for some number $y \in \mathbb{R}$, the corresponding β-tail distribution for a given confidence level β is

$$\Psi_\beta(y) = \begin{cases} 0, & \text{if } y < \mathrm{VaR}_\beta, \\ \frac{\Psi(y)-\beta}{1-\beta}, & \text{if } y \geq \mathrm{VaR}_\beta, \end{cases} \tag{10.1}$$

where $\mathrm{VaR}_\beta = \inf\{z \mid \Psi(y) \geq \beta\}$. The CVaR of the loss function $f(\mathbf{x}, \mathbf{R})$ is then given by

$$\mathrm{CVaR}[f(\mathbf{x}, \mathbf{R})] := \int_{f(\mathbf{x}, \mathbf{R}) \geq \mathrm{VaR}_\beta} f(\mathbf{x}, \mathbf{R}) d\Psi_\beta(y), \tag{10.2}$$

where the integration should be understood as a summation when $\mathbf{R}$ is a discrete random vector. Note that the above definition of CVaR is for the general distribution function of the loss function $f(\mathbf{x}, \mathbf{R})$, see, e.g., Rockafellar and Uryasev (2002) for some subtle difference on the definition of the CVaR between the cases of discrete random variable and continuous random variable. Rockafellar and Uryasev (2000) and Rockafellar and Uryasev (2002) showed that the $\mathrm{CVaR}[f(\mathbf{x}, \mathbf{R})]$ can be computed by solving a simple convex optimization problem.

Lemma 2.1 *The CVaR of the loss $f(\mathbf{x}, \mathbf{R})$ of the terminal wealth can be computed as follows:*

$$CVaR[f(\mathbf{x}, \mathbf{R})] = \min_{\alpha} \left\{ \alpha + \frac{1}{1-\beta} \mathrm{E}\big[(f(\mathbf{x}, \mathbf{R}) - \alpha)^{+}\big] \right\},$$

where α is an auxiliary variable and $(y)^{+} := \max y, 0.$

Let $\mathbf{D} = \{\mathbf{r}_1, \mathbf{r}_2, \cdots, \mathbf{r}_m\}$ be the data set of the historical returns, where $\mathbf{r}_i \in \mathbb{R}^n$ is the i-th sample of the returns and m is the number of the samples we can observe. Without loss of generality, we assume $\mathbf{r}_i$ and $\mathbf{r}_j$ to be independent for any $i, j \in \{1, \cdots, m\}$. The data set $\mathbf{D}$ can also be regarded as m realizations of the random return $\mathbf{R}$. From Lemma 2.1, if we fix the loss function as $f(\mathbf{R}, \mathbf{x}) = b - \mathbf{R}'\mathbf{x}$, the data driven mean-CVaR portfolio optimization model is given as follows:

$$(\mathcal{P}_1) \ \min_{\mathbf{x}} \ \alpha + \frac{1}{m(1-\beta)} \sum_{i=1}^{m} (b - \mathbf{r}_i'\mathbf{x})^{+} \tag{10.3}$$

$$\text{Subject to}: \sum_{i=1}^{n} x_i = 1, \tag{10.4}$$

$$\frac{1}{m} \sum_{i=1}^{m} \mathbf{r}_i'\mathbf{x} \geq d, \tag{10.5}$$

where d is a pre-given target return level. By introducing some auxiliary variables, problem $(\mathcal{P}_1)$ can be reformulated as a linear programming problem. To overcome the instability of the out-of-sample performance of the DDMC model $(\mathcal{P}_1)$, we propose to use the following model $(\mathcal{P}_2(\omega))$ with some given weighting vector $\omega \in \mathbb{R}^n$,

$$(\mathcal{P}_2(\omega)): \ \min_{\mathbf{x}} \ \alpha + \frac{1}{m(1-\beta)} \sum_{i=1}^{m} (b - \mathbf{r}_i'\mathbf{x})^{+} + \|\mathbf{x}\|_1^{\omega}, \tag{10.6}$$

$$\text{Subject to}: \ \mathbf{x} \text{ satisfies (10.4) and (10.5)},$$

where $\omega = (\omega_1, \cdots, \omega_n)'$ with $\omega_i \geq 0$, for $i = 1, \cdots, n$ and

$$\|\mathbf{x}\|_1^{\omega} := \sum_{i=1}^{n} \omega_i |x_i|.$$

When ω is a unit vector with all elements being 1, the weighted l_1-norm formulation becomes the l_1-norm formulation, which is denoted by $\|\mathbf{x}\|_1$. Using the l_1 norm as the penalty for the sparsity of the solution is a standard routine in data analysis. The ideal penalty of the sparsity of the solution is l_0 norm, which is defined as

$\|\mathbf{x}\|_0 = \sum_{i=1}^{n} |\text{Sign}(x_i)|$ with $\text{Sign}(a) = 1$ if $a > 0$, $\text{Sign}(a) = -1$ if $a < 0$ and $\text{Sign}(a) = 0$ if $a = 0$. However, the l_0 norm is highly nonconvex and hard to be optimized directly. It has been proved that the l_1 norm of $\mathbf{x}$, $\|\mathbf{x}\|_1$, is the convex hull of $\|\mathbf{x}\|_0$(see Zhao and Li 2012). Thus, it is reasonable to use l_1 norm as the surrogate of l_0 norm to penalize the sparsity. In model $(\mathcal{P}_2(\omega))$, we prefer to use the formulation of weighted-l_1 norm, which further enhances the sparsity by varying the choice of vector ω. Note that problem $(\mathcal{P}_2(\omega))$ can be reformulated as a linear programming problem,

$$
\begin{aligned}
(\bar{\mathcal{P}}_2(\omega)) : \quad & \min_{\mathbf{x},\tau,\phi} \ \alpha + \frac{1}{m(1-\beta)} \sum_{i=1}^{m} \tau_i + \sum_{j=1}^{n} \phi_j, \\
\text{Subject to :} \quad & \tau_i \geq 0, \ i = 1, \cdots, m, \\
& b - \mathbf{r}_i'\mathbf{x} \leq \tau_i, \ \ i = 1, \cdots, m, \\
& \omega_j x_j \leq \phi_j, \ \ j = 1, \cdots, n, \\
& \omega_j x_j \geq -\phi_j, \ \ j = 1, \cdots, n, \\
& \sum_{j}^{n} x_j = 1, \\
& \frac{1}{m} \sum_{j}^{n} \mathbf{r}_j'\mathbf{x} \geq d,
\end{aligned}
$$

where τ_i for $i = 1, \cdots, m$ and ϕ_j for $j = 1, \cdots, n$ are auxiliary decision variables.

In this work, we also consider to integrate the variance term of the portfolio return in model $(\mathcal{P}_2(\omega))$ to further enhance the stability of the out-of-sample performance, i.e.,

$$
(\mathcal{P}_3(\omega)) : \quad \min_{\mathbf{x}} \ \alpha + \frac{1}{m(1-\beta)} \sum_{i=1}^{m} (b - \mathbf{r}_i'\mathbf{x})^+ + \|\mathbf{x}\|_1^{\omega} + \rho \mathbf{x}' F \mathbf{x} \tag{10.7}
$$

$$
\text{Subject to : } \mathbf{x} \text{ satisfies (10.4) and (10.5)},
$$

where $F \in \mathbb{R}^{n \times n}$ is the sample covariance matrix of the asset returns. Note that, similar to problem $(\mathcal{P}_2(\omega))$, problem $(\mathcal{P}_3(\omega))$ can be reformulated as a convex quadratic programming formulation, which can be solved efficiently by a commercial solver like IBM CPLEX (IBM 2015).

10.3 Evaluation and Discussion

10.3.1 Evaluation Methods

To evaluate the out-of-sample performance of the three portfolio optimization models $(\mathcal{P}_1)$, $(\mathcal{P}_2(\omega))$ and $(\mathcal{P}_3(\omega))$, we mainly adopt the simulation approach with all parameters being estimated from the real historical price data of some stock index. The main reason of using this approach is as follows. The number of the historical data of the monthly return is very limited in real portfolio management. Thus, it is hard to carry on various tests by solely using the true market historical data. On the other hand, by using the simulation approach, different types of test data sets can be generated, which provides us more freedom to evaluate the performances of the three models under different situations. More specifically, we adopt the following procedures.

(a) **Data Generation**: Generate a data set of returns $\mathbf{D}_{\text{sample}} = \{\mathbf{r}_1, \cdots, \mathbf{r}_m\}$ with a sample size being m according to some distributions of the returns.[1] For example, if we assume the random returns follow a mixed distribution of multivariate normal distribution and exponential distribution with given mean vector and covariance matrix, we then generate m samples of the returns according to this distribution.
(b) **Optimization**: Solve all three problems $(\mathcal{P}_1)$, $(\mathcal{P}_2(\omega))$ and $(\mathcal{P}_3(\omega))$ according to the data set $\mathbf{D}_{\text{sample}}$ to generate the portfolio decisions $\mathbf{x}^1$, $\mathbf{x}^2$ and $\mathbf{x}^3$, respectively. If it is necessary, we can vary the target return level d in three models to achieve the portfolio policy $\mathbf{x}^i(d)$, $i = 1, 2, 3$, for different level of d.
(c) **Evaluation**: Generate 50 data set $\mathbf{D}^{(i)}_{\text{test}}$, $i = 1, \cdots, 50$ according to the similar distribution used in step **Data Generation** with the size of the each data set $\mathbf{D}^{(i)}_{\text{test}}$ being m. For each test set $\mathbf{D}^{(i)}_{\text{test}}$, we implement the portfolio policy $\mathbf{x}^i(d)$, $i = 1, 2, 3$ and compute the corresponding empirical expected return and CVaR.

In step **Evaluation**, we actually perform 50 trials of out-of-sample tests and the resulted empirical sample expected return and CVaR are recorded. In each iteration, we use the IBM CPLEX (IBM 2015) as the solver to solve the corespondent linear programming and convex quadratic programming problems of $(\mathcal{P}_1)$, $(\mathcal{P}_2(\omega))$ and $(\mathcal{P}_3(\omega))$.

[1]The detailed discussion of the distribution is given in Sect. 10.3.2.

10.3.2 Data Generation

In this paper, we use the 48 industry portfolios constructed by Fama and Frech as the basic data set for our test.[2] We estimate the mean return vector and covariance matrix of monthly return by using the historical monthly returns from Jan 1998 to Dec 2015. Note that there are only 216 samples of the returns, however, we need to estimate 1176 unknown parameters in the covariance matrix,[3] which implies that using the sample covariance matrix method may generate a singular matrix. To overcome this difficulty, we adopt the shrinkage estimation method for the covariance matrix proposed by Ledoit and Wolf (2003) by setting the shrinkage coefficient to 0.1. After we have achieved the sample mean vector of the returns, $\hat{\mathbf{R}} := (\hat{R}_1, \cdots, \hat{R}_n)'$ and the estimation of the covariance matrix $\hat{\Sigma} := \{\Sigma_{i,j}\}_{i=1,j=1}^{n,n}$, we then use the following method to generate the samples. Adopting a similar setting given by Lim et al. (2011), we construct a hybrid distribution combining the multivariate normal distribution and the exponential distribution. Let $B(\eta)$ be the Bernoulli random variable with parameter η, i.e., $B(\eta) = 1$ with probability η and $B(\eta) = 0$ with probability $1 - \eta$. Let z be the exponential random variable with the probability distribution function being

$$\mathbb{P}(z < a) = \int_0^a \lambda e^{\lambda s} ds.$$

In this paper, we simply fix $\lambda = 10$. Suppose the random vector $Y \in \mathbb{R}^n$ follows the multivariate normal distribution with mean and covariance matrix being $\hat{R}$ and $\hat{\Sigma}$, respectively. We assume the random return is captured by the hybrid distribution as follows:

$$\mathbf{R} \sim -B(\eta)\big(z\mathbf{1} + \mathbf{l}\big) + \big(1 - B(\eta)\big)Y,$$

where $\mathbf{c} := (c_1, \cdots, c_n)'$ with $c_i := \hat{R}_i - \sqrt{\Sigma_{ii}}$ for $i = 1, \cdots, n$ and Σ_{ii} is the i-th diagonal element of Σ. Note that the parameter η controls the tail-loss of the distribution, i.e., the larger the η is, the heavier tail of the distribution will be. Figure 10.1 gives the distribution of one entry of $\mathbf{R}$ for different η.

[2]The data of 48 industry portfolio can be found in http://mba.tuck.dartmouth.edu/pages/faculty/ken.french/data.

[3]Since the covariance matrix is symmetrical, we only need to estimate the upper triangle of the matrix. Thus, the total number of unknown parameters is $(48 + 1) \times 48/2 = 1176$.

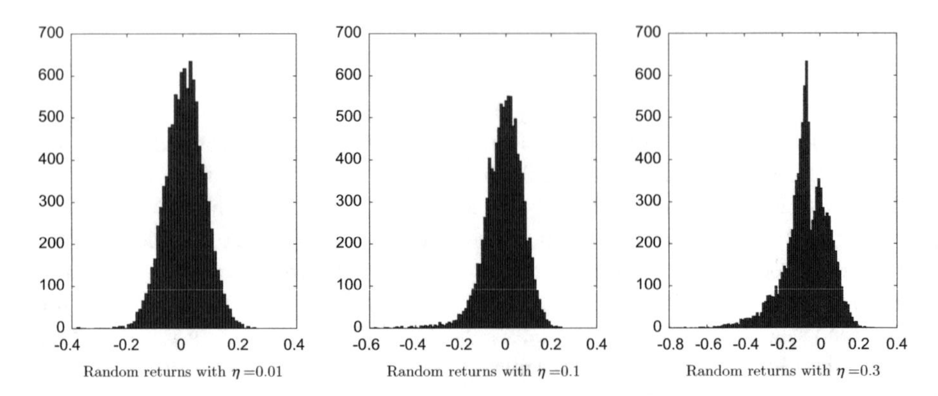

Fig. 10.1 The empirical distribution of hybrid random returns R_1 with different value of η

10.3.3 Re-Weighted Method for Sparse Solution

In portfolio optimization models $(\mathcal{P}_2(\omega))$ and $(\mathcal{P}_3(\omega))$, we use the weighted-l_1 norm to penalize the sparsity of the solution. However, since the objective function is a weighted summation of the CVaR and the weighted-l_1 norm of the portfolio weight, we need to choose the weighting parameter ω carefully. If $\|\omega\|$ is too large, the optimality of the CVaR will be jeopardized. On the other hand, if $\|\omega\|$ is too small, then the resulted solution will be not sparse enough. To overcome this difficulty, we adopt the iterative reweighted method of the l_1 norm to enhance the sparsity of the solution(see, e.g., Zhao and Li 2012). More specifically, we apply the following iterative procedure to change the weighting parameter ω dynamically and adaptively. Let $\omega^{(k)} \in \mathbb{R}^n$ and $\mathbf{x}^{(k)}$ be the weighting vector and portfolio decision vector in k-th iteration, respectively. We repeat the following steps.

(1) For any given $\omega^{(k)}$, solve the problem $\mathcal{P}_2(\omega^{(k)})$(or problem $(\mathcal{P}_3(\omega^{(k)}))$), which gives the solution $\mathbf{x}^{(k)}$. If the stopping criteria is satisfied, e.g., the sparsity of $\mathbf{x}^k$ does not change any more, we stop the iteration. Otherwise, go to step II.
(2) Use $\mathbf{x}^{(k)}$ to construct the new weighting parameter $\omega^{(k+1)}$ and let $k = k + 1$. Go to step 1.

There are several ways to construct the new weighting vector $\omega^{(k+1)} = \big(\omega_1^{(k+1)}, \omega_2^{(k+1)}, \cdots, \omega_n^{(k+1)}\big)'$ by using the information of $\mathbf{x}^{(k)} = \big(x_1^{(k)}, \cdots, x_n^{(k)}\big)'$. Motivated by Zhao and Li (2012) and based on our numerical experiments, we select the following three methods which perform relatively better than the others. Let $\epsilon > 0$ be a small positive number.

(a) Method I: Let $\omega_j^{(k+1)} = 1/(|x_j^{(k)}| + \epsilon)$ for $j = 1, \cdots, n$.
(b) Method II: Let $\omega_j^{(k+1)} = 1/(|x_j^{(k)}| + \epsilon)^{(1-p)}$, for $j = 1, \cdots, n$ and $p \in (0, 1)$.
(c) Method III: Let $\omega_j^{(k+1)} = (p + (|x_i^k| + \epsilon)^{1-p})/((|x_i^{(k)}| + \epsilon)^{1-p}\big[|x_i^{(k)}| + \epsilon + (|x_j^k| + \epsilon)^p\big])$ for $j = 1, \cdots, n$ with $p \in (0, 1)$.

It is not hard to see that when x_i^k is a small number then the corresponding weighting coefficient ω_i^{k+1} will be large, which will drive x_i^{k+1} to be even smaller in the next round of optimization.

10.3.4 Comparison of the Global Mean-CVaR Portfolio

In this section, we compare the out-of-sample performance of the three models $(\mathcal{P}_1)$, $(\mathcal{P}_2(\omega))$ and $(\mathcal{P}_3(\omega))$ for the special case of finding the *global minimum CVaR* portfolio. More specifically, we consider the problems with ignoring the constraint (10.5) in all three models $(\mathcal{P}_1)$, $(\mathcal{P}_2(\omega))$ and $(\mathcal{P}_3(\omega))$. Following the evaluation procedure illustrated in Sect. 10.3.1, we generate one data set $\mathbf{D}_{\text{sample}}$ to compute the correspondent portfolio weights and apply such portfolio decision in 50 testing data sets $\mathbf{D}_{\text{test}}^{(j)}$ for $j = 1, \cdots, 50$ as the out-of-sample tests. We check three different types of size of $\mathbf{D}_{\text{sample}}$ and $\mathbf{D}_{\text{test}}^{(j)}$ as $m = 200$, $m = 300$ and $m = 400$.

Figures 10.2, 10.3 and 10.4 plot the empirical mean value and CVaR of the global minimum CVaR portfolio return generated from 50 out-of-sample tests. We can observe that the empirical mean and CVaR pair spread in a quite large range for model $(\mathcal{P}_1)$. However, by using our proposed models $(\mathcal{P}_2(\omega))$ and $(\mathcal{P}_3(\omega))$, we can see that the range of the resulted empirical mean and CVaR pair are significantly reduced. Table 10.1 records the detail of the above experiments. The column 'min', 'max' and 'range' show the minimum value, the maximum value and the range(i.e., 'max'-'min') of the corresponding data set, respectively. For the case $m = 200$, the minimum and maximum value of resulted mean and CVaR of model $(\mathcal{P}_1)$ is from -0.0037 to 0.541 and 0.2634 to 0.5943, respectively. That is to say, the relative difference of the out-of-sample CVaR and mean value are 0.33 and 0.0578 for model $(\mathcal{P}_1)$. In the same row of $m = 200$, we can observe that this range is reduced to 0.1756 and 0.0343, respectively, for model $(\mathcal{P}_2(\omega))$ and reduce to 0.1394 and

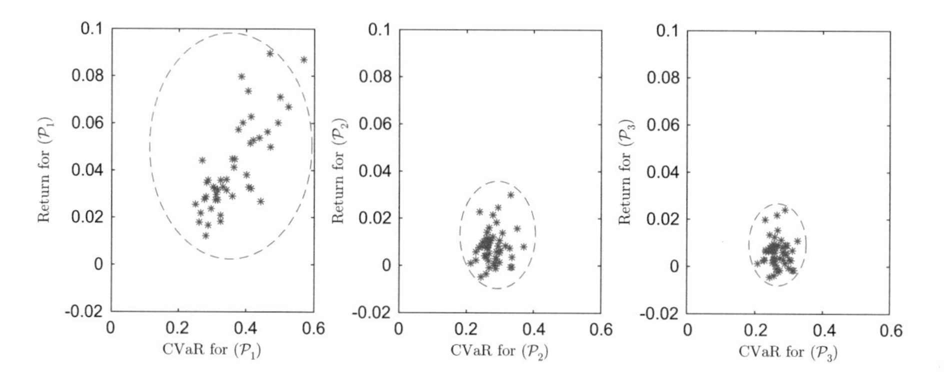

Fig. 10.2 The out-of-sample performance of three models with sample size $m = 200$ and $\epsilon = 0.1$

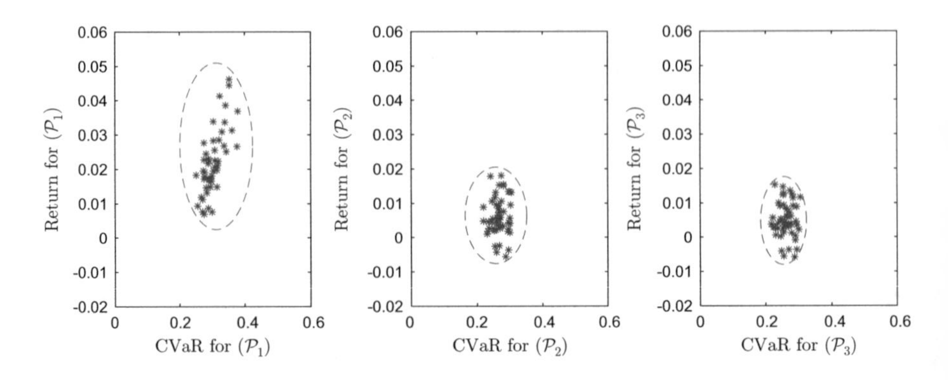

Fig. 10.3 The out-of-sample performance of three models with sample size $m = 300$ and $\epsilon = 0.1$

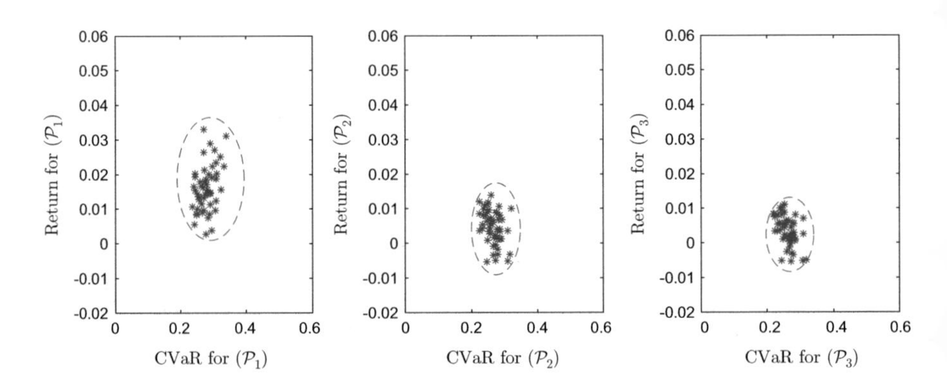

Fig. 10.4 The out-of-sample performance of three models with sample size $m = 400$ and $\epsilon = 0.1$

Table 10.1 The empirical mean value and CVaR of portfolio returns for global minimum CVaR problem generated by different models with $\eta = 0.1$

	$(\mathcal{P}_1)$			$\mathcal{P}_2(\omega)$			$\mathcal{P}_3(\omega)$		
m	Min	Max	Range	Min	Max	Range	Min	Max	Range
CVaR ($\times 10^{-2}$)									
$m = 200$	26.34	59.43	33.09	20.61	38.17	17.56	19.30	33.24	13.94
$m = 300$	25.25	38.00	12.75	21.80	30.43	8.63	20.63	28.85	8.22
$m = 400$	23.70	34.03	10.33	22.30	32.09	9.79	20.75	30.24	9.48
Exp return ($\times 10^{-2}$)									
$m = 200$	−0.37	5.41	5.78	−1.09	2.34	3.43	−1.13	1.79	2.92
$m = 300$	−0.32	3.93	4.26	−0.57	1.80	2.37	−0.55	1.45	2.00
$m = 400$	−0.35	2.56	2.91	−0.55	1.38	1.93	−0.50	1.05	1.55

Table 10.2 The empirical mean value and CVaR of portfolio returns for global minimum CVaR problem generated by different models with $\eta = 0.2$

	$(\mathcal{P}_1)$			$\mathcal{P}_2(\omega)$			$\mathcal{P}_3(\omega)$		
m	Mean	Max	Range	Min	Max	Range	Min	Max	Range
CVaR ($\times 10^{-2}$)									
$m = 200$	33.57	118.67	85.10	25.67	46.31	20.65	24.23	40.95	16.72
$m = 300$	30.64	48.43	17.79	27.20	36.74	9.54	23.78	34.89	11.11
$m = 400$	26.99	39.05	12.05	25.62	33.03	7.41	23.89	30.88	6.99
Exp return ($\times 10^{-2}$)									
$m = 200$	−0.79	14.62	15.41	−0.50	3.56	4.06	−0.49	3.07	3.56
$m = 300$	0.24	5.22	4.99	−0.54	3.19	3.73	−0.68	2.43	3.11
$m = 400$	0.19	5.10	4.90	−0.27	2.01	2.27	−0.42	1.75	2.18

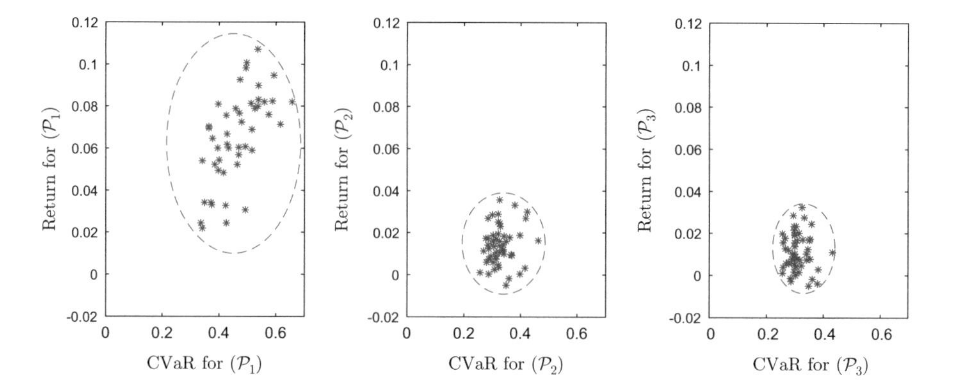

Fig. 10.5 The out-of-sample performance of three models with sample size $m = 200$ and $\epsilon = 0.2$

0.0292, respectively, for model $(\mathcal{P}_3(\omega))$. From Table 10.1, we can see that, as the size of the sample increases, e.g., the case of $m = 300$ and $m = 400$, the variation of the resulted empirical mean and CVaR of model $(\mathcal{P}_1)$, $(\mathcal{P}_2(\omega))$ and $(\mathcal{P}_3(\omega))$ are reduced. However, the performance of the models $(\mathcal{P}_2(\omega))$ and $(\mathcal{P}_3(\omega))$ is better than model $(\mathcal{P}_1)$.

Table 10.2 and Figs. 10.5, 10.6 and 10.7 show the detailed results of the comparison between the three models when $\eta = 0.2$. As we have illustrated in Sect. 10.3.2, the parameter η controls the shape of the tail distribution of the random returns. Under this case, the stock returns have heavier tails comparing with the previous case with $\eta = 0.1$. However, a similar pattern can be observed that the formulation $(\mathcal{P}_2(\omega))$ and $(\mathcal{P}_3(\omega))$ can better control the variation of the empirical mean return and CVaR.

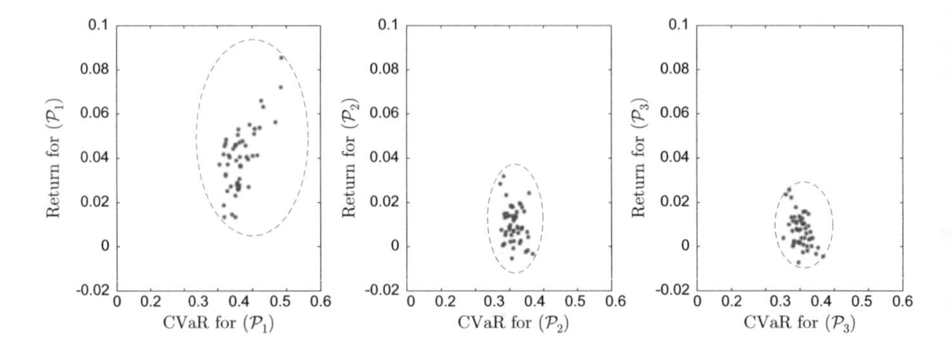

Fig. 10.6 The out-of-sample performance of three models with sample size $m = 300$ and $\epsilon = 0.2$

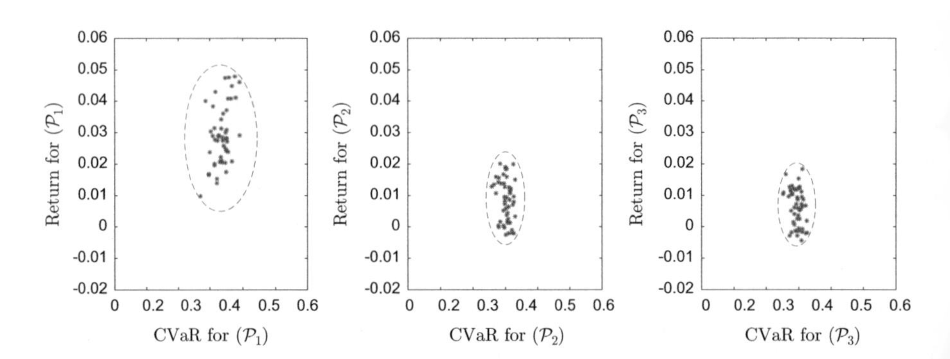

Fig. 10.7 The out-of-sample performance of three models with sample size $m = 400$ and $\epsilon = 0.2$

10.3.5 Comparison of the Empirical Efficient Frontiers

In this section, we compare the mean-CVaR efficient frontiers generated by three models $(\mathcal{P}_1)$, $(\mathcal{P}_2(\omega))$ and $(\mathcal{P}_3(\omega))$. The efficient frontiers are generated by varying the target return d from 0.01 to 0.1 in all these models. Figures 10.8, 10.9 and 10.10 plot the out-of-sample empirical mean-CVaR efficient frontier for 50 trials of simulations with $\eta = 0.1$. Table 10.3 shows the detailed statistics of the comparison. In Table 10.3, the columns 'min dev', 'max dev' and 'mean dev' represent the minimum deviation, maximum deviation and average deviation of the out-of-sample CVaR and expected return.[4] Note that the minimum, maximum and average deviation is computed for all different value of d in 50 trials of simulation.

[4]Given some random samples $a_1, \cdots, a_m$, the maximum, minimum and average deviation is defined as $\max\{|a_i - \bar{a}| \mid i = 1, \cdots, m\}$, $\min\{|a_i - \bar{a}| \mid i = 1, \cdots, m\}$ and $\frac{1}{m}\sum_{i=1}^{m}(|a_i - \bar{a}|)$, where $\bar{a} = \frac{1}{m}\sum_{i=1}^{m} a_i$.

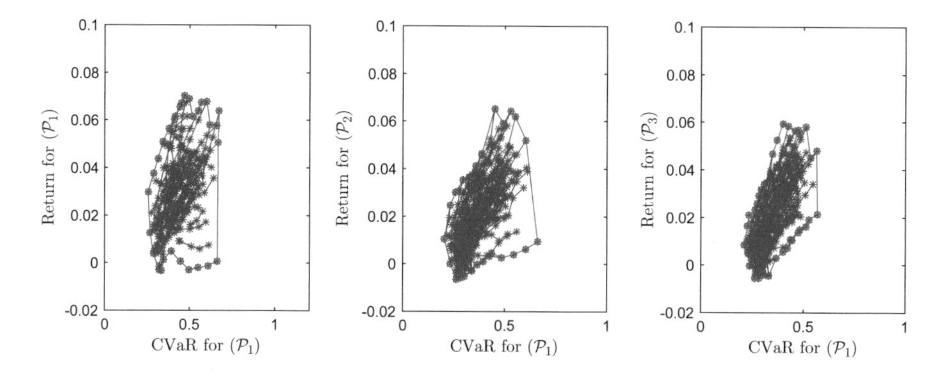

Fig. 10.8 The out-of-sample performance of three models with sample size $m = 200$ and $\epsilon = 0.2$

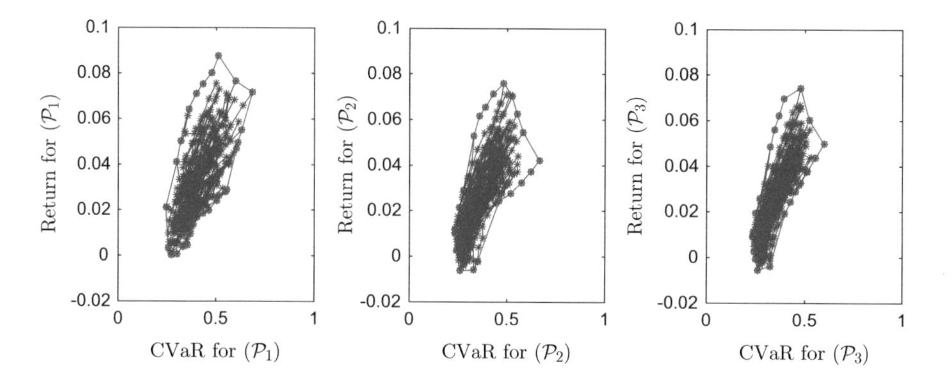

Fig. 10.9 The out-of-sample performance of three models with sample size $m = 300$ and $\epsilon = 0.2$

For all of these tests, we can observe that the proposed formulations $(\mathcal{P}_2(\omega))$ and $(\mathcal{P}_3(\omega))$ perform better than the traditional model $(\mathcal{P}_1)$. For example, in the row of $m = 200$ in Table 10.3, the maximum deviation of three models are 19.98%, 17.35% and 12.54%, respectively. The average deviation of three models are 5.04%, 2.99%, and 2.68%, respectively. Similar pattern can be observed when we increase the tail part of the distribution of the random return. Figures 10.11, 10.12, and 10.13 and Table 10.4 provide the detail of the improvement under this case.

10.4 Conclusion

In this work, we proposed some methods to reduce the instability issue of the out-of-sample performance for mean-CVaR portfolio optimization model. More specifically, we suggest to add the weighted l_1 norm as a penalty of the sparsity

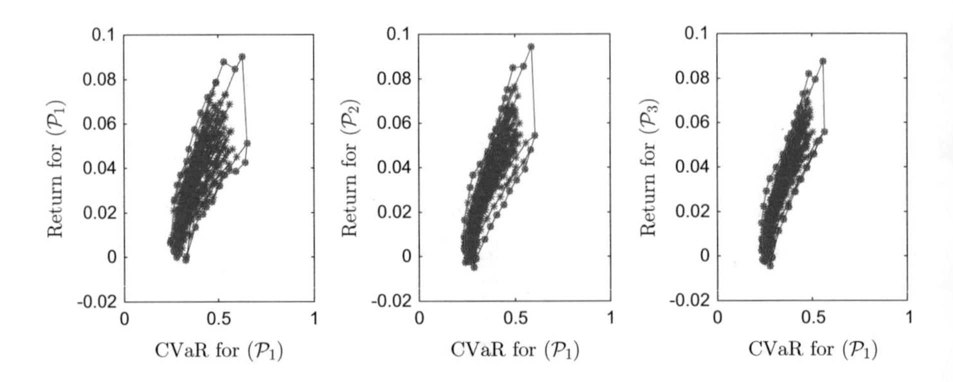

Fig. 10.10 The out-of-sample performance of three models with sample size $m = 400$ and $\epsilon = 0.2$

Table 10.3 Comparison of mean-CVaR efficient frontiers for different models with $\eta = 0.1$

m	$(\mathcal{P}_1)$ Min dev	Max dev	Mean dev	$\mathcal{P}_2(\omega)$ Min dev	Max dev	Mean dev	$\mathcal{P}_3(\omega)$ Min dev	Max dev	Mean dev
CVaR ($\times 10^{-2}$)									
$m = 200$	0.04	19.98	5.04	0.01	17.35	2.99	0.00	12.54	2.68
$m = 300$	0.01	16.14	3.52	0.01	19.15	2.82	0.00	15.68	2.48
$m = 400$	0.01	13.95	2.30	0.00	13.33	2.12	0.00	12.49	2.04
Exp return ($\times 10^{-2}$)									
$m = 200$	0.00	4.56	1.11	0.00	3.21	0.83	0.00	2.87	0.80
$m = 300$	0.00	3.72	0.85	0.00	2.94	0.67	0.00	3.12	0.64
$m = 400$	0.00	3.36	0.67	0.00	4.04	0.55	0.00	3.68	0.51

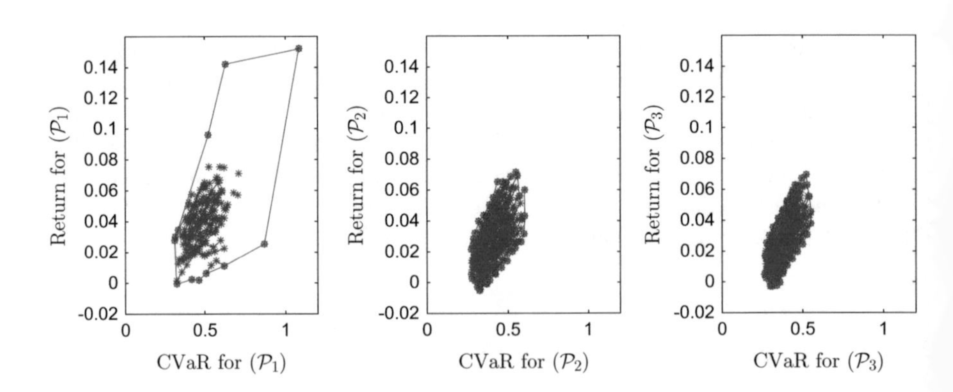

Fig. 10.11 The out-of-sample performance of three models with sample size $m = 200$ and $\epsilon = 0.2$

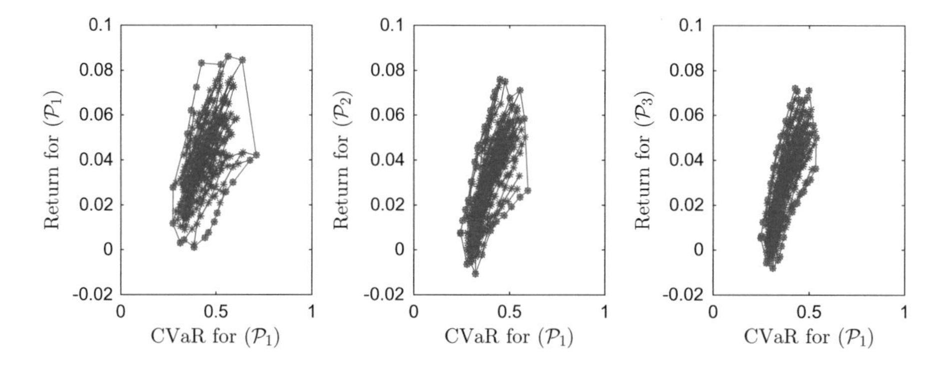

Fig. 10.12 The out-of-sample performance of three models with sample size $m = 300$ and $\epsilon = 0.2$

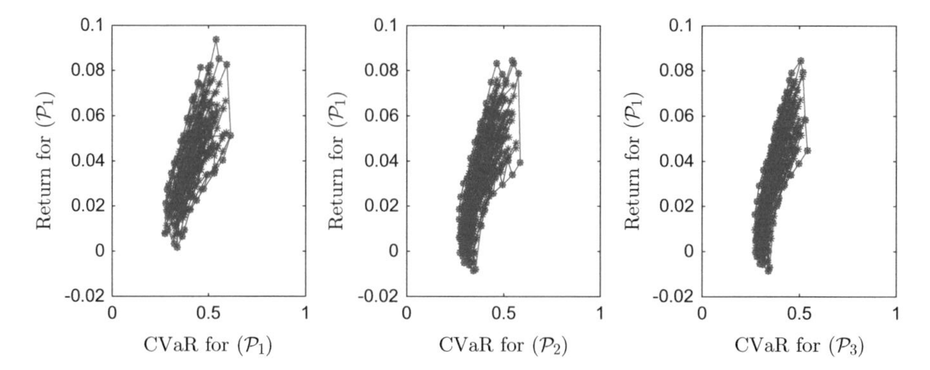

Fig. 10.13 The out-of-sample performance of three models with sample size $m = 400$ and $\epsilon = 0.2$

of the portfolio decision and add the variance term in the objective function to control the total variation in mean-CVaR portfolio formulation. In order to balance the sparsity and optimality of the solution, the reweighted l_1 norm method is adopted to adjust the weighting coefficients. Our simulation based experiments show that the proposed methods reduce the variation of the empirical mean value and the CVaR of the portfolio return in out-of-sample test significantly. However, observing from our experiment, the proposed methods still have some limitations. When the size of the portfolio is large, e.g., when $n = 500$, solely using our methods may not control the variation of the out-of-sample test to a desired level. A possible solution for this case is to increase the number of the samples by using some statistical sampling methods like bootstrap. Another important issue is the computational burden of the proposed methods when n and m are large. For example, for problem $(\mathcal{P}_2(\omega))$, the linear programming formulation (given in Sect. 10.2) has almost $m + 2n$ decision variables and $2(m + n)$ constraints. In the literature, Kunzi-Bay and Janos (2006) have showed that using the dual formulation and decomposition approach may

Table 10.4 Comparison of mean-CVaR efficient frontiers for different models with $\eta = 0.2$

	$(\mathcal{P}_1)$			$\mathcal{P}_2(\omega)$			$\mathcal{P}_3(\omega)$		
m	Min dev	Max dev	Mean dev	Min dev	Max dev	Mean dev	Min dev	Max dev	Mean dev
CVaR ($\times 10^{-2}$)									
$m = 200$	0.01	58.24	9.41	0.01	12.73	3.90	0.00	9.46	3.00
$m = 300$	0.01	17.88	4.06	0.00	11.02	2.66	0.01	9.67	2.42
$m = 400$	0.01	10.49	2.68	0.00	10.32	2.54	0.01	8.71	2.31
Exp return ($\times 10^{-2}$)									
$m = 200$	0.00	10.63	1.83	0.00	3.03	0.91	0.00	2.98	0.81
$m = 300$	0.00	3.13	0.93	0.00	2.97	0.71	0.00	2.31	0.68
$m = 400$	0.00	3.25	0.73	0.00	2.96	0.69	0.00	2.30	0.66

enhance the efficiency of the solution procedure. All the models considered in this work belong to the static portfolio optimization formulation, which gives the buy-and-hold type of portfolio policy. Studying the stability issue of the out-of-sample test for multiperiod mean-CVaR portfolio optimization problem is an interesting and challenging topic.

Acknowledgements This research work was partially supported by National Natural Science Foundation of China under grant 71201102 and 61573244.

References

S. Alexander, T. Coleman, Y. Li, Minimizing cvar and var for portfolio of derivatives. J. Bank. Financ. **30**, 583–605 (2006)

F. Andersson, H. Mausser, D. Rosen, S. Uryasev, Credit risk optimization with conditional value at risk criterion. Math. Program. Ser. B **89**, 273–291 (2001)

P. Artzner, F. Delbaen, J. Eber, D. Heath, Coherent measure of risk. Math. Financ. **9**, 203–228 (1999)

V. Demiguel, L. Garlappi, F.J. Nogales, R. Uppal, A generalized approach to portfolio optimization: improving performance by constraining portfolio norms. Manag. Sci. **55**(5), 798–812 (2009a)

V. Demiguel, L. Garlappi, R. Uppal, Optimal versus naive diversification: how inefficient is the 1/n portfolio strategy? Rev. Financ. Stud. **22**(5), 1915–1953 (2009b)

J. Gao, D. Li, Optimal cardinality constrained portfolio selection, Oper. Res. **61**, 745–761 (2013)

J. Gotoh, A. Takeda, On the role of norm constraints in portfolio selection. Comput. Manag. Sci. **8**, 323–353 (2011)

J. Gotoh, K. Shinozaki, A. Takeda, Robust portfolio techniques for mitigating the fragile of cvar minimization and generalization to coherent risk measures. Quant. Financ. **13**(10), 1621–1635 (2013)

IBM, Reference Manual CPLEX 12.5, IBM, USA, (2015)

P.N. Kolm, R. Tütüncü, F. Fabozzi, 60 years of portfolio optimization: practical challenges and current trends. Eur. J. Oper. Res. **234**(2), 356–371 (2014)

I. Kondor, S. Pafka, G. Nagy, Noise sensitivity of portfolio selection under various risk measures. J. Bank. Financ. **31**, 1545–1573 (2007)

A. Kunzi-Bay, M. Janos, Computational aspects of minimizing conditional value-at-risk. Comput. Manag. Sci. **3**, 3–27 (2006)

O. Ledoit, M. Wolf, Improved estimation of the covariance matrix of stock rreturn with an application to portfolio selection. J. Empir. Financ. **10**, 603–621 (2003)

D. Li, X. Sun, J. Wang, Optimal lot solution to cardinality constrained mean-variance formulation for portfolio selecton. Math. Financ. **16**, 83–101 (2006)

A. Lim, J.G. Shanthikumar, G. Vahn, Conditional value-at-risk in portfolio optimization: chorent but fragile. Oper. Res. Lett. **39**, 163–171 (2011)

H. Markowitz, Portfolio selection. J. Financ. **7**, 77–91 (1952)

R.C. Merton, On estimating the expected return on the market: an exploratory investigation. J. Fianc. Econ. **8**, 43–54 (1980)

G. Pflug, Some remarks on the value-at-risk and conditonal value-at-risk, in *Probabilistic Constrained optimization: Methodology and Applications* (Springer, Boston, 2000), pp. 272–281

R. Rockafellar, S. Uryasev, Optimization of conditional value-at-risk. J. Risk **2**, 21–41 (2000)

R. Rockafellar, S. Uryasev, Conditional value-at-risk for general loss distributions. J. Bank. Financ. **26**, 1443–1471 (2002)

A. Takeda, T. Kanamori, A robust approach based on conditional value-at-risk measure to statistical learning problems. Eur. J. Oper. Res. **198**(1), 287–296 (2009)

Y. Zhao, D. Li, Reweighted l1-norm minimization for sparse solutions to underdetermined lineaer systems. SIAM J. Optim. **22**(3), 1065–1088 (2012)

S.S. Zhu, M. Fukushima, Worst-case conditional value-at-risk with application to robust portolio management. Oper. Res. **57**, 1155–1168 (2009)

Chapter 11
Multistage Optioned Portfolio Selection: Mean-Variance Model and Target Tracking Model

Jianfeng Liang

Abstract Options form an indispensable part of the modern financial markets. One reason for this phenomenon is the versatile payoff structures of options, which can serve to form investment portfolios with desirable risk profiles. This chapter introduces mean-variance models and develops target tracking model for optioned portfolio selection problem in both static and dynamic formulations. We focus on the rich properties of the payoff functions and the solution methodologies. Two different solution techniques for multistage mean-variance model are discussed: one is based on stochastic programming and optimality conditions, and the other one is based on stochastic control and dynamic programming. In addition, tracking-error-variance optimization models are proposed and solved by dynamic programming. It turns out that the optimal tracking portfolio holds mean-variance efficiency. Close form relationships between the mean-variance model and the tracking model are proved, which bring new insights to dynamically solve the classical multistage mean-variance model. Throughout the chapter, numerical examples with real life data are used to illustrate and validate the results.

Keywords Portfolio selection • Index options • Multistage mean-variance model • Multistage tracking model • Scenario tree • Dynamic programming • Stochastic control

11.1 Literature Review

Options on stocks were first traded on an organized exchange in 1973. Since then there has been a dramatic growth in the options markets. Options are now traded on many exchanges throughout the world. Huge volumes of options are also

---

J. Liang (✉)
Department of Finance, Lingnan (University) College, Sun Yat-sen University, Guangzhou, People's Republic of China
e-mail: jfliang@mail.sysu.edu.cn

T.-M. Choi et al. (eds.), *Optimization and Control for Systems in the Big-Data Era*, International Series in Operations Research & Management Science 252,
DOI 10.1007/978-3-319-53518-0_11

traded over the counter by banks and other financial institutions. Options form an indispensable part of the modern financial markets. One reason for this phenomenon is the versatile payoff structures of options, which can serve to form investment portfolios with desirable risk profiles (Hull 1999; Mcmillan 2002). It makes sense to investigate how an individual investor should select his/her derivative optioned portfolios in a multistage investment framework, and to study the payoff patterns as a result of different market environments.

The mean-variance analysis of Markowitz (1952, 1959) plays a key role in the theory of portfolio selection, which quantifies the return and the risk in computable terms. In the presence of short-selling, the model is analytically solvable (Merton 1972). The mean-variance model was later extended to the multistage dynamic case. Research on dynamic portfolio selection problem had been dominated by maximizing expected utility function of the terminal wealth (Mossion 1968; Samuleson 1969; Hakansson 1971; Merton 1971; Elton and Gruber 1974a,b; Dumas and Luciano 1991; Grauer and Hakansson 1993). However, it was only until 2000 when an analytical formulations of the optimal portfolio for the multistage mean-variance model along with an expression of the mean-variance efficient frontier were derived, due to Li and Ng (2000), by means of an embedding scheme and a stochastic control strategy. The multistage mean-variance analysis is further developed with more considerations, such as bankruptcy consideration, uncertain investment horizon, and cardinality constraints (see Zhu et al. 2004; Yi et al. 2008; Chiu and Li 2006; Li et al. 2006). A generalized mean-variance model involving options is proposed by Morard and Naciri (1990), which aims to optimize the hedging ratios. The hedging is implemented with the covered call writing strategy. The empirical results showed that the use of covered calls improved the performance of stock portfolios. Isakov and Morard (2001) point out that in the case of incomplete hedging, the mean-variance formula could be applied for the optioned portfolio selection problems, since the hedged return does not necessarily have a non-symmetric distribution. It turns out that the mean-variance criteria is a reasonable choice for the optioned portfolio selection problem.

Parallel to this development, stochastic programming has found wide range of applications in financial planning. It is agreed that the discrete decision framework is more prone to numerical implementations than the continuous setting. Scenario tree constitutes a generic, relatively simple approach to represent future states of the world in stochastic optimization problems. In finance, such trees have been used in numerous modes, both in computational and in theoretical frameworks (Koskosidis and Duarte 1997; Odenkamp 1999; Rockafellar and Uryasev 2000; Berkelaar et al. 2002; King 2002; Kallio and Ziemba 2007). Naturally, stochastic linear programming models for optioned portfolio selection have been proposed in the literature. The models are mostly proposed to maximize the expected return under some desired constraints (Dert and Oldenkamp 2000; Berkelaar et al. 2002, 2005). Carr and Madan (2001) discuss the optimal payoffs of optioned portfolio in a single-stage setting by adopting expected utility-maximization model. Liang et al. (2008) studies mean-variance optioned portfolio selection models based on multistage scenario tree structure. It studies the individual investor's payoff patterns analytically, which lends more insights on the relationship between the optioned portfolio payoffs and that of the portfolios in the security market. This paper also empowers the decision-maker by offering an alternative solution method, based on stochastic programming. Schyns et al. (2010) present a multistage optioned portfolio selection model under VaR (Value-at-Risk) constraints. The model contains several features, like the consideration of transaction costs, bid-ask spreads, and the possibility to rebalance the portfolio with options introduced at the start of each period. The resulting mixed integer programming model can be near-optimized by a standard branch-and-cut solver or by a specialized heuristic.

Target tracking models have been widely used in recent years in the investment industry. It enables the investor to follow a performance benchmark closely by holding a portfolio involving a few assets, including derivatives. Minimization of tracking error has become an important criterion for assessing overall manager performance. The classical tracking problem focuses on minimizing the tracking error from a benchmark portfolio under some restrictions. There are many different definitions of tracking error and as a consequence different models. Roll (1992) makes a mean-variance analysis of tracking error, and defines the tracking error variance (TEV) criterion, where the tracking error is measured by the square of the difference between the performance of the portfolio and a benchmark. Solution methods of the portfolio selection under TEV criterion are investigated in literatures (see Ammann and Zimmermann 2001; Clarke et al. 1994; Fang and Zhang 2006; Jorion 2003; Ma and Tang 2001), which almost concentrate on static framework. Rohweder (1998) offers portfolio segmentation as an alternative to tracking error optimization, and discusses the problem with transaction cost by simulation method. Wang (1999) considers the problem of tracking multiple targets with multiple portfolios. For a relation between tracking error models and tactical asset allocation, see Ammann and Zimmermann (2001) and Clarke et al. (1994).

Static models usually assume a backward perspective, which allows to find the portfolio that best tracks the performance of given benchmark during a past period and then keep it for a subsequent period. Introducing scenarios in a static model improves this approach since the optimal portfolio is composed at the beginning of the period considering different possible future realizations (scenarios) and not only past history. For a scenario approach in static models, Dembo and Rosen (1999), Dempster and Thompson (2002), and Ziemba (2003). This approach is an improvement on the static models based on past history since it allows forecasts of future realizations and blending of forecasts and subjective views. This also allows use of some more general distributions and non-linear instruments such as options. Cesari and Cremonini (2003) propose a comparison of benchmark with other asset allocation strategies in Monte Carlo simulation framework. Barror and Canestrelli (2009) formulate and solve a multistage tracking error model in stochastic programming framework based on scenario tree. They consider an increasing number of scenarios and assets and show the superior performance of the dynamically optimization tracking portfolio over static strategies. Liang (2009) and Liang and Liu (2009) investigate the tracking problems by applying portfolio of options under TEV measure, and analyze the efficiency of the optimal portfolio of the tracking model. A double tracking error portfolio model is proposed by Barror and Canestrelli (2009), which combines the goals of replicating the performance of a benchmark and controlling downside risk. The choice of a proper measure for downside risk leads to different problem formulations and investment strategies that reflect different attitudes towards risk. The proposed model is test through a set of out-of-sample rolling simulation in different market conditions.

This chapter introduces the recent developments in multistage optioned portfolio selection problems derived in the literature, and also proposes new insights in solving methods. Specifically, we discuss the mean-variance model and the target tracking model of optioned portfolio selection, in both problem formulations and solving methods. The rest part of this chapter is organized as follows. Section 11.2 reviews the static optioned portfolio selection problems, in mean-variance and tracking formulations. It focuses on characterizations of optioned portfolio payoff and the efficiency of optimal tracking portfolio. Section 11.3 describes the dynamic mean-variance formulations and the solving methods from both quadratic optimization and stochastic control viewpoints. Multistage tracking model analysis is presented in Sect. 11.4, which brings in new insights in solving the mean-variance model. Summary follows.

11.2 The Static Models

We first consider the static mean-variance model and tracking model for portfolio selection of a portfolio of options in this section. Holding portfolio involving options allows traders to shape the future payoff of their portfolios, thanks to the intrinsic properties of options. We restrict our portfolio to the case of options linked to a same financial index. Models with several underlying securities and their options could also be handled by a similar approach but they require more parameters. With one underlying index, clear conclusions are easier to draw. The static models and solving methods are straightforward. The main purpose to discuss this case is to deliver the rich properties of the optimal optioned portfolio payoff, and also analyze the relations between two types of models.

11.2.1 Statement and Notations

Consider the investment problem based on a single-stage scenario tree structure. There is a stock index, a set of m European call options on the stock index, and a risk-free asset. The options have the same expiration, and their strike prices are $K_1 < K_2 < \cdots < K_m$. The decision horizon is the same as the options' expiration, and r is the gross risk-free return rate in this period. The decision variables are thus denoted to be X and x, where X is the number of shares of stock index and its options to hold ($X = (x_0, x_1, \cdots, x_m) \in \Re^{m+1}$), and x is the dollar invested in the risk-free asset ($x \in \Re$).

As for the decision tree. There are n scenarios, with p_i the probability of the ith scenario occurring, $i = 1, 2, \cdots, n$, and so $\sum_{i=1}^{n} p_i = 1$. Other data includes the price vector of risky assets (stock index and its options) in the beginning, u, and the per-share payoff vector of the risky assets at the ith scenario, v_i. We denote $\bar{v} = \sum_{i=1}^{n} p_i v_i$ as the average per-share payoff vector, and $A = \sum_{i=1}^{n} p_i (v_i - \bar{v})(v_i - \bar{v})'$ as the covariance matrix of risky assets.

Assumption 2.1 *Assume that the scenario tree is well generated in the following sense. There are in total at least $m + 2$ scenarios, and for each given interval, $I_0 = (0,\ K_1)$, $I_i = (K_i,\ K_{i+1})$, $i = 1, 2, \cdots, m-1$, and $I_m = (K_m,\ +\infty)$, there is at least one scenario. Moreover, the prices of the options are generated so as to ensure that there is no arbitrage opportunity on the scenario tree.*

As a consequence of the above assumption, this structure makes sure that the estimated covariance matrix A is positive definite. Remark that in practice Assumption 2.1 can be easily verified, for instance, by solving a linear programming model. In numerical experience, the theoretical Black–Scholes option prices automatically satisfy the no arbitrage property if the scenario tree is generated in the way as described in Assumption 2.1.

Furthermore, let us introduce the following quantities which are helpful to present the models in clear formulations.

$$
\begin{aligned}
\widetilde{A} &= A + \bar{v}\bar{v}', \\
\alpha &= u'A^{-1}u, \\
\beta &= \bar{v}'A^{-1}u, \\
\gamma &= \bar{v}'A^{-1}\bar{v}, \\
\delta &= r^2\alpha - 2r\beta + \gamma = (\bar{v} - r \cdot u)'A^{-1}(\bar{v} - r \cdot u), \\
\rho &= \tfrac{r}{\delta+1}.
\end{aligned}
$$

Following, we will introduce the static mean-variance model and tracking model of optioned portfolio selection problem.

11.2.2 The Mean-Variance Model

Given the initial wealth B, the terminal wealth W, and the expected return R, the objective is to construct a optioned portfolio with minimum final payoff volatility. The single-stage mean-variance model is to minimize the variance Var (W), subject to the constraints on expected return $\mathsf{E}(W)$ and budget. Based on the scenario tree just introduced, we have

$$\mathsf{E}(W) = \sum_{i=1}^{n} p_i v_i' X + rx = \bar{v}' X + rx \tag{11.1}$$

$$\text{Var}\,(W) = \mathsf{E}\left[(v_i' X + rx - R)^2\right] = \begin{pmatrix} X \\ x \end{pmatrix}' \begin{pmatrix} \widetilde{A} & r\bar{v} \\ r\bar{v}' & r^2 \end{pmatrix} \begin{pmatrix} X \\ x \end{pmatrix} - R^2 \tag{11.2}$$

So the deterministic equivalent of the single-stage mean-variance model is

$$(M_1)\ \min\ \frac{1}{2} \begin{pmatrix} X \\ x \end{pmatrix}' \begin{pmatrix} \widetilde{A} & r\bar{v} \\ r\bar{v}' & r^2 \end{pmatrix} \begin{pmatrix} X \\ x \end{pmatrix} - \frac{1}{2} R^2 \tag{11.3}$$

$$\text{s.t.}\ \ u' X + x = B \tag{11.4}$$

$$\bar{v}' X + rx = R. \tag{11.5}$$

Under Assumption 2.1, (M_1) is a strictly convex quadratic optimization problem, and its solution is given in the following theorem.

Theorem 2.1 *The single-stage mean-variance model* (M_1) *has the following unique primal-dual solutions:*

$$X = -\frac{\lambda}{r} A^{-1} (\bar{v} - ru),$$

$$x = \frac{1}{r} \left[\frac{\lambda}{r} (\gamma - r\beta + 1) + \mu \right],$$

$$\lambda = \frac{r}{\delta + 1} (rB - \mu) \equiv \rho (rB - \mu),$$

$$\mu = r \frac{R - \rho B}{r - \rho},$$

where λ *and* μ *are the Lagrangian multipliers related to the constraints (11.4) and (11.5), respectively, and the associated risk is*

$$Var\,(W) = \frac{\rho}{r - \rho} (R - rB)^2.$$

The proof of the theorem is straightforward, by using the optimality condition, and is thus omitted here. The main purpose of discussing this special case is to highlight the properties of the payoff structures of the optimal optioned portfolio.

Proposition 2.1 *The optimal payoff curve is piecewise linear with respect to the underlying index value. The slopes of the line segments are steeper for a larger target* R *value for all* $R > rB$. *At any breakpoint where the slope of the curve changes, the index value must equal to one of the strike prices of the options.*

Furthermore, by defining $\psi := \frac{\rho r}{r-\rho}$ *and* $\theta := \frac{\psi}{r}A^{-1}(\bar{v} - ru)$*, a scenario* $S = K_j$ *is a local maximum point for the payoff function iff* $\sum_{i=0}^{j-1} \theta_i > 0$*, and* $\sum_{i=0}^{j} \theta_i < 0$*, and it is a local minimum point iff* $\sum_{i=0}^{j-1} \theta_i < 0$*, and* $\sum_{i=0}^{j} \theta_i > 0$*.*

Proof Denote $\psi := \frac{\rho r}{r-\rho}$ and $\theta := \frac{\psi}{r}A^{-1}(\bar{v} - ru)$, and rewrite the optimal solutions of (M_1) to be

$$\begin{aligned} \lambda &= \psi(rB - R), \\ X &= \theta(R - rB), \\ x &= B - u'\theta(R - rB) \equiv B + a(R - rB), \end{aligned}$$

where $\theta = (\theta_0, \theta_1, \cdots, \theta_m)'$, and $a := -u'\theta$. For convenience, scenario S represents a fact that the index value is S in this scenario. Denote the payoff at scenario S to be $P(S)$. Since the m European call options have strike prices $K_1 < K_2 < \cdots < K_m$, for $K_j < S < K_{j+1}$,

$$P(S) = x_0 S + \sum_{i=1}^{j} x_i(S - K_i) + rx.$$

Let two scenarios S_1 and S_2 be between K_j and K_{j+1}, i.e., $K_j < S_1 < S_2 < K_{j+1}$, then

$$P(S_2) - P(S_1) = (S_2 - S_1)x_0 + \sum_{i=1}^{j} x_i(S_2 - S_1) = (S_2 - S_1)\sum_{i=0}^{j} x_i.$$

Since $x_i = \theta_i(R - rB)$, $i = 0, 1, \ldots, m$, the slope of the payoff function between K_j and K_{j+1} is

$$\frac{P(S_2) - P(S_1)}{S_2 - S_1} = (R - rB)\sum_{i=0}^{j} \theta_i. \tag{11.6}$$

Therefore, for a fixed R this slope is constant, and the payoff is linear between two neighboring strike prices. Also because θ_i is independent of R, for any $R \geq rB$, the value of $\sum_{i=0}^{j} \theta_i$ is fixed and so the payoff function is linear in R. Moreover, for a larger R value with $R > rB$, the slopes of line segments between any two neighboring strike prices will become steeper.

Equation (11.6) further leads that a scenario $S = K_j$ is a local maximum point for the payoff function iff $\sum_{i=0}^{j-1} \theta_i > 0$, and $\sum_{i=0}^{j} \theta_i < 0$, and it is a local minimum point iff $\sum_{i=0}^{j-1} \theta_i < 0$, and $\sum_{i=0}^{j} \theta_i > 0$. □

Corollary 2.1 *There are scenarios where the payoffs of the optimal portfolio are constantly rB, regardless the value of R. In any scenario S, for all* $R > rB$*, if* $P(S) > rB$*, then the payoff of a higher R dominates that of a lower R; else, if* $P(S) < rB$*, then the payoff of a higher R is dominated by that of a lower R.*

The above corollary depicts an essential structure of the optimal payoff curve. For different target expected return R, the optimal payoff curves form piecewise line segments with an invariant set of breakpoints and varying degree of slopes. The following specific example helps to visualize the picture.

Example 1 We consider the model using the market data of call options on the S and P 500 index, which are listed on the CBOE. The prices are drawn from the CBOE web page in the morning of Aug. 9, 2006. The horizon is equal to the expiration date of the options, which is Sep. 16, 2006. The investment horizon is 38 days. In this example, we simply use the mid prices as the short/long option prices. The scenario tree is generated under the assumption that the index value at the horizon is lognormally distributed with an expected annualized growth rate $\mu = 13.24\%$ and annualized standard deviation of $\sigma = 16.25\%$, which are listed in the webpage of the Standard and Poor's. The risk-free return rate for the investment horizon is set to be $r = 1 + 0.5\%$. The initial wealth is $B = \$10,000$. We assume three different target expected payoffs $R_1 = rB = \$10{,}050$, $R_2 = \$10{,}100$, and $R_3 = \$10{,}200$. The information of the assets and the optimal solutions is shown in Table 11.1. The third column shows the θ value in the optimal portfolio, which is independent of the target R value. We have shown in Proposition 2.1 that the slope of the $(i+1)$th line segment of the payoff curve is $(R-rB)\sum_{j=0}^{i}\theta_j$. Therefore from the values given in the fourth column, we can tell if a breakpoint is a local maximum or minimum point of the payoff curve, which is shown in the fifth column. Figure 11.1 shows the optimal payoff curves for R_1, R_2, and R_3 by the flat red solid line, the more fluctuate blue dash line, and the most fluctuate green dash-dot line, respectively.

Table 11.1 The assets and the optimal solutions of Example 1

	Mid price	The variable	θ_i^*	$\sum_{j=0}^{i}\theta_j^*$	Local max/min of the breakpoint on the strike price
S and P 500	1265.95	x_0	0.0043	0.0043	
Call 1230	50.30	x_1	0.1728	0.1771	Neither
Call 1240	42.60	x_2	−0.7919	−0.6148	Max
Call 1245	38.80	x_3	0.2316	−0.3832	Neither
Call 1250	35.20	x_4	1.2239	0.8407	Min
Call 1260	28.65	x_5	−2.9942	−2.1535	Max
Call 1265	25.30	x_6	3.4876	1.3341	Min
Call 1270	22.30	x_7	−1.6472	−0.3131	Max
Call 1280	16.80	x_8	0.6515	0.3384	Min
Call 1285	14.40	x_9	−0.4816	−0.1432	Max
Call 1290	11.40	x_{10}	0.3079	0.1647	Min
Call 1295	10.20	x_{11}	−0.7173	0.5526	Neither
Call 1300	8.40	x_{12}	1.1943	0.6417	Neither
Call 1305	6.90	x_{13}	−0.7389	−0.0972	Max 7.40
Call 1315	4.35	x_{14}	0.1204	0.0232	Min 4.70

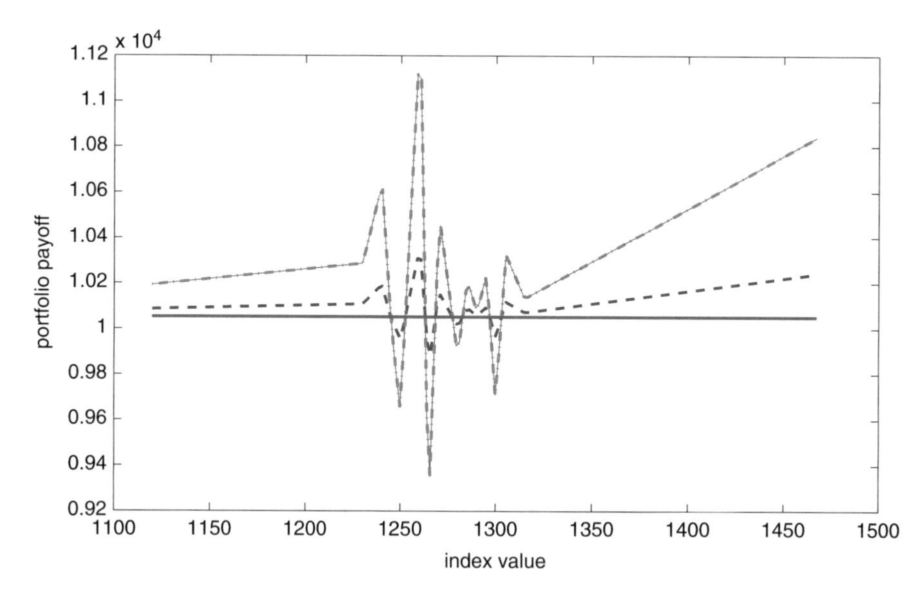

Fig. 11.1 The optimal payoff curves of Example 1

11.2.3 The Target Tracking Model

The classical tracking error problem focuses on minimizing the deviations from a benchmark under some restrictions. Let W be the final investment payoff, and denote R as a given target payoff, then the tracking error variance (TEV) is defined as the expectation of the square of the difference between W and R, that is, TEV $=$ $\mathsf{E}[(W-R)^2]$. Given initial wealth B, the investment object is to construct a portfolio with minimum TEV . Applying the same assets and notation in the previous section, the static tracking problem is presented as follows:

$$\begin{aligned} \min \ & \frac{1}{2}\mathsf{E}[(W-R)^2] \\ \text{s.t.} \ & u'X + x_f = B, \end{aligned}$$

In the given scenario tree structure, we deliver the explicit formulation of the tracking error variance $\mathsf{E}[(W-R)^2]$:

$$\begin{aligned} \mathsf{E}[(W-R)^2] &= \mathsf{E}[(v'X + rx_f - R)^2] \\ &= \mathsf{E}[X'vv'X + r^2x_f^2 + R^2 + 2rx_fu'X - 2Rv'X - 2Rrx_f] \\ &= X'(A + \bar{v}\bar{v}')X + r^2x_f^2 + 2rx_f\bar{v}'X - 2R\bar{v}'X - 2Rrx_f + R^2 \end{aligned}$$

$$= \begin{pmatrix} X \\ x_f \end{pmatrix}' \begin{pmatrix} \widetilde{A} & r\bar{v} \\ r\bar{v}' & r^2 \end{pmatrix} \begin{pmatrix} X \\ x_f \end{pmatrix} - 2R \begin{pmatrix} \bar{v} \\ r \end{pmatrix}' \begin{pmatrix} X \\ x_f \end{pmatrix} + R^2 \tag{11.7}$$

The deterministic equivalent form of the static tacking model is

(M_2)

$$\min \frac{1}{2} \begin{pmatrix} X \\ x_f \end{pmatrix}' \begin{pmatrix} \widetilde{A} & r\bar{v} \\ r\bar{v}' & r^2 \end{pmatrix} \begin{pmatrix} X \\ x_f \end{pmatrix} - R \begin{pmatrix} \bar{v} \\ r \end{pmatrix}' \begin{pmatrix} X \\ x_f \end{pmatrix} + \frac{1}{2} R^2 \tag{11.8}$$

$$\text{s.t.} \quad u'X + x_f = B \tag{11.9}$$

Theorem 2.2 *The static target tracking model* (M_2) *has a unique primal-dual solution*

$$X = -\frac{\lambda}{r} A^{-1} (\bar{v} - ru)$$

$$x_f = \frac{1}{r} \left[\frac{\lambda}{r} (\gamma - r\beta + 1) + R \right]$$

$$\lambda = \frac{r}{\delta + 1} (rB - R) \equiv \rho(rB - R),$$

with the associated optimal tracking error variance TEV^*:

$$TEV^* = \frac{\rho}{r} (R - rB)^2,$$

where λ is the Lagrangian multiplier associated with the budget constraint (11.9).

Review Theorems 2.1 and 2.2, it shows that the solution formulations in both models are close to each other. Especially, the solutions for tracking model are a special case of those for the mean-variance model with assigning $\mu = R$. Now it is interesting to further investigate the efficiency of the optimal tracking portfolio, given by the following proposition.

Proposition 2.2 *The optimal portfolio of a tracking model holds mean-variance efficiency under the same scenario tree structure and same set of constraints.*

Proof We discuss the models in general case, and both models hold the same set of budget constraints and other given constraints.

The mean-variance problem with expected payoff R^M can be shown as:

(MV)

$$\begin{aligned} \min\ & \frac{1}{2}\text{Var} \\ \text{s.t.}\ & \textit{Expected payoff constraints} \\ & \textit{Budget constraints} \\ & \textit{Other given constraints.} \end{aligned}$$

Review (11.2) and (11.7) leads the following relations between the tracking error variance TEV and the variance Var

$$\text{TEV} = \text{Var} + (Mean - R^T)^2 \tag{11.10}$$

where *Mean* represents the expectation of the portfolio payoff, and R^T is the tracking target payoff. Therefore the general tracking model is reformulated as:

(TEV)

$$\begin{aligned} \min\ & \frac{1}{2}\text{Var} + \frac{1}{2}(Mean - R^T)^2 \\ \text{s.t.}\ & \textit{Budget constraints} \\ & \textit{Other given constraints.} \end{aligned}$$

Suppose that (X, x_f) is an optimal portfolio for **(TEV)** with given tracking target R^T. Denote $Mean(X, x_f)$ as the expected payoff of this portfolio. If (X, x_f) is not optimal for the model **(MV)** with a given $R^M = Mean(X, x_f)$, then there must exist another portfolio, $(Y, y_f) \neq (X, x_f)$, which is on the efficient frontier with the same mean R^M, that is,

$$\begin{cases} Mean(Y, y_f) = Mean(X, x_f) = R^M \\ \text{Var}\ (Y) < \text{Var}\ (X). \end{cases}$$

Therefore

$$\text{TEV}\ (Y, y_f) = \frac{1}{2}\text{Var}\ (Y) + \frac{1}{2}[Mean(Y, y_f) - R^T]^2$$

$$< \frac{1}{2}\text{Var}\ (X) + \frac{1}{2}[Mean(X, x_f) - R^T]^2 = \text{TEV}\ (X, x_f)$$

this contradicts to the optimality of (X, x_f) for **(TEV)**. Thus, the optimal tracking portfolio must be mean-variance efficient. □

Proposition 2.2 tells that the optimal tracking portfolio is on the MV efficient frontier. It is a natural question to fix the optimal portfolio point on the efficient frontier for the tracking problem. This is realized by deriving the relations between the tracking target R^T and the expected payoff R^M, with which both models hold the same optimal portfolio. Now we present Proposition 2.3 for details.

Proposition 2.3 *Given both mean-variance model and target tracking model based on a same static scenario tree, they share the same set of optimal solutions iff* $R^T - rB = h(R^M - rB)$*, where* R^T *is the tracking target,* R^M *is the expected return, and* $h := \frac{r}{r-\rho}$.

Proof For the given two models on a same scenario tree, when they share the same optimal portfolio, Eq. (11.10) shows that the following relations must be realized,

$$\text{TEV}^* = \text{Var}^* + (R^M - R^T)^2$$

where the optimal objective values are proposed in Theorems 2.1 and 2.2 as:

$$\begin{aligned} \text{TEV}^* &= \frac{\rho}{r}(R^T - rB)^2, \\ \text{Var}^* &= \frac{\rho}{r-\rho}(R^M - rB)^2 \end{aligned}$$

Therefore,

$$\begin{aligned} &\text{TEV}^* = \text{Var}^* + (R^M - R^T)^2 \\ \Rightarrow\ &\frac{\rho}{r}(R^T - rB)^2 = \frac{\rho}{r-\rho}(R^M - rB)^2 + (R^M - R^T)^2 \\ \Rightarrow\ &\frac{\rho - r}{r}(R^T - rB)^2 = \frac{r}{r-\rho}(R^M - rB)^2 - 2(R^T - rB)(R^M - rB) \\ \Rightarrow\ &[(R^T - rB) - \frac{r}{r-\rho}(R^M - rB)]^2 = 0 \\ \Rightarrow\ &(R^T - rB) = h(R^M - rB) \end{aligned}$$

where the last equation is delivered with $h := \frac{r}{r-\rho}$. □

This section introduces the static mean-variance model of optioned portfolio selection, shows the properties of the optimal optioned portfolio. A static target tracking model is also proposed, and proved for its mean-variance efficiency. We derive a deterministic transformation between these two models in static case, by which they share the same optimal portfolio. In the following context, we will discuss the models in multistage case and further specify the relationships in between.

11.3 The Multistage Mean-Variance Model

Consider the multistage mean-variance model for optioned portfolio selection. We focus on the problem formulations and the solution methods in both mathematical programming and stochastic control perspectives. Explicit solutions are proved in both cases, and the transformations between two sets of optimal solutions are delivered.

11.3.1 A Mathematical Programming Resolution

There are $T + 1$ stages denoted from stage 0 to T in the multistage scenario tree. Denote N_t to be the index set of the scenarios at stage t, and S_{nt} as the nth scenario at stage t, for $n \in N_t,\ t = 0, 1, \cdots, T$. For those data at this scenario, the price vector of the risky assets is denoted by u_{nt}, and the payoff vector of the risky assets (scaled to be the rates of returns) is denoted by v_{nt}. We denote the wealth at this scenario to be W_{nt}. The decision variables at this scenario are X_{nt} and x_{nt}. X_{nt} is the vector of number of shares for risky assets to hold on S_{nt}, and x_{nt} is the dollar invested in the risk-free asset. Moreover, let $S_{a(n),t-1}$ be the ancestor of S_{nt}, and $S_{c(n),t+1}$ be the set of immediate children of S_{nt}. The conditional probability distribution of S_{nt}, given $S_{a(n),t-1}$, is p_{nt}. Finally, the gross risk-free return rate is r_t for stage t.

For other notations, B is the initial wealth, W_T is the wealth at the end of the last stage, and R is the given level of the expected final payoff.

The multistage mean-variance model is in the following formulation:

$$\begin{aligned}
\min\ & \tfrac{1}{2}\mathrm{Var}\,(W_T)\\
\text{s.t.}\ & u_0'X_0 + x_0 = B\\
& v_{nt}'X_{a(n),t-1} + r_{t-1}x_{a(n),t-1} = u_{nt}'X_{nt} + x_{nt},\ t = 1, 2, \cdots, T,\ n \in N_t,\\
& \mathsf{E}(W_T) = R,
\end{aligned}$$

Now define some constants in order to get an explicit formulation of the model:

$$\begin{aligned}
\alpha_{nt} &= u_{nt}'A_{nt}^{-1}u_{nt},\\
\beta_{nt} &= \bar{v}_{nt}'A_{nt}^{-1}u_{nt},\\
\gamma_{nt} &= \bar{v}_{nt}'A_{nt}^{-1}\bar{v}_{nt},\\
\delta_{nt} &= (\bar{v}_{nt} - r_t u_{nt})'A_{nt}^{-1}(\bar{v}_{nt} - r_t u_{nt}),\\
q_{nt} &= \textstyle\sum_{i \in c(n)} \frac{q_{i,t+1}}{1+\delta_{i,t+1}}, \qquad t = 0, 1, \cdots, T-1\\
w_{nt} &= \frac{q_{nt}/(1+\delta_{nt})}{q_{a(n),t-1}},\\
\widetilde{v_{nt}} &= q_{nt}\bar{v}_{nt},\\
\widetilde{A_{nt}} &= q_{nt}(A_{nt} + \bar{v}_{nt}\bar{v}_{nt}'),\\
\widetilde{r_{nt}} &= q_{nt}r_t r_{t+1}\cdots r_T,\\
r &= r_0 r_1 \cdots r_T,\\
\tilde{r} &= q_0 r,\\
\rho &= \frac{\tilde{r}}{\delta_0+1} \equiv \frac{q_0 r}{\delta_0+1}.
\end{aligned}$$

Review the definitions of w_{nt} and q_{nt}, w_{nt} represents a modified conditional probability distribution, and for the last stage there are

$$w_{n,T+1} = p_{n,T+1}$$
$$q_{nT} = p_{nT} \cdot p_{a(n),T-1} \cdot \cdots \cdot p_{a(\cdots a(n)),1}$$

In the above formulas, $\bar{v}_{nt}$ is defined as the expected per-share payoff of the single-stage subtree derived from scenario S_{nt} with the modified probability distribution, and A_{nt} is the conditional covariance matrix of this subtree with the modified probability distribution. In this multistage case, we also assume that the scenario tree is well generated as proposed in Assumption 2.1, that is, all of these covariance matrixes are positive definite, and no arbitrage opportunity exists.

Denoting X_T and x_T as the solutions at the beginning of the final stage, $\mathsf{E}(W_T)$ and Var (W_T) are expressed as follows:

$$\mathsf{E}(W_T) = \sum_{n\in N_T} (\widetilde{v_{nT}}' X_{nT} + \widetilde{r_{nT}} x_{nT})$$

$$\text{Var}\,(W_T) = \sum_{n\in N_T} \begin{pmatrix} X_{nT} \\ x_{nT} \end{pmatrix}' \begin{pmatrix} \widetilde{A_{nT}} & r_T\widetilde{v_{nT}} \\ r_T\widetilde{v_{nT}}' & r_T\widetilde{r_{nT}} \end{pmatrix} \begin{pmatrix} X_{nT} \\ x_{nT} \end{pmatrix} - R^2$$

Thus, the deterministic equivalent mathematical programming formulation for the multistage mean-variance model shows in (M_3):

$$(M_3)\ \min \frac{1}{2} \sum_{n\in N_T} \begin{pmatrix} X_{nT} \\ x_{nT} \end{pmatrix}' \begin{pmatrix} \widetilde{A_{nT}} & r_T\widetilde{v_{nT}} \\ r_T\widetilde{v_{nT}}' & r_T\widetilde{r_{nT}} \end{pmatrix} \begin{pmatrix} X_{nT} \\ x_{nT} \end{pmatrix} - \frac{1}{2}R^2 \tag{11.11}$$

$$\text{s.t.}\ u_0' X_0 + x_0 = B \tag{11.12}$$

$$v_{nt}' X_{a(n),t-1} + r_{t-1} x_{a(n),t-1} = u_{nt}' X_{nt} + x_{nt}, \quad n \in N_t,\ t = 1, 2, \cdots, T \tag{11.13}$$

$$\sum_{n\in N_T} \left(\widetilde{v_{nT}}' X_{nT} + \widetilde{r_{nT}} x_{nT}\right) = R \tag{11.14}$$

Assume that the multistage scenario tree is well generated, that is, all of these covariance matrixes are positive definite, and no arbitrage opportunity exists. The multistage model (M_3) is a strictly convex quadratic programming problem. It can be solved analytically as shown in Theorem 3.1. An important feature of (M_3) is that it is flexible. If other complicating constraints are added, then the model may not admit an explicit solution as stipulated in Theorem 3.1; however, the model can still be solved very efficiently in the numerical sense (see, e.g., Berkelaar et al. 2005).

Theorem 3.1 *The multistage mean-variance model* (M_3) *has the following unique primal-dual solutions:*

$$
\begin{aligned}
X_{nt} &= -\frac{\lambda_{nt}}{r_{t+1}r_{t+2}\cdots r_T\widetilde{r_{nt}}}A_{nt}^{-1}(\bar{v}_{nt} - r_t u_{nt}),\\
x_{nt} &= \frac{1}{r_t r_{t+1}\cdots r_T}\left[\frac{\lambda_{nt}}{\widetilde{r_{nt}}}(\gamma_{nt} - r_t\beta_{nt} + 1) + \mu\right],\\
\lambda_{nt} &= \frac{\widetilde{r_{nt}}}{\delta_{nt} + 1}\left[r_t r_{t+1}\cdots r_T(W_{nt}) - \mu\right],\\
\mu &= \frac{r}{r-\rho}(R - \rho B),
\end{aligned}
$$

where λ_0, λ_{nt}, *and* μ *are the Lagrangian multipliers of the model. The associated risk is*

$$
Var\,(W_T) = \frac{\rho}{r-\rho}(R - rB)^2.
$$

Proof The Lagrangian function for (M_3) is

$$
\begin{aligned}
L = {} & \frac{1}{2}\sum_{n\in N_T}\begin{pmatrix}X_{nT}\\ x_{nT}\end{pmatrix}'\begin{pmatrix}\widetilde{A_{nT}} & r_T\widetilde{v_{nT}}\\ r_T\widetilde{v_{nT}}' & r_T\widetilde{r_{nT}}\end{pmatrix}\begin{pmatrix}X_{nT}\\ x_{nT}\end{pmatrix} - \frac{1}{2}R^2\\
& -\lambda_0(u_0'X_0 + x_0 - B)\\
& -\sum_{t=1}^{T}\left\{\sum_{n\in N_t}\lambda_{nt}[u_{nt}'X_{nt} + x_{nt} - v_{nt}'X_{a(n),t-1} - r_{t-1}x_{a(n),t-1}]\right\}\\
& -\mu\left[\sum_{n\in N_T}\left(\widetilde{v_{nT}}'X_{nT} + \widetilde{r_{nT}}x_{nT}\right) - R\right].
\end{aligned}
\tag{11.15}
$$

We prove the theorem by induction. First check the last-stage solutions. As $\frac{\partial L}{\partial x_{nT}} = 0$, we have

$$
r_T\widetilde{v_{nT}}'X_{nT} + r_T\widetilde{r_{nT}}x_{nT} - \lambda_{nT} - \mu\widetilde{r_{nT}} = 0
$$

leading to

$$
x_{nT} = \frac{1}{r_T}\left[-\bar{v}_{nT}'X_{nT} + \frac{\lambda_{nT}}{\widetilde{r_{nT}}} + \mu\right].
\tag{11.16}
$$

Substituting (11.16) into $\frac{\partial L}{\partial X_{nT}} = 0$ yields

$$
\widetilde{A_{nT}}X_{nT} + r_T\widetilde{v_{nT}}x_{nT} - \lambda_{nT}u_{nT} - \mu\widetilde{v_{nT}} = 0.
$$

Therefore,

$$p_{nT} \cdot p_{a(n),T-1} \cdots p_{a(\cdots a(n)),1} \cdot A_{nT} X_{nT} + \frac{\lambda_{nT}}{r_T}(\bar{v}_{nT} - r_T u_{nT}) = 0.$$

Solving the above for X_{nT}, and substituting it back to (11.16) gives an expression for x_{nT},

$$X_{nT} = -\frac{\lambda_{nT}}{\widetilde{r_{nT}}} A_{nT}^{-1}(\bar{v}_{nT} - r_T u_{nT}) \tag{11.17}$$

$$x_{nT} = \frac{1}{r_T}\left[\frac{\lambda_{nT}}{\widetilde{r_{nT}}}(\gamma_{nT} - r_T \beta_{nT} + 1) + \mu\right]. \tag{11.18}$$

The condition $\frac{\partial L}{\partial \lambda_{nT}} = 0$ is equivalent to

$$W_{nT} = u'_{nT} X_{nT} + x_{nT} = \frac{\lambda_{nT}}{r_T} \cdot \frac{\delta_{nT} + 1}{\widetilde{r_{nT}}} + \frac{\mu}{r_T}$$

yielding

$$\lambda_{nT} = \frac{\widetilde{r_{nT}}}{\delta_{nT} + 1}\left[r_T(W_{nT}) - \mu\right].$$

Therefore, the last-stage solutions satisfy the KKT conditions. Next we shall apply induction to the stage index t. Suppose that the following formulas hold true for the solutions at the tth stage, that is,

$$\begin{aligned} X_{nt} &= -\frac{\lambda_{nt}}{r_{t+1} r_{t+2} \cdots r_T \widetilde{r_{nt}}} A_{nt}^{-1}(\bar{v}_{nt} - r_t u_{nt}), \\ x_{nt} &= \frac{1}{r_t r_{t+1} \cdots r_T}\left[\frac{\lambda_{nt}}{\widetilde{r_{nt}}}(\gamma_{nt} - r_t \beta_{nt} + 1) + \mu\right], \\ \lambda_{nt} &= \frac{\widetilde{r_{nt}}}{\delta_{nt} + 1}\left[r_t r_{t+1} \cdots r_T(W_{nt}) - \mu\right]. \end{aligned}$$

We shall now prove that they also hold for stage $t-1$. By $\frac{\partial L}{\partial x_{n,t-1}} = 0$, we have

$$-\lambda_{n,t-1} + \sum_{l \in c(n)} \lambda_{lt} r_{t-1} = 0$$

implying

$$-\lambda_{n,t-1} + \sum_{l \in c(n)} \frac{r_{t-1}\widetilde{r_{lt}}}{\delta_{lt} + 1}\left[r_t r_{t+1} \cdots r_T(v'_{lt} X_{n,t-1} + r_{t-1} x_{n,t-1}) - \mu\right] = 0$$

i.e.,

$$-\lambda_{n,t-1} + r_{t-1}(r_t \cdots r_T)^2 \widetilde{v_{n,t-1}}' X_{n,t-1} + r_{t-1} \cdots r_T \widetilde{r_{n,t-1}} x_{n,t-1} - \mu \widetilde{r_{n,t-1}} = 0.$$

Finally,

$$x_{n,t-1} = \frac{1}{r_{t-1}\cdots r_n}\left[-r_t\cdots r_T\bar{v}'_{n,t-1}X_{n,t-1} + \frac{\lambda_{n,t-1}}{\widetilde{r_{n,t-1}}} + \mu\right]. \tag{11.19}$$

Substituting (11.19) into $\frac{\partial L}{\partial X_{n,t-1}} = 0$ yields

$$-\lambda_{n,t-1}u_{n,t-1} + \sum_{l\in c(n)} \lambda_{lt}v_{lt} = 0.$$

Thus,

$$-\lambda_{n,t-1}u_{n,t-1} + (r_t\cdots r_T)^2\widetilde{A_{n,t-1}}X_{n,t-1} + r_{t-1}(r_t\cdots r_T)^2\widetilde{v_{n,t-1}}x_{n,t-1} - \mu(r_t\cdots r_T)\widetilde{v_{n,t-1}} = 0$$

and so,

$$(r_t\cdots r_T)^2 q_{n,t-1}A_{n,t-1}X_{n,t-1} + \frac{\lambda_{n,t-1}}{r_{t-1}}(\bar{v}_{n,t-1} - r_{t-1}u_{n,t-1}) = 0,$$

which gives rise to $X_{n,t-1}$, as shown below,

$$X_{n,t-1} = -\frac{\lambda_{n,t-1}}{r_t\cdots r_T\widetilde{r_{n,t-1}}}A^{-1}_{n,t-1}(\bar{v}_{n,t-1} - r_{t-1}u_{n,t-1}), \tag{11.20}$$

and substituting this back into (11.19) gives

$$x_{n,t-1} = \frac{1}{r_{t-1}\cdots r_T}\left[\frac{\lambda_{n,t-1}}{\widetilde{r_{n,t-1}}}(\gamma_{n,t-1} - r_{t-1}\beta_{n,t-1} + 1) + \mu\right]. \tag{11.21}$$

Thus, the condition $\frac{\partial L}{\partial \lambda_{n,t-1}} = 0$ is equivalent to

$$\begin{aligned} W_{n,t-1} &= u'_{n,t-1}X_{n,t-1} + x_{n,t-1} \\ &- \frac{\lambda_{n,t-1}}{r_{t-1}\cdots r_T}\cdot\frac{\delta_{n,t-1}+1}{\widetilde{r_{n,t-1}}} + \frac{\mu}{r_{t-1}\cdots r_T} \end{aligned}$$

and so

$$\lambda_{n,t-1} = \frac{\widetilde{r_{n,t-1}}}{\delta_{n,t-1}+1}\left[r_{t-1}\cdots r_n(W_{n,t-1}) - \mu\right]. \tag{11.22}$$

Therefore, if the expressions in Theorem 3.1 are correct for the tth stage, then they are also correct for stage $t-1$. By induction, we have proven the theorem on the part of the primal solutions. Next we shall show that the expression for the dual variable μ and the objective function are also correct.

It follows from the solution forms of X, x, and λ just derived, in the single-stage subtree at stage $t-1$ with the root node $S_{n,t-1}$, we have the following two recursive equations:

Equation 1

$$\sum_{l\in c(n)}\left(\frac{\lambda_{lt}}{r_t\cdots r_T}+p_{lt}\cdot p_{n,t-1}\cdot p_{a(n),t-2}\cdots p_{a(\cdots a(n)),1}\cdot\mu\right)=\frac{\lambda_{n,t-1}}{r_{t-1}\cdots r_T}+p_{n,t-1}\cdot p_{a(n),t-2}\cdots p_{a(\cdots a(n)),1}\cdot\mu$$

Equation 2

$$\sum_{l\in c(n)} q_{lt}\left[\left(\frac{\lambda_{lt}}{\widetilde{r_{lt}}}\right)^2(\delta_{lt}+1)+2\mu\frac{\lambda_{lt}}{\widetilde{r_{lt}}}\right]=q_{n,t-1}\left[\left(\frac{\lambda_{n,t-1}}{\widetilde{r_{n,t-1}}}\right)^2(\delta_{n,t-1}+1)+2\mu\frac{\lambda_{n,t-1}}{\widetilde{r_{n,t-1}}}\right].$$

Then, based on Eq. (1), and the condition $\frac{\partial L}{\partial\mu}=0$, we have

$$\begin{aligned}
R &= \textstyle\sum_{n\in N_T}\left(\widetilde{v_{nT}}'X_{nT}+\widetilde{r_{nT}}x_{nT}\right)\\
&= \textstyle\sum_{n\in N_T} p_{nT}\cdot p_{a(n),T-1}\cdots p_{a(\cdots a(n)),1}\left[\frac{\lambda_{nT}}{\widetilde{r_{nT}}}+\mu\right]\\
&= \textstyle\sum_{n\in N_T}\left[\frac{\lambda_{nT}}{r_T}+p_{nT}\cdot p_{a(n),T-1}\cdots p_{a(\cdots a(n)),1}\cdot\mu\right]\\
&= \textstyle\sum_{n\in N_{T-1}}\left[\frac{\lambda_{n,T-1}}{r_{T-1}r_T}+p_{n,T-1}\cdot p_{a(n),T-2}\cdots p_{a(\cdots a(n)),1}\cdot\mu\right]\\
&\ \ \vdots\\
&= \tfrac{\lambda_0}{r}+\mu\\
&= \tfrac{\rho}{r}(rB-\mu)+\mu.
\end{aligned}$$

This implies

$$\mu=\frac{r}{r-\rho}(R-\rho B).$$

The expression for μ is thus proven.

By Eq. (2), we derive the optimal variance as

$$\begin{aligned}
\mathrm{Var}\,(W_T)+R^2 &= \textstyle\sum_{n\in N_T}\begin{pmatrix}X_{nT}\\ x_{nT}\end{pmatrix}'\begin{pmatrix}\widetilde{A_{nT}} & r_T\widetilde{v_{nT}}\\ r_T\widetilde{v_{nT}}' & r_T\widetilde{r_{nT}}\end{pmatrix}\begin{pmatrix}X_{nT}\\ x_{nT}\end{pmatrix}\\
&= \textstyle\sum_{n\in N_T} q_{nT}\left[\left(\frac{\lambda_{nT}}{\widetilde{r_{nT}}}\right)^2(\delta_{nT}+1)+2\mu\frac{\lambda_{nT}}{\widetilde{r_{nT}}}\right]+\mu^2\\
&= \textstyle\sum_{n\in N_{T-1}} q_{n,T-1}\left[\left(\frac{\lambda_{n,T-1}}{\widetilde{r_{n,T-1}}}\right)^2(\delta_{n,T-1}+1)+2\mu\frac{\lambda_{n,T-1}}{\widetilde{r_{n,T-1}}}\right]+\mu^2\\
&\ \ \vdots\\
&= q_0\left[\left(\tfrac{\lambda_0}{r}\right)^2(\delta_0+1)+2\mu\tfrac{\lambda_0}{r}\right]+\mu^2\\
&= r\rho B^2+\mu^2\left(1-\tfrac{\rho}{r}\right).
\end{aligned}$$

Hence, $\mathrm{Var}\,(W_T)=\frac{\rho}{r-\rho}(R-rB)^2$, as desired. □

Theorem 3.1 tells a way to solve the multistage mean-variance model explicitly. One can first compute all the quantities in a backwards fashion, and then get the explicit solutions forwards. Starting from the last stage T, the modified conditional probabilities can be firstly calculated (for the last stage they are equal to the real probabilities). Applying these modified probabilities in the last stage, the conditional expected per-share payoffs and the covariance matrices for each single-stage subtrees and also α, β, γ, δ at each decision node are computed. The following job is to go backward and carry out similar computations on stage $T - 1$. This process is repeated until stage 0, and μ could be calculated to finish the backward process. The backward process gathers all the required parameters and information. Then we start from the first stage to reap the solution. Applying the formulations in Theorem 3.1, optimal solutions for each decision node can be computed in a forward fashion using the data and quantities that have already been computed.

Example 2 We solve a two-stage problem with options on the S and P 500 index, which are listed on the CBOE. The prices are drawn from the CBOE web page in the morning of March 27, 2015, shown in Table 11.2. The portfolios can be constructed in the beginning of the investment and can also be reorganized on April 17, 2015. The investment horizon is May 15, 2015, on which the options can be exercised. The whole investment horizon is 48 days, with the first stage 20 days, and the second stage 28 days. The scenario tree is generated under the assumption that the index value at the horizon is lognormally distributed with an expected annualized growth rate $\mu = 14.43\%$ and an annualized standard deviation of $\sigma = 13.02\%$, which are listed in the Standard and Poor's company's web page. The annual risk-free return rates for both stages are $r_1 = r_2 = 1 + 0.02\%$. In this example, we simply use the mid prices as the initial prices for both buying and selling. The options prices at the end of first stage are calculated using the Black–Scholes formula with the volatility value σ given above. The initial wealth is \$10,000, and we assign an expected final payoffs $R =$ \$10,600.

Figure 11.2 shows the optimal payoff surface. We observe that the payoff curves for any one subtree in the second stage are piecewise linear and hold the properties described in Propositions 2.2–2.3.

Table 11.2 The data of the S and P 500 index and the options of Example 2

	Expiration	Last	Bid	Ask
S and P 500		2056.15	2056.15	2056.15
Call 2010	May 15	76.20	72.30	75.70
Call 2050	May 15	45.70	44.90	47.90
Call 2070	May 15	32.50	34.00	34.70
Call 2090	May 15	24.40	23.60	24.30
Call 2110	May 15	15.05	15.10	15.70
Call 2130	May 15	8.95	8.70	9.10

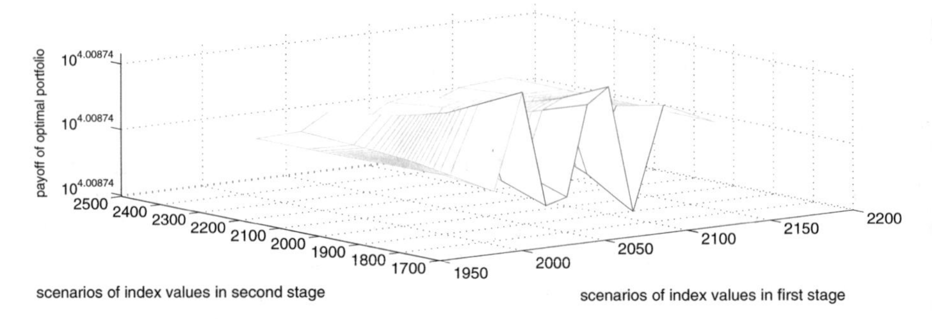

Fig. 11.2 The optimal payoff surfaces of Example 2

11.3.2 A Stochastic Control Perspective

The preceding section presents an easy computable procedure for solving the discrete multistage mean-variance model. In this section, we introduce an alternative approach. These two methods have distinctive features, and complement each other in many ways. The method in question lends itself from stochastic control. For a pure mean-variance model without derivatives, the first such analysis was due to Li and Ng (2000). We shall adopt that approach to accommodate options portfolio selection in an discrete problem statement. An important advantage of the stochastic control approach is its strength in dealing with extensions of the model to the continuous case (in time and in the space of scenarios), while it renders computational difficulties if inequality constraints are included in the model. In contrast to this, the stochastic programming approach can easily handle inequality constraints in the numerical sense, while it cannot work with the continuous extension of the model.

Let us briefly review the main points in Li and Ng (2000). In order to enable the stochastic control approach, we reformulate the multistage model as follows:

$$(M(\omega))\ \max\ \mathsf{E}(W_T) - \omega \text{Var}\,(W_T) \tag{11.23}$$

$$\text{s.t.}\ \ u_0' X_0 + x_0 = B \tag{11.24}$$

$$v_{nt}' X_{a(n),t-1} + r_{t-1} x_{a(n),t-1} = u_{nt}' X_{nt} + x_{nt}$$
$$n \in N_t,\ \ t = 1, 2, \cdots, T, \tag{11.25}$$

where $\omega \in [0,\ \infty)$ is a tradeoff factor between the variance and expected payoff, given by the decision-maker. Let I^t be an information set available at time t and $I^{t-1} \subset I^t$, $\forall t$. While the expectation operator satisfies the smoothing property: $\mathsf{E}[\mathsf{E}(\cdot|I^j)|I^k] = \mathsf{E}(\cdot|I^k)$, $\forall j > k$, the variance operator does not: Var [Var $(\cdot|I^j)|I^k] \neq$

Var $(\cdot|I^k)$, $\forall j > k$. Thus dynamic programming is not directly applicable to solve $(M(\omega))$. So we need an auxiliary problem as follows:

$$(M(\omega,\lambda)) \max -\omega \mathsf{E}(W_T^2) + \lambda \mathsf{E}(W_T) \tag{11.26}$$

$$\text{s.t. } u_0' X_0 + x_0 = B \tag{11.27}$$

$$v_{nt}' X_{a(n),t-1} + r_{t-1} x_{a(n),t-1} = u_{nt}' X_{nt} + x_{nt}$$
$$n \in N_t, \quad t = 1, 2, \cdots, T. \tag{11.28}$$

The auxiliary problem is of a separable structure suitable for dynamic programming. The objective function of $(M(\omega,\lambda))$ is in a quadratic form while the system dynamic is in a linear form. The following theorem points out the relationship between these two problems, which was established by Li and Ng (2000).

Theorem 3.2 (Li and Ng 2000) *Let $(X^*,\ x^*)$ be an optimal solution of $(M(\omega,\lambda^*))$. If $(X^*,\ x^*)$ is optimal for $(M(\omega))$ when $\lambda^* = 1 + 2\omega \mathsf{E}(W_T)|_{(X^*,\ x^*)}$.*

Since our model on a well-generated scenario tree is strictly convex, it has a unique optimal solution. Therefore the necessary condition established in Theorem 3.2 is also sufficient for our model. The exact implementation to apply this result is as follows. First, solve the optimal solution of the auxiliary problem as functions of λ. Then, we shall find the optimal λ^* by using the condition established in Theorem 3.2. Finally, we substitute this λ^* back to the optimal solutions of the auxiliary problem to get the solution for the original problem.

In order to simplify the notations in the multistage framework, we introduce the modified probabilities:

$$\begin{aligned} q_{nt} &= \sum_{l\in c(n)} \tfrac{p_{l,t+1}\cdot q_{l,t+1}}{\delta_{l,t+1}+1}, \quad t = 0, 1, \ldots, T, \\ w_{nt} &= \tfrac{p_{nt}\cdot q_{nt}/(\delta_{nt}+1)}{q_{a(n),t-1}}, \end{aligned}$$

where w_{nt} can be understood as a modified conditional probability distribution in each subtree. We have $w_{n,T+1} = p_{n,T+1}$, and $q_{nT} = 1$. Comparing the current method with the previous one in Sect. 11.3.1, and observe that w_{nt} are the same for both cases, therefore the basic quantities derived from these modified probability distributions are the same in both cases.

We now apply the stochastic control approach to solve our problem with the discrete scenario structure. Using the same notations as Sect. 11.3.1, we first solve the auxiliary problem $(M(\omega,\lambda))$ using dynamic programming.

Lemma 3.1 *The auxiliary problem $(M(\omega,\ \lambda))$ can be solved analytically. In particular, at stage t, the objective value is*

$$\begin{aligned} J_{nt} = &-\omega \left\{ q_{nt}(r_{t+1}\cdots r_T)^2 [X_{nt}' A_{nt} X_{nt} + (\bar{v}_{nt}' X_{nt} + r_t x_{nt})^2] + \tfrac{\lambda^2}{4\omega^2}(1-q_{nt}) \right\} \\ &+\lambda \left\{ q_{nt}(r_{t+1}\cdots r_T)(\bar{v}_{nt}' X_{nt} + r_{nt} x_{nt}) + \tfrac{\lambda}{2\omega}(1-q_{nt}) \right\}, \end{aligned}$$

and the optimal primal-dual solution is

$$\begin{aligned}
X_{nt} &= -\frac{\epsilon_{nt}}{2\omega(r_{t+1}r_{t+2}\cdots r_n)\widetilde{r}_{nt}} A_{nt}^{-1}(\bar{v}_{nt} - r_t u_{nt}),\\
x_{nt} &= \frac{1}{2\omega\widetilde{(r_t r_{t+1}\cdots r_T)}}\left[\frac{\epsilon_{nt}}{\widetilde{r}_{nt}}(\gamma_{nt} - r_t\beta_{nt} + 1) + \lambda\right],\\
\epsilon_{nt} &= \frac{r_{nt}}{\delta_{nt}+1}\left[2\omega(r_t r_{t+1}\cdots r_T)(W_{nt}) - \lambda\right].
\end{aligned}$$

The proof of the lemma is similar with that of Theorem 3.1.

Now we have obtained the analytical solutions of the auxiliary problem. After that we can apply Theorem 3.2 to get the solutions for our original problem. Before that, we introduce one more lemma.

Lemma 3.2 *Let* $\nu = \frac{r-\rho}{2r}$, *and* $\eta = \frac{\lambda}{\omega}$. *For the multistage auxiliary problem* $(M(\omega,\lambda))$, *we have* $E(W_T) = \rho B + \nu\eta$ *and* $E(W_T^2) = r\rho B^2 + \frac{\nu}{2}\eta^2$, *where* W_T *is the final payoff of the optimal portfolio for* $(M(\omega,\lambda))$. *Further, Problem* $(M(\omega,\lambda^*))$ *and* $(M(\omega))$ *have the same optimal solutions if and only if* $\lambda^* = \frac{1+2\omega\rho B}{1-2\nu} = 2\omega r B + \frac{r}{\rho}$.

The lemma can be easily proved using Theorem 3.2 with $\lambda^* = 1 + 2\omega \mathsf{E}(W_T)$.

Using these lemmas, we substitute λ^* in the optimal solutions of the auxiliary problem to get the optimal solution for the original problem $(M(\omega))$, leading to the expression of the efficient frontier. The results are given as follows.

Theorem 3.3 *The original multistage problem* $(M(\omega))$ *has the optimal solutions given as follows:*

$$\begin{aligned}
X_{nt} &= -\frac{\epsilon_{nt}}{2\omega(r_{t+1}r_{t+2}\cdots r_T)\widetilde{r}_{nt}} A_{nt}^{-1}(\bar{v}_{nt} - r_t u_{nt}),\\
x_{nt} &= \frac{1}{2\omega\widetilde{(r_t r_{t+1}\cdots r_T)}}\left[\frac{\epsilon_{nt}}{\widetilde{r}_{nt}}(\gamma_{nt} - r_t\beta_{nt} + 1) + \lambda\right],\\
\epsilon_{nt} &= \frac{r_{nt}}{\delta_{nt}+1}\left[2\omega(r_t r_{t+1}\cdots r_T)(W_{nt}) - \lambda\right],\\
\lambda &= 2\omega r B + \frac{r}{\rho}.
\end{aligned}$$

The associated expected payoff and the variance are, respectively,

$$\mathsf{E}(W_T) = \frac{r-\rho}{2\omega\rho} + rB$$

$$Var\,(W_T) = \frac{\rho}{r-\rho}[\mathsf{E}(W_T) - rB]^2.$$

Theorem 3.3 presents solutions for a tradeoff factor ω. The theorem also displays the relationship between the tradeoff factor ω and the expected return $\mathsf{E}(W_T)$ of the optimal portfolio. Finally, the mean-variance efficient frontier is computed. Comparing Theorem 3.3 with Theorem 3.1, we see that these two methods end up with the same mean-variance efficient frontier. If we express ω in terms of R using the formulas in Theorem 3.3, then we can get an expression of the efficient frontier. As expected, one can see that the solutions of (M_3) and $(M(\omega))$ are indeed identical.

So far we introduced two different approaches to solve the portfolio selection problem for a multistage scenario tree structure. These two approaches need almost the same amount of computational efforts, and eventually reach the same efficient frontier. However, they shed lights on the problem from two very different angles, both are important for different reasons. On the one hand, the stochastic programming approach is confined itself in discrete time for a finite scenario structure, while the stochastic control approach can be applied for the problems with continuous scenario or in a continuous time setting (as shown in Zhou and Li 2000). On the other hand, the stochastic programming approach can be used to deal with almost any complicating constraints, which cannot be dealt with by the stochastic control approach.

11.4 The Multistage Target Tracking Model

11.4.1 The Model and Solutions

Consider the tracking model in a multistage case. With final payoff denoting as W and a given target payoff R, the tracking strategy is to minimize the tracking error variance (TEV), $\mathsf{E}[(W-R)^2]$, on the investment horizon. We apply the same scenario tree as in Sect. 11.3. The general multistage tracking model is

$$(M_4)\ \min\ \mathsf{E}[(W-R)^2] \tag{11.29}$$

$$\text{s.t.}\ u_0'X_0 + x_0 = B \tag{11.30}$$

$$v_{nt}'X_{a(n),t-1} + r_{t-1}x_{a(n),t-1} = u_{nt}'X_{nt} + x_{nt},\ t = 1, 2, \cdots, T,\ n \in N_t, \tag{11.31}$$

Now define constants on the scenario S_{nt} in order to get an explicit formulation of the model:

$$\begin{aligned}
\alpha_{nt} &= u_{nt}'A_{nt}^{-1}u_{nt}, \\
\beta_{nt} &= \bar{v}_{nt}'A_{nt}^{-1}u_{nt}, \\
\gamma_{nt} &= \bar{v}_{nt}'A_{nt}^{-1}\bar{v}_{nt}, \\
\delta_{nt} &= (\bar{v}_{nt} - r_t u_{nt})'A_{nt}^{1}(\bar{v}_{nt} - r_t u_{nt}), \\
\rho_{nt} &= \tfrac{r_t}{\delta_{nt}+1},
\end{aligned}$$

$$\begin{aligned}
C_{nt} &= \textstyle\sum_{i\in c(n),t+1} p_{i,t+1}\rho_{i,t+1}C_{i,t+1}, \quad C_{nT} = 1, \\
\tilde{p}_{c(n),t+1} &= \frac{p_{c(n),t+1}\rho_{c(n),t+1}C_{c(n),t+1}}{\sum_{i\in c(n),t+1} p_{i,t+1}\rho_{i,t+1}C_{c(n),t+1}}, \quad \tilde{p}_{n,T+1} = p_{n,T+1}, \\
r &= r_0 r_1 \cdots r_T,
\end{aligned}$$

where $\tilde{p}$ can be understood as a modified conditional probability distribution. $\bar{v}_{nt}$ is the expected per-share payoff of the single-stage subtree derived from scenario S_{nt} with the modified probability distribution $\tilde{p}_{c(n),t+1}$, and A_{nt} is the conditional covariance matrix of this subtree with the modified probability distribution $\tilde{p}_{c(n),t+1}$.

Different from the mean-variance model, the multistage tracking model (M_4) is separable and can be solved dynamically. The analytical solutions are presented in the following Theorem 4.1.

Theorem 4.1 *The multistage target tracking model* (M_4) *can be solved dynamically, and on the scenario* S_{nt}*, the optimal solution is*

$$X^*_{nt} = -\frac{\lambda_{n,t}}{r_t} A^{-1}_{n,t}(\bar{v}_{n,t} - r_t u_{n,t}) \tag{11.32}$$

$$x^*_{nt} = \frac{1}{r_t}\left[\frac{\lambda_{n,t}}{r_t}(\gamma_{n,t} - r_t\beta_{n,t} + 1) + R_{t+1}\right] \tag{11.33}$$

$$\lambda_{nt} = \frac{r_t}{\delta_{n,t} + 1}(r_t W_{nt} - R_{t+1}) \equiv r_t \rho_{nt}(W_{nt} - R_t) \tag{11.34}$$

with the associated optimal TEV on this scenario:

$$TEV^*_{nt} = r_{t+1}\cdots r_T C_{nt}\frac{\rho_{nt}}{r_t}(R_{t+1} - r_t W_{n,t})^2 = r_t \cdots r_T C_{n,t}\rho_{n,t}(W_{n,t} - R_t)^2 \tag{11.35}$$

where

$$R_t := \frac{R_{t+1}}{r_t} = \frac{R}{r_t\cdots r_T}, \quad \text{and } R_T := \frac{R}{r_T}$$

and for the overall optimal TEV^**, we have*

$$TEV^* = rC_0\rho_0(B - R_0)^2 = \frac{C_0\rho_0}{r}(R - rB)^2$$

Proof We prove by induction. Firstly derive the expressions of TEV, i.e., $\mathsf{E}[(W - R)^2]$. Denoting X_T and x_T as the solutions at the beginning of the final stage, we have the final tracking error variance as

$$\begin{aligned}
\text{TEV} &= \mathsf{E}[\mathsf{E}[(W-R)^2|S_{\cdot,1}]] \\
&= \mathsf{E}[\mathsf{E}[\mathsf{E}[(W-R)^2|\{S_{\cdot,1},\ S_{\cdot,2}\}]]] \\
&\vdots \\
&= \mathsf{E}\left\{\mathsf{E}[\cdots\mathsf{E}[\mathsf{E}[(W-R)^2|\{S_{\cdot,1},\cdots,S_{\cdot,T}\}]]]\right\}
\end{aligned} \tag{11.36}$$

Start from the scenario S_{nT} in the last stage T, for the single-stage subtree with root node S_{nT}, we have the tracking model in a static formulation:

$$\begin{aligned}
\min\ & \mathsf{E}[(W-R)^2] \\
\text{s.t.}\ & u'X_{nT} + x_{nT} = W_{nT},
\end{aligned}$$

Theorem 2.2 presents the optimal solutions for this static problem as:

$$
\begin{aligned}
X_{nT}^* &= -\frac{\lambda_{nT}}{r_T} A_{nT}^{-1} (\bar{v}_{nT} - r_T u_{nT}) \\
x_{nT}^* &= \frac{1}{r_T} \left[\frac{\lambda_{nT}}{r_T} (\gamma_{nT} - r_T \beta_{nT} + 1) + R \right] \\
\lambda_{nT} &= \frac{r_T}{\delta_{nT} + 1} (r_T W_{nT} - R) \equiv r_T \rho_{nT} (W_{nT} - R_T)
\end{aligned}
$$

and the optimal tracking error variance (TEV_{nT}^*)

$$
\text{TEV}_{nT}^* = \frac{\rho_{nT}}{r_T} (R - r_T W_{nT})^2 = r_T \rho_{nT} (W_{nT} - R_T)^2
$$

where $R_T := \frac{R}{r_T}$. Solutions (11.32)–(11.35) hold true for the last stage.

Next we apply induction to the stage index t. Suppose that the solutions (11.32)–(11.35) hold true at the tth stage on the scenario S_{nt}. Now prove that they also hold for stage $t-1$. Firstly, Substitute (11.35) into (11.36) leads the tracking error variance on scenario $S_{n,t-1}$:

$$
\begin{aligned}
\text{TEV}_{n,t-1} &= \mathsf{E}[\text{TEV}_{i,t}^* | i \in \{c(n), t\}] \\
&= \mathsf{E}[r_t \cdots r_T C_{i,t} \rho_{i,t} (W_{i,t} - R_t)^2 | i \in \{c(n), t\}]
\end{aligned}
$$

The problem on the scenario $S_{n,t-1}$ shows

$$
\begin{aligned}
\min\ & \mathsf{E}[r_t \cdots r_T C_{i,t} \rho_{i,t} (W_{i,t} - R_t)^2 | i \in \{c(n), t\}] \\
\text{s.t. }\ & u_{n,t-1}' X_{n,t-1} + x_{n,t-1} = W_{n,t-1} \\
& v_{i,t}' X_{n,t-1} + r_{t-1} x_{n,t-1} = W_{i,t}, \quad i \in \{c(n), t\}
\end{aligned}
$$

Define the modified conditional probability

$$
\tilde{p}_{c(n),t} = \frac{p_{c(n),t} \rho_{c(n),t} C_{c(n),t}}{\sum_{i \in c(n),t} p_{i,t} \rho_{i,t} C_{c(n),t}}
$$

and the constant $C_{n,t-1}$,

$$
C_{n,t-1} = \sum_{i \in c(n),t} p_{i,t} \rho_{i,t} C_{i,t}
$$

Than the problem on $S_{n,t-1}$ is reformulated as

$$
\begin{aligned}
\min\ & r_t \cdots r_T C_{n,t-1} \mathsf{E}[(W_{i,t} - R_t)^2] \\
\text{s.t. }\ & u_{n,t-1}' X_{n,t-1} + x_{n,t-1} = W_{n,t-1} \\
& v_{i,t}' X_{n,t-1} + r_{t-1} x_{n,t-1} = W_{i,t}, \quad i \in \{c(n), t\}
\end{aligned}
$$

where the $\mathsf{E}[\cdot]$ is the expectation under the modified probability $\tilde{p}$. Therefore the problem is once again presented as a static tracking problem, the only difference is that the distribution probabilities are modified. So for this problem on scenario $S_{n,t-1}$, the optimal solutions are

$$\begin{aligned} X_{n,t-1} &= -\frac{\lambda_{n,t-1}}{r_{t-1}} A_{n,t-1}^{-1} (\bar{v}_{n,t-1} - r_{t-1} u_{n,t-1}) \\ x_{n,t-1} &= \frac{1}{r_{t-1}} \left[\frac{\lambda_{n,t-1}}{r_{t-1}} (\gamma_{n,t-1} - r_{t-1}\beta_{n,t-1} + 1) + R_t \right] \\ \lambda_{n,t-1} &= \frac{r_{t-1}}{\delta_{n,t-1} + 1} (r_{t-1} W_{n,t-1} - R_t) \equiv r_{t-1}\rho_{n,t-1} (W_{n,t-1} - R_{t-1}) \end{aligned}$$

with the associated tracking error variance ($\text{TEV}^{\,*}_{n,t-1}$)

$$\begin{aligned} \text{TEV}^{\,*}_{n,t-1} &= r_t \cdots r_T C_{n,t-1} \frac{\rho_{n,t-1}}{r_{t-1}} (R_t - r_{t-1} W_{n,t-1})^2 \\ &= r_{t-1} \cdots r_T C_{n,t-1} \rho_{n,t-1} (W_{n,t-1} - R_{t-1})^2 \end{aligned}$$

Therefore, the solutions (11.32)–(11.35) are also correct for stage $t-1$. □

Theorem 4.1 presents a way to dynamically solve the multistage tracking model explicitly. Start from the last stage, the modified conditional probabilities can be firstly calculated (for the last stage they are equal to the real probabilities). Applying these modified probabilities, the constants of α, β, γ, δ, and ρ are computed. And based on Theorem 4.1, the optimal solutions are delivered on each decision node in the last stage. Then, one can go backward to the previous stage and carry out the same process to get optimal solutions. This process is repeated until back to the root of the whole tree, and the overall ($\text{TEV}^{\,*}$) is finally obtained.

11.4.2 Relations Between Models

In the previous context, we have established the mean-variance efficiency of the optimal tracking portfolio. Review (M_3) and solutions in Theorem 3.1, we find that the solutions of both multistage models (i.e., the tracking model and the mean-variance model) hold similar formulas, expected for the definitions of a few quantities. What's more, we have derived out the condition on which the static model shares the same set of optimal portfolio in Proposition 4.1. It makes sense to verify whether this holds for the multistage case.

We now consider both models based on the same multistage scenario tree, and for the same sets of assets. Comparing solutions of both models in Theorems 3.1 and 4.1, they are presented in quite similar formulas, expected for the expressions of the modified probabilities.

Review the modified probability distributions $\tilde{p}$ and w. For the tracking model (M_4) we have

$$C_{n,t} = \sum_{i \in c(n),t+1} p_{i,t+1}\rho_{i,t+1}C_{i,t+1}, \quad C_{nT} = 1 \tag{11.37}$$

$$\tilde{p}_{n,t} = \frac{p_{nt}\rho_{nt}C_{nt}}{\sum_{i \in n,t} p_{it}\rho_{i,t}C_{i,t}}, \quad \tilde{p}_{n,T+1} = p_{n,T+1} \tag{11.38}$$

and for the mean-variance model (M_3),

$$q_{nt} = \sum_{i \in c(n)} \frac{q_{i,t+1}}{1+\delta_{i,t+1}}, \qquad q_{nT} = p_{nT} \cdot p_{a(n),T-1} \cdot \cdots \cdot p_{a(\cdots a(n)),1}; \tag{11.39}$$

$$w_{nt} = \frac{q_{nt}/(1+\delta_{nt})}{q_{a(n),t-1}}, \qquad w_{n,T+1} = p_{n,T+1} \tag{11.40}$$

We state the relations between the modified probabilities in both models in the following lemma.

Lemma 4.1 *For both problems* (M_3) *and* (M_4) *based on a same scenario tree, the probability distributions are modified in a same way during solving processes. That is, on any given scenario* S_{nt}*, we have* $w_{nt} = \tilde{p}_{nt}$*, and* $q_{nt} = p_{nt} \cdots p_{a(\cdots a(n)),1} \cdot \frac{C_{nt}}{r_{t+1} \cdots r_T}$.

Proof Start from the last stage T on a given scenario S_{nT}, with the definitions of $\rho_{nT} = \frac{r_T}{1+\delta_{nT}}$ and $C_{nT} = 1$, (11.38)–(11.40) give

$$w_{nT} = \frac{q_{nT}/(1+\delta_{nT})}{\sum[q_{iT}/(1+\delta_{iT})]} = \frac{p_{nT}\rho_{nT}}{\sum p_{iT}\rho_{iT}} = \tilde{p}_{nT}$$
$$q_{nT} = p_{nT} \cdots p_{a(\cdots a(n)),1} \cdot C_{nT}$$

Lemma 4.1 holds true in stage T. Next we prove by induction. Suppose the following relations hold for stage t, we now prove they hold for stage $t-1$ and finish the proof by induction.

$$q_{nt} = p_{nt} \cdots p_{a(\cdots a(n)),1} \cdot \frac{C_{nt}}{r_{t+1} \cdots r_T} \tag{11.41}$$

$$w_{nt} = \tilde{p}_{nt} \tag{11.42}$$

Substitute (11.41) to (11.39) for stage $t-1$ gives

$$\begin{aligned} q_{n,t-1} &= \sum_{i \in c(n)} \frac{q_{it}}{1+\delta_{it}} \\ &= p_{n,t-1} \cdots p_{a(\cdots a(n)),1} \cdot \sum_{i \in c(n)} p_{it} \frac{C_{it}}{r_{t+1} \cdots r_T} \frac{\rho_{it}}{r_t} \\ &= p_{n,t-1}, \cdots, p_{a(\cdots a(n)),1} \cdot \frac{C_{n,t-1}}{r_t, \cdots r_T} \end{aligned} \tag{11.43}$$

further substitute (11.43) to (11.40) for stage $t-1$, together with (11.38) we get

$$\begin{aligned} w_{n,t-1} &= \frac{q_{n,t-1}/(1+\delta_{n,t-1})}{\sum[q_{i,t-1}/(1+\delta_{i,t-1})]} \\ &= \frac{p_{n,t-1}C_{n,t-1}\rho_{n,t-1}}{\sum[p_{i,t-1}C_{i,t-1}\rho_{i,t-1}]} \\ &= \tilde{p}_{n,t-1} \end{aligned}$$

□

Lemma 4.1 states that the modified distributions are same for both multistage models in the solving process. This is helpful to verify Proposition 4.1 in the multistage case. We modify the Proposition 4.1 as follows.

Proposition 4.1 *Given both mean-variance model and target tracking model based on a same multistage scenario tree, they share the same set of optimal solutions iff* $R^T - rB = h(R^M - rB)$, *where* R^T *is the final tracking target,* R^M *is the final expected return, and* $h := \frac{r}{r-\rho}$.

Proof The only different from this Proposition 4.1 is the multistage case. In the multistage case, the optimal objective values for both problems (M_4) and (M_3) are proposed in Theorems 4.1 and 3.1 as:

$$\begin{aligned} \text{TEV}^* &= \frac{C_0\rho_0}{r}(R^T - rB)^2, \\ \text{Var}^* &= \frac{\rho}{r-\rho}(R^M - rB)^2 \end{aligned}$$

and Lemma 4.1 gives

$$q_0 = \frac{C_0}{r_1, \cdots r_T},$$

leading

$$C_0\rho_0 = \frac{rq_0}{r_0}\frac{r_0}{1+\delta_0} = \frac{rq_0}{1+\delta_0} = \rho$$

therefore we reformulate the optimal objectives in multistage as:

$$\begin{aligned} \text{TEV}^* &= \frac{\rho}{r}(R^T - rB)^2, \\ \text{Var}^* &= \frac{\rho}{r-\rho}(R^M - rB)^2 \end{aligned}$$

which are same as the forms in static case. Thus we can follow the same proof as that for Proposition 2.3 in Sect. 11.2. □

It brings us a new insight to solve a multistage mean-variance model. A classical mean-variance problem could be transformed to a target tracking problem, and further be separated and dynamically solved. The necessary work is to modify the expected final payoff R^M to a final tracking target R^T, by applying the rule of $R^T - rB = h(R^M - rB)$ in Proposition 4.1, where h is a constant defined based on the scenario tree structure.

Let's compare with Theorem 3.2, where a relationship between R^M and the tradeoff factor ω is proposed. And the classical mean-variance model is solved by applying an auxiliary problem in optimal control perspective. We now find out that these two methodologies are different approaches but equally satisfactory results in solving the multistage mean-variance model.

Finally, consider the equations stated in Theorem 3.2 and Proposition 2.3,

$$R^M = \frac{r - \rho}{2\omega\rho} + rB, \tag{11.44}$$

$$R^T - rB = h(R^M - rB). \tag{11.45}$$

Combine (11.44) and (11.45) together, and it is straightforward to get:

$$R^T = \frac{r}{2\omega\rho} + rB \tag{11.46}$$

Equation (11.46) presents a relationship between the tradeoff factor ω in model $(M(\omega))$ and the tracking target R^T in model (M_4), which guarantees both models achieve optimality by same optimal portfolio.

We have discussed three methods in solving the multistage mean-variance model, described in Theorems 3.1, 3.2, and 4.1, and Proposition 2.3, respectively. These methods are closely linked to each other by the key factors, i.e., the expected return R^M for (M_3), the tradeoff factor ω for $(M(\omega))$, and the target return R^T for (M_4). It is proved that with the transformations presented in (11.44)–(11.46), the problems in these three formulations share the same optimal portfolio.

We notice that these three approaches need almost the same amount of computational efforts, and eventually reach the same efficient frontier. In this problem, estimating the covariance and inverse covariance matrices of the returns of the assets in the portfolio plays an important role. As the number of assets or the investment stages grows, those parameters to be estimated quickly become massive. Even if we could estimate each individual parameter accurately, the cumulated error of the whole estimation can be large under matrix norms. Big data size brings new challenges to the mode, and this requires new statistical procedures on estimating large covariance matrices and their inverse.

11.5 Conclusions

This chapter introduces the optioned portfolio selection models on discrete scenario tree framework. We present the models and derive the analytical optimal solutions. Attentions are paid to the properties of the optimal payoff in static case and the solution methods in multistage case. Two different solution techniques for multistage mean-variance model are firstly proposed, which are, respectively, based on stochastic programming and stochastic control. We also develop a target tracking models. It turns out that the optimal tracking portfolio holds mean-variance efficiency, which further provides an alternative method to dynamically solve the classical multistage mean-variance problem. Future research can be carried out on the model statements and analysis. Solution methods for models with random tracking target and uncertainty investment horizon are under research. And new estimation methods for large size of parameters will be applied to improve the accuracy and efficiency of models.

References

M. Ammann, H. Zimmermann, Tracking error and tactical asset allocation. Financ. Anal. J. **57**(2), 32–43 (2001)

D. Barror, E. Canestrelli, Tracking error: a multistage portfolio model. Ann. Oper. Res. **165**, 47–66 (2009)

A. Berkelaar, C. Dert, B. Odenkamp, S. Zhang, A primal-dual decomposition-based interior point approach to two-stage stochastic linear programming. Oper. Res. **50**, 904–915 (2002)

A. Berkelaar, J. Gromicho, R. Kouwenberg, S. Zhang, A primal-dual decomposition algorithm for multistage stochastic convex programming. Math. Program. **104**, 153–177 (2005)

P. Carr, D. Madan, Optimal positioning in derivative securities. Quant. Financ. **1**, 19–37 (2001)

R. Cesari, D. Cremonini, Benchmarking portfolio insurance and technical analysis: a Monte Carlo comparison of dynamic strategies of asset allocation. J. Econ. Dyn. Control. **27**, 987–1011 (2003)

M.C. Chiu, D. Li, Asset and liability management under a continuous-time mean-variance optimization framework. Insur. Math. Econ. **39**, 330–355 (2006)

R.G. Clarke, S. Krase, M. Statman, Tracking errors, regret, and tactical asset allocation. J. Portf. Manag. **20**(3), 16–25 (1994)

R. Dembo, D. Rosen, The practice of portfolio replication, a practical overview of forward and inverse problems. Ann. Oper. Res. **85**, 267–284 (1999)

M.A.H. Dempster, G.W.P. Thompson, Dynamic portfolio replication using stochastic programming, in *Risk Management: Value at Risk and Beyond*, ed. M.A.H. Dempster (Cambridge University Press, Cambridge, 2002), pp. 100–128

C. Dert, B. Oldenkamp, Optimal guaranteed return portfolio and the casino effect. Oper. Res. **48**, 768–775 (2000)

B. Dumas, E. Luciano, An exact solution to a dynamic portfolio choice problem under transaction cost. J. Financ. **46**, 577–595 (1991)

E.J. Elton, M.J. Gruber, The multi-period consumption investment problem and single period analysis. Oxf. Econ. Pap. **26**(2), 289–301 (1974)

E.J. Elton, M.J. Gruber, On the optimality of some multiperiod portfolio selection criteria. J. Bus. **47**, 231–243 (1974)

Y. Fang, Y.S. Zhang, Risk control mechanism of active portfolio investment with tracking error constraints. Chin. J. Manag. Sci. **14**(4), 19–24 (2006)
R.R. Grauer, N.H. Hakansson, On the use of mean-variance and quadratic approximations in implementing dynamic investment strategies: a comparison of return and investment policies. Manag. Sci. **39**, 856–971 (1993)
N.H. Hakansson, On optimal myopic portfolio policy, with and without serial correlation of yields. J. Bus. **44**, 324–334 (1971)
J.C. Hull, *Options, Futures, and Other Derivatives* (Prentice Hall, Upper Saddle River, 1999)
D. Isakov, B. Morard, Improving portfolio performance with option strategies: evidence from Switzerland. Eur. Financ. Manage. **7**(1), 73–91 (2001)
P. Jorion, Portfolio optimization with tracking-error constraints. Financ. Anal. J. **59**(5), 70–82 (2003)
M. Kallio, W.T. Ziemba, Using Tucker's theorem of the alternative to simplify, review and expand discrete arbitrage theory. J. Bank. Financ. **31**, 2281–2302 (2007)
A.J. King, Duality and martingales: a stochastic programming perspective on contingent claims. Math. Program. Ser. B **91**, 543–562 (2002)
Y.A. Koskosidis, A.M. Duarte, A scenario-based approach to active asset allocation. J. Portf. Manag. **23**(2), 74–85 (1997)
D. Li, W.L. Ng, Optimal dynamic portfolio selection: multi-period mean-variance formulation. Math. Financ. **10**, 387–406 (2000)
D. Li, X.L. Sun, J. Wang, Optimal lot solution to cardinality constrained mean-variance formulation for portfolio selection. Math. Financ. **16**(1), 83–101 (2006)
J.F. Liang, Multistage tracking models: solutions and analysis, in *Electronic Commerce and Business Intelligence* (IEEE Computer Society, Los Alamitos, 2009)
J.F. Liang, J.J. Liu, Tracking error analysis of optioned portfolio optimization, in *Business Intelligence: Artificial Intelligence in Business, Industry and Engineering* (IEEE Computer Society, Los Alamitos, 2009)
J.F. Liang, S.Z. Zhang, D. Li, Optioned portfolio selection: models and analysis, Math. Financ. **18**(4), 569–593 (2008)
Y.K. Ma, X.W. Tang, A study on the model of portfolio investment decision based on tracking error. Syst. Eng. Theory Appl. **12**, 11–16 (2001)
H.M. Markowitz, Portfolio selection. J. Financ. **7**, 77–91 (1952)
H.M. Markowitz, *Portfolio Selection: Efficient Diversification of Investment* (Wiley, New York, 1959)
L.G. Mcmillan, *Options as a Strategic Investment* (New York Institute of Finance, New York, 2002)
R.C. Merton, Optimum consumption and portfolio rules in a continuous time model. J. Econ. Theory **3**, 373–413 (1971)
R.C. Merton, An analytical derivation of the efficient portfolio frontier. J. Financ. Quant. Anal. **7**, 1851–1872 (1972)
B. Morard, A. Naciri, Options and investment strategies. J. Futur. Mark. **10**, 505–517 (1990)
J. Mossion, Optimal multiperiod portfolio policy. J. Bus. **41**, 215–229 (1968)
B. Odenkamp, Derivatives in Portfolio Management. Ph.D. thesis, Erasmus University Rotterdam, Thesis Publishers, Amsterdam, The Netherlands, 1999
R.T. Rockafellar, S. Uryasev, Optimization of conditional value-at-risk. J. Risk **2**, 21–41 (2000)
H.C. Rohweder, Implementing stock selection ideas: does tracking error optimization do any good? J. Portf. Manag. **24**(3), 49–59 (1998)
R. Roll, A mean/variance analysis of tracking error. J. Portf. Manag. **18**(4), 13–22 (1992)
P.A. Samuleson, Lifetime portfolio selection by dynamic stochastic programming. Rev. Econ. Stat. **50**, 239–246 (1969)
M. Schyns, Y. Crama, G. Hubner, Optimal selection of a portfolio of options under Value-at-Risk constraints: a scenario approach. Ann. Oper. Res. **181**, 683–708 (2010)
M.Y. Wang, Multiple-benchmark and multiple-portfolio optimization. Financ. Anal. J. **51**(1), 63–72 (1999)

L. Yi, Z.F. Li, D. Li, Multi-period portfolio selection for asset-liability management with uncertain investment horizon. J. Ind. Manag. Optim. **4**(3), 535–552 (2008)

W.T. Ziemba, The research foundation of AIMR. in *The Stochastic Programming Approach to Asset, Liability and Wealth Management* (AIMR, Charlottesville, 2003)

X.Y. Zhou, D. Li, Continuous-time mean-variance portfolio selection: a stochastic LQ framework. Appl. Math. Optim. **42**, 19–33 (2000)

S.S. Zhu, D. Li, S.Y. Wang, Risk control over bankruptcy in dynamic portfolio selection: a generalized mean-variance formulation. IEEE Trans. Autom. Control **49**(3), 447–457 (2004)

Chapter 12
Multi-Period Portfolio Selection with Stochastic Investment Horizon

Lan Yi

Abstract It is often the case that some unexpected events may force an investor to terminate her investment and exit the financial market. In this work, the mean-variance formulation of multi-period portfolio optimization with stochastic investment horizon is considered. Given the distribution of the uncertain investment horizon, the problem under investigation can be formulated as a nonseparable dynamic problem. By making use of the embedding technique of Li and Ng (Math Financ 4(2):387–406, 2000), an analytical optimal strategy and an analytical expression of the mean-variance efficient frontier for the mean-variance formulation of the problem are achieved. Two special cases are also discussed in this work.

Keywords Multi-period • Mean-variance portfolio optimization • Stochastic investment horizon • Embedding technique • Dynamic programming

12.1 Introduction

Portfolio theory deals with the question of how to find an optimal distribution of the wealth among various assets. Mean-variance analysis and expected utility formulation are two different tools for dealing with portfolio selections. The mean-variance formulation proposed by Markowitz (1959) provides a fundamental basis for portfolio allocation in a single period. Analytical expression of the mean-variance efficient frontier in single-period portfolio selection was derived by Merton (1972). Extending the single-period portfolio selection problem to multi-period one is an important development of this research area. However, multi-period portfolio selection model has been dominated by the results of maximizing expected utility functions of the terminal wealth for years, because of the nonseparability of the mean-variance objective function in the sense of dynamic programming. Recently, by using the embedding techniques, Li and Ng (2000) solved analytically the multi-period portfolio selection problem under the mean-variance framework.

L. Yi (✉)
Management School, Jinan University, Guangzhou, People's Republic of China
e-mail: tlyi@jnu.edu.cn

T.-M. Choi et al. (eds.), *Optimization and Control for Systems in the Big-Data Era*,
International Series in Operations Research & Management Science 252,
DOI 10.1007/978-3-319-53518-0_12

The embedding technique introduced by Li and Ng (2000) overcomes the bottleneck of research on multi-period mean-variance portfolio selection problems, and a lot of interesting research have done based on Li and Ng (2000)'s work (see Çelikyurt and özekici 2007; Chiu and Li 2006; Costa and Araujo 2008; Cui et al. 2012, 2013, 2012; Guo and Hu 2005; Josa-fombellida and Rincón-zapatero 2008; Li et al. 2010; Liang et al. 2008; Yi et al. 2008; Zhou and Li 2000; Zhu et al. 2011). Mean-field framework is another method to tackle the issue of nonseparability which was recently introduced by Cui et al. (2013). It offers an efficient modeling tool and a tractable solution scheme in deriving the optimal policies analytically of the multi-period mean-variance-type portfolio selection problems (see Cui et al. 2013; Yi et al. 2014).

An assumption often taken for granted in general portfolio selection models is that the investment horizon is deterministic, which implies that an investor knows with certainty the exit time at the beginning of her investment. However, an investment horizon, in the real world, is always unknown when an investor starts her investment. There are many exogenous and endogenous factors that can drive the exit strategy of an investor. Sudden huge consumption, serious illness, retirement, etc. are market-unrelated exogenous reasons to force an investor to exit the financial market. At the same time, there also exit some market-related exogenous reasons, e.g., an anticipation for long-turn depression of financial market could make some investors to exit market earlier. While the exogenous reasons are independent of the investor's investment policy, endogenous factors are policy-dependent. For example, the investor may decide to exit the market once her wealth hits her investment target, or the investor carefully searches for a stopping time to maximize the expected utility of her terminal wealth. In such situations, the exit time is determined endogenously.

Recognizing a clear gap between theory and practice, it seems sympathetic to relax the restrictive assumption that the investment horizon is pre-fixed with certainty. Research on this subject was actually pioneered by Yaari (1965), who deals with the problem of optimal consumption for an individual with uncertain date of death, under a pure deterministic investment environment. Other related works include (Hakansson 1969; Merton 1971; Karatzas and Wang 2000; Browne 2000; Guo and Hu 2005; Martellini and Urosevic 2006). Karatzas and Wang (2000) address the optimal dynamic investment problem in a complete market with an assumption that the uncertain investment horizon is a stopping time of asset price filtration. A different problem of minimizing the expected time to beat a benchmark is addressed in Browne (2000), where the exit time is a random variable related to the portfolio. The uncertain exit time concerned in these two works is endogenous. Martellini and Urosevic (2006) analyze a static mean-variance portfolio selection problem for both the situations where exit time is independent and dependent of asset returns. Exogenous and endogenous exit times are considered, respectively, in these two different cases. Multi-period mean-variance portfolio optimization problem with uncertain exit time is studied in Guo and Hu (2005), where the uncertain exit time is exogenous. Yi et al. (2008) also considered an exogenous uncertain exit time in a multi-period mean-variance asset-liabilities optimization problem, and derived the analytical solution by using embedding technique.

Yi et al. (2014) reconsidered the exogenous uncertain exit time in multi-period mean-variance portfolio selection problem by using the mean-field framework. Although the exogenous exit time has been investigated in the investment literature since (Yaari 1965), the only case concerned about is market-independent exit time. That means, the probability of the exit time is independent of the financial market. Market-dependent exogenous exit time is considered in Blanchet-Scaillet et al. (2005), which applies the uncertain time horizon into dynamic asset pricing theory. Blanchet-Scalliet et al. (2008) incorporate an uncertain time horizon into a continuous-time optimal portfolio selection problem.

In this work, a market-dependent exogenous exit time is introduced into the multi-period mean-variance portfolio selection problem. By introducing the uncertain exit time, there are two kinds of uncertainties in this portfolio model, return risk and exit risk. Both analytical optimal policy and the efficient frontier of the mean-variance portfolio selection problem can be derived by adopting the embedding technique in Li and Ng (2000). Furthermore, this work will illustrate that the state-independent exit time (Li and Ng (2000)'s work) is a special case of the state-dependent one. By comparing this case with cases with certain exit time, it is found that the condition of uncertain exit time increases the investment risk.

This work is organized as follows. After an introduction of the time uncertainty, the mean-variance portfolio selection model with uncertain exit time is described in Sect. 12.3. In Sect. 12.4 the analytical solution is derived by using dynamic programming, and the efficient frontier is obtained. Section 12.5 discusses a special case when the uncertain exit time is state-independent, and also compares cases with uncertain or certain exit time by means of examples. Finally, Sect. 12.6 gives the conclusion.

12.2 Exit-Time Uncertainty

Assume that the investor's investment time horizon is a positive discrete random variable $\tau \in \mathcal{N}$, rather than a positive constant T.

Denote by $\mathscr{F} = \{\mathcal{F}_0, \mathcal{F}_1, \cdots, \mathcal{F}_t \cdots\}$ the filtration reflecting financial market information, $\mathscr{T} = \{\mathcal{T}_1, \cdots, \mathcal{T}_t \cdots\}$, with $\mathcal{T}_t := \sigma(\tau \wedge t)$ the information about whether the exit has occurred or not. Let the filtration $\mathscr{G} = (\mathcal{G}_0, \mathcal{G}_1, \cdots, \mathcal{G}_t, \cdots)$ represent the total information (may not completely available to investor), which is generated by filtrations $\mathscr{F}$ and $\mathscr{T}$. Denote $\mathscr{A} := \{\mathcal{A}_t\}$ as an enlargement filtration of $\mathscr{F}$, and $\mathcal{F}_t \subseteq \mathcal{A}_t \subseteq \mathcal{G}_t$. Filtration $\mathscr{A}$ presents all the available information to investors.

Notice that the assets' prices at time t are $\mathcal{F}_t$-measurable, hence $\mathcal{A}_t$-measurable and $\mathcal{G}_t$-measurable. The event $\{\tau > t\}$ is $\mathcal{G}$-measurable, but may not be $\mathcal{A}$-measurable. If τ is an $\mathscr{F}$-stopping time, we have $\mathcal{G}_t = \mathcal{A}_t = \mathcal{F}_t$. However, in this study, we suppose that $\mathcal{A}_t \subset \mathcal{G}_t$, that is, the random variable τ is not an $\mathscr{A}$-stopping time, so the event $\{\tau > t\}$ is not $\mathcal{A}_t$-measurable, which means we cannot imply whether or not the exit has occurred by time t under the σ-algebra $\mathcal{A}_t$.

Suppose that the probability of the event $\{t < \tau\}$ is $\mathcal{A}_t$-measurable, which is pre-given. Denote the conditional probability of $\{\tau \leq t\}$ as $P_t = P(\tau \leq t|\mathcal{A}_t)$. Assume that $P_t = P(\tau \leq t|\mathcal{A}_t)$ is an increasing process with respect to t. A sufficient condition for this assumption is that $P(\tau \leq t|\mathcal{A}_t) = P(\tau \leq t|\mathcal{A}_\infty)$.

Assumption 2.1

$$P(\tau > t|\mathcal{A}_t) = P(\tau > t|\mathcal{A}_\infty). \tag{12.1}$$

To understand the above definitions, let us consider the following example. While an investor invests her money in the financial market, she is waiting at the same time for a gold mining opportunity. Once the opportunity is available and is more profitable than the market portfolio, she will exit the market and invest all her money on the gold mining project. However, whether the gold mining opportunity will be available at time t is unknown under the information $\mathcal{F}_t$. Assume that the availability of gold mining is described as a Poisson process with density $\widehat{\lambda}_t$, which is $\mathcal{A}_t$-measurable random variable. Let $\{M_t\}$ be the return process of market portfolio, which is $\mathcal{F}_t$-measurable. So the probability that $\{\tau \leq t\}$ happens can be determined by

$$P_t = P(\tau \leq t \mid \mathcal{A}_t) = 1 - \exp\{-\sum_{s=1}^{t} \lambda_s\}, \tag{12.2}$$

where $\lambda_t := f(\widehat{\lambda}_t, M_t)$ is determined by $\widehat{\lambda}_t$ and the return of market portfolio at time t. Actually, λ_t can be thought as the average failure rate (exit occurrence) during the interval $(t-1, t]$. Notice that λ_t is $\mathcal{A}_t$-measurable, so is P_t. A more specific example will be given in Example 2.1.

Assumption 2.2 The random time τ is finite almost surely, i.e., $P(\tau < \infty) = 1$.

Given a constant T, we define a stochastic process ξ_t as follows,

$$\xi_t := P(\tau = t|\mathcal{A}_t) = \begin{cases} P_1 & t = 1; \\ P_t - P_{t-1} & t = 2, \cdots, T-1; \\ 1 - P_{T-1} & t = T. \end{cases} \tag{12.3}$$

It is easy to check that $\sum_{t=1}^{T} \xi_t = 1$ and ξ_t is $\mathcal{A}_t$-measurable.

Remark 2.1 In the above example, if the density $\widehat{\lambda}_t$ is constant, then

$$P_t = P(\tau \leq t \mid \mathcal{F}_t) = 1 - \exp\{-\sum_{s=1}^{t} \lambda_s\}.$$

λ_t is $\mathcal{F}_t$-measurable, so is P_t and ξ_t. Specifically, if the investor draw her money out of financial market once the gold mining project is available, no matter it is more profitable than the market portfolio or not, λ_t will be $\mathcal{F}_t$-independent, and $\lambda_t = \widehat{\lambda}_t$.

Remark 2.2 To model the probability of uncertain exit time, we could use the historical data of financial market and the related information which will influence investor's exit time. To estimate λ_t in previous example, we could use historical data of the S and P 500 index to estimate $\{M_t\}$, and use the published statistical data of gold mining industries to estimate $\widehat{\lambda}_t$.

12.3 Problem Formulation

Consider a financial market with T trading dates (indexed by $0, 1, \cdots, T-1$), and a finite time horizon T. Uncertainty of the economy is described through a probability space $(\Omega, \mathscr{A}, P)$. Without lose of generality, let us suppose $\mathscr{A} = \mathscr{F}$. The obtained results can be easily generalized to situations where $\mathscr{A} \supset \mathscr{F}$. There are $(n+1)$ risky securities $S_1, \cdots, S_{n+1}$. An investor enters the financial market with an initial wealth v_0. The investor can allocate her wealth among the $(n+1)$ assets. The wealth can be reallocated among the $(n+1)$ assets at the beginning of each of the following T consecutive time periods until she exits the market. The investor plans to invest her wealth at most for T periods. However, she will exit the market at some random time τ by some reasons related to the financial market. Hence the exiting time is $T \wedge \tau$.

The rate of return of the risky security S_{n+1} between time periods t and $t+1$ within the planning horizon is denoted by r_t^0, and those of the other risky assets are denoted by a vector $r_t = (r_t^1, \cdots, r_t^n)'$, where r_t^i is the random return for security i between time periods t and $t+1$. It is assumed in this work that vectors $\widetilde{r}_t = [r_t^0, r_t']'$, $t = 0, 1, \cdots, T-1$, are statistically independent and return $\widetilde{r}_t$ has a know mean $E(\widetilde{r}_t) = [E(r_t^0), E(r_t^1), \cdots, E(r_t^n)]'$ and a known covariance

$$Cov(\widetilde{r}_t) = \begin{bmatrix} \sigma_{t,00} & \cdots & \sigma_{t,0n} \\ \vdots & \ddots & \vdots \\ \sigma_{t,0n} & \cdots & \sigma_{t,nn} \end{bmatrix}.$$

Denote $R_t := r_t - r_t^0\mathbf{e}$ where $\mathbf{e} = (1, 1, \cdots, 1)'$. It is reasonable to assume that $E(\widetilde{r}_t\widetilde{r}_t')$ is positive definite for all time periods, i.e.,

$$E(\widetilde{r}_t\widetilde{r}_t') = \begin{bmatrix} E((r_t^0)^2) & E((r_t^1 r_t^0)) & \cdots & E((r_t^n r_t^0)) \\ E((r_t^0 r_t^1)) & E((r_t^1)^2) & \cdots & E((r_t^n r_t^1)) \\ \cdots & \cdots & \cdots & \cdots \\ E((r_t^0 r_t^n)) & E((r_t^1 r_t^n)) & \cdots & E((r_t^n)^2) \end{bmatrix} > 0, \quad \forall\, t = 0, 1, \cdots, T-1.$$

Suppose that τ is a discrete random processes defined in Sect. 12.2. Hence, the exit probability is $\xi_t (t = 1, 2, \cdots, T)$, where ξ_t and R_{t-1} can be dependent.

Let V_t be the wealth of the investor at the beginning of the tth period, and let π_t^i be the amount of wealth invested in the ith risky asset at the beginning of the tth period. Let the vector $\pi_t = (\pi_t^1, \cdots, \pi_t^n)'$. So $V_t - \sum_{i=1}^n \pi_i^t$ is the amount of wealth invested in the asset S_{n+1}. The relationship between wealth of periods t and $t+1$ is

$$V_{t+1} = V_t r_t^0 + \pi_t' R_t, \quad t = 0, 1, \cdots, T-1. \tag{12.4}$$

The investor is seeking a best investment strategy $\pi_t = (\pi_t^1, \cdots, \pi_t^n)'$ for $t = 0, 1, \cdots, T-1$, such that (1) the expected value of the uncertain terminal wealth $V_{T\wedge\tau}$ is maximized while the variance of the terminal wealth is not greater than a preselected risk level,

$$(P1(\sigma)) \begin{cases} \max\limits_{\pi} & E(V_{T\wedge\tau}) \\ s.t. & Var(V_{T\wedge\tau}) \le \sigma \text{ and } (12.4), \end{cases}$$

for $\sigma \ge 0$, or (2) the variance of the uncertain terminal wealth $V_{T\wedge\tau}$ is minimized while the expected terminal wealth is not smaller than a preselected level,

$$(P2(\epsilon)) \begin{cases} \min\limits_{\pi} & Var(V_{T\wedge\tau}) \\ s.t. & E(V_{T\wedge\tau}) \ge \epsilon \text{ and } (12.4), \end{cases}$$

for $\epsilon \ge 0$.

By varying the value of σ in $(P1(\sigma))$ or the value of ϵ in $(P2(\epsilon))$, the set of efficient multi-period portfolio policies can be generated, which are the same for both problem $(P1(\sigma))$ and $(P2(\epsilon))$.

Using the Lagrangian approach, either problem $(P1(\sigma))$ or $(P2(\epsilon))$ can be expressed equivalently as

$$(P3(\omega)) \begin{cases} \max\limits_{\pi} E(V_{T\wedge\tau}) - \omega Var(V_{T\wedge\tau}) \\ s.t. \qquad (12.4), \end{cases}$$

where $\omega \in [0, \infty)$, which represents a trade-off between the expected terminal wealth and the associated risk. Actually, $(P3(\omega))$ generates the same set of multi-period portfolio policies as problem $(P1(\sigma))$ and $(P2(\epsilon))$. If π^* solves $(P3(\omega))$, then π^* solves $(P1(\sigma))$ with $\sigma = Var(V_{T\wedge\tau})|_{\pi^*}$ and π^* solves $(P2(\epsilon))$ with $\epsilon = E(V_{T\wedge\tau})|_{\pi^*}$ In this work, we concentrate on problem $(P3(\omega))$.

12.4 Analytical Solution to Multi-Period Mean-Variance Formulation with Exit-Time Uncertainty

12.4.1 Construction of Auxiliary Problem

Since the mean-variance formulation is nonseparable in the sense of dynamic programming, we use the embedding technique of Li and Ng (2000) to analyze our problem. It will be proved in the following that the embedding technique still works when the exit time is uncertain. Let us introduce an alternative optimization problem $(P4(\lambda,\omega))$:

$$(P4(\lambda,\omega)) \begin{cases} \max\limits_{\pi} E(\lambda V_{T\wedge\tau} - \omega V_{T\wedge\tau}^2) \\ s.t. \qquad (12.4), \end{cases}$$

Define $\Phi_A(\lambda,\omega)$ to be the set of optimal solutions of problem $(P4(\lambda,\omega))$ and $\Phi_P(\omega)$ to be the set of optimal solutions of problem $(P3(\omega))$, i.e.,

$\Phi_A(\lambda,\omega) = \{\phi \mid \phi \text{ is an optimal solution to } (P4(\lambda,\omega))\}$,

$\Phi_P(\omega) = \{\phi \mid \phi \text{ is an optimal solution to } (P3(\omega))\}$.

Denote a new variable $d(\phi,\omega)$ as a function of ϕ and ω, i.e.,

$$d(\phi,\omega) = 1 + 2\omega E(V_{T\wedge\tau}) \mid_{\phi} . \tag{12.5}$$

The following two theorems will show the relationship between the original problem $(P3(\omega))$ and the auxiliary problem $(P4(\lambda,\omega))$.

Theorem 1 *For any* $\phi^* \in \Phi_P(\omega)$, $\phi^* \in \Phi_A(d(\phi^*,\omega),\omega)$.

Proof If ϕ^* is a solution of $(P3(\omega))$, but not a solution to $(P4(d(\phi^*,\lambda),\omega))$, there exists a ϕ such that

$$-\omega E(V_{T\wedge\tau}^2(\phi)) + d(\phi^*,\omega)E(V_{T\wedge\tau}(\phi)) > -\omega E(V_{T\wedge\tau}^2(\phi^*)) + d(\phi^*,\omega)E(V_{T\wedge\tau}(\phi^*)),$$

that is

$$(-\omega, d(\phi^*,\omega)) \begin{pmatrix} E(V_{T\wedge\tau}^2(\phi)) \\ E(V_{T\wedge\tau}(\phi)) \end{pmatrix} > (-\omega, d(\phi^*,\omega)) \begin{pmatrix} E(V_{T\wedge\tau}^2(\phi^*)) \\ E(V_{T\wedge\tau}(\phi^*)) \end{pmatrix}. \tag{12.6}$$

Let

$$\begin{aligned} U &= E(V_{T\wedge\tau}(\phi)) - \omega Var(V_{T\wedge\tau}(\phi)) \\ &= E(V_{T\wedge\tau}(\phi)) - \omega[E(V_{T\wedge\tau}^2(\phi)) - E^2(V_{T\wedge\tau}(\phi))]. \end{aligned} \tag{12.7}$$

As U is convex with respect to $E(V_{T\wedge\tau}(\phi))$ and $E(V_{T\wedge\tau}^2(\phi))$, we have

$$\begin{aligned} & U[E(V_{T\wedge\tau}^2(\phi)), E(V_{T\wedge\tau}(\phi))] - U[E(V_{T\wedge\tau}^2(\phi^*)), E(V_{T\wedge\tau}(\phi^*))] \\ & \geq \left(\frac{\partial U}{\partial E(V_{T\wedge\tau}^2(\phi))}, \frac{\partial U}{\partial E(V_{T\wedge\tau}(\phi))}\right)|_{\phi^*} \begin{pmatrix} E(V_{T\wedge\tau}^2(\phi)) - E(V_{T\wedge\tau}^2(\phi^*)) \\ E(V_{T\wedge\tau}(\phi)) - E(V_{T\wedge\tau}(\phi^*)) \end{pmatrix} \\ & = (-\omega, d(\phi^*,\omega)) \begin{pmatrix} E(V_{T\wedge\tau}^2(\phi)) - E(V_{T\wedge\tau}^2(\phi^*)) \\ E(V_{T\wedge\tau}(\phi)) - E(V_{T\wedge\tau}(\phi^*)) \end{pmatrix} > 0, \end{aligned}$$

which is a contradiction.

Theorem 2 *Assume $\phi^* \in \Phi_A(\lambda^*, \omega)$. A necessary condition for $\phi^* \in \Phi_P(\omega)$ is $\lambda^* = 1 + 2\omega E(V_{T\wedge\tau}) \mid_{\phi^*}$.*

Proof For fixed ω, the set of all solutions to $((P4(\lambda, \omega))$ can be parameterized by λ. If ϕ^* is an optimal solution of $(P3(\omega))$, then $\phi^* \in \bigcup_\lambda \Phi_A(\lambda, \omega)$. Hence $(P3(\omega))$ is equivalent to the following problem:

$$\begin{aligned} &\max\nolimits_\lambda U[E(V^2_{T\wedge\tau}(\lambda, \omega)), E^2(V_{T\wedge\tau}(\lambda, \omega)] \\ &= \max\nolimits_\lambda \{E(V_{T\wedge\tau}(\lambda, \omega)) - \omega[E(V^2_{T\wedge\tau}(\lambda, \omega)) - E^2(V_{T\wedge\tau}(\lambda, \omega))]\}, \end{aligned} \tag{12.8}$$

The necessary condition for optimal λ^* is $\frac{\partial U}{\partial \lambda} \mid_{\lambda^*} = 0$, that is

$$\frac{\partial E(V_{T\wedge\tau}(\lambda^*, \omega))}{\partial \lambda}[1 + 2\omega E(V_{T\wedge\tau}(\lambda^*, \omega))] - \omega \frac{\partial E(V^2_{T\wedge\tau}(\lambda^*, \omega)}{\partial \lambda} = 0. \tag{12.9}$$

On the other hand, because $\phi^* \in \Phi_A(\lambda, \omega)$, the optimality condition for $((P4(\lambda, \omega))$ gives rise,

$$\lambda^* \frac{\partial E(V_{T\wedge\tau}(\lambda^*, \omega))}{\partial \lambda} - \omega \frac{\partial E(V^2_{T\wedge\tau}(\lambda^*, \omega))}{\partial \lambda} = 0. \tag{12.10}$$

These two conditions, (12.9) and (12.10), yield

$$\lambda^* = 1 + 2\omega E(V_{T\wedge\tau}(\lambda^*, \omega)) = [1 + 2\omega E(V_{T\wedge\tau}(\lambda, \omega))] \mid_{\phi^*} .$$

Based on these two theorems, we can get the optimal solution to the original problem by solving the auxiliary problem $((P4(\lambda, \omega))$. The objective function of $((P4(\lambda, \omega))$ can be reformulated by using the definition of exit probability ξ_t.

Proposition 4.1 *The auxiliary problem $(P4(\lambda, \omega))$ is equivalent to*

$$\begin{cases} \max\limits_{\pi} & E[\sum_{t=1}^{T}(\lambda V_t - \omega V_t^2)\xi_t] \\ s.t. & V_{t+1} = V_t r_t^0 + \pi_t' R_t \ \ for \ \ t = 0, 1, \cdots, T-1. \end{cases} \tag{12.11}$$

Proof Using the property of conditional probability, we can derive the following:

$$\begin{aligned} &E(\lambda V_{T\wedge\tau} - \omega V^2_{T\wedge\tau}) \\ &= E[E[(\lambda V_{T\wedge\tau} - \omega V^2_{T\wedge\tau})\mathbf{1}_{\{\tau=1\}} \mid \mathcal{F}_1]] + E[E[(\lambda V_{T\wedge\tau} - \omega V^2_{T\wedge\tau})\mathbf{1}_{\{\tau>1\}} \mid \mathcal{F}_1]] \\ &= E[(\lambda V_1 - \omega V_1^2)E[\mathbf{1}_{\{\tau=1\}} \mid \mathcal{F}_1]] + E[(\lambda V_{T\wedge\tau} - \omega V^2_{T\wedge\tau})\mathbf{1}_{\{\tau>1\}}] \\ &= E[(\lambda V_1 - \omega V_1^2)\xi_1] + E[E[(\lambda V_{T\wedge\tau} - \omega V^2_{T\wedge\tau})\mathbf{1}_{\{\tau>1\}} \mid \mathcal{F}_2]] \\ &= E[(\lambda V_1 - \omega V_1^2)\xi_1] + E[E[(\lambda V_{T\wedge\tau} - \omega V^2_{T\wedge\tau})\mathbf{1}_{\{\tau=2\}} \mid \mathcal{F}_2]] \\ &\quad + E[E[(\lambda V_{T\wedge\tau} - \omega V^2_{T\wedge\tau})\mathbf{1}_{\{\tau>2\}} \mid \mathcal{F}_2]] \end{aligned}$$

$$= E[(\lambda V_1 - \omega V_1^2)\xi_1] + E[(\lambda V_2 - \omega V_2^2)\xi_2] + E[(\lambda V_{T\wedge\tau} - \omega V_{T\wedge\tau}^2)\mathbf{1}_{\{\tau>2\}}]$$

$$\vdots$$

$$= \sum_{i=1}^{T} E[(V_t - \lambda V_t^2)\xi_t],$$

which proves the equivalence between (12.11) and $(P4(\lambda, \omega))$.

12.4.2 Analytical Form of the Optimal Dynamic Portfolio Policy

The optimal solution to the auxiliary problem can be derived analytically by using dynamic programming. In the following, we denote $E_t(\cdot) := E(\cdot \mid \mathcal{F}_t)$ for our convenience.

Theorem 3 *The optimal solution of the auxiliary problem* $(P4(\lambda, \omega))$ *at each time period* t *is of the following form:*

$$\pi_t^*(V_t, \gamma) = \frac{\gamma}{2} u_t(\gamma) - K_t V_t, \tag{12.12}$$

where

$$\gamma = \frac{\lambda}{\omega}, \tag{12.13}$$

$$u_t = E_t[R_t R_t'(\xi_{t+1} + B_{t+1})]^{-1} E_t[R_t(\xi_{t+1} + A_{t+1})], \tag{12.14}$$

$$K_t = E_t[R_t R_t'(\xi_{t+1} + B_{t+1})]^{-1} E_t[r_{t+1}^0 R_t(\xi_{t+1} + B_{t+1})], \tag{12.15}$$

$$\begin{aligned} A_t = {} & E_t[r_t^0(\xi_{t+1} + A_{t+1})] \\ & - E_t[R_t(\xi_{t+1} + A_{t+1})]' E_t(R_t R_t'(\xi_{t+1} + B_{t+1}))^{-1} E_t[r_t^0 R_t(\xi_{t+1} + B_{t+1})], \end{aligned} \tag{12.16}$$

$$\begin{aligned} B_t = {} & E_t[(r_t^0)^2(\xi_{t+1} + B_{t+1})] \\ & \times E_t[r_t^0 R_t(\xi_{t+1} + B_{t+1})]' E_t[R_t R_t'(\xi_{t+1} + B_{t+1})]^{-1} E_t[r_t^0 R_t(\xi_{t+1} + B_{t+1})], \end{aligned} \tag{12.17}$$

$$\begin{aligned} C_t = {} & C_{t+1} + \\ & \frac{\lambda^2}{4\omega} E_t[R_t(\xi_{t+1} + A_{t+1})]' E_t[R_t R_t'(\xi_{t+1} + B_{t+1})]^{-1} E_t[R_t(\xi_{t+1} + A_{t+1})], \end{aligned} \tag{12.18}$$

with the following boundary conditions:

$$\begin{aligned} u_{T-1} =& E_{T-1}(R_{T-1}R'_{T-1}\xi_T)^{-1}E_{T-1}(R_{T-1}\xi_T), \\ K_{T-1} =& E_{T-1}(R_{T-1}R'_{T-1}\xi_T)^{-1}E_{T-1}(r^0_{T-1}R_{T-1}\xi_T), \\ A_{T-1} =& E_{T-1}(r^0_{T-1}\xi_T) \\ &- E_{T-1}(R_{T-1}\xi_T)'E_{T-1}(R_{T-1}R'_{T-1}\xi_T)^{-1}E_{T-1}(r^0_{T-1}R_{T-1}\xi_T), \\ B_{T-1} =& E_{T-1}[(r^0_{T-1})^2\xi_T] \\ &- E_{T-1}(r^0_{T-1}R_{T-1}\xi_T)'E_{T-1}(R_{T-1}R'_{T-1}\xi_T)^{-1}E_{T-1}(r^0_{T-1}R_{T-1}\xi_T), \\ C_{T-1} =& \frac{\lambda^2}{4\omega}E_{T-1}(R_{T-1}\xi_T)'E_{T-1}(R_{T-1}R'_{T-1}\xi_T)^{-1}E_{T-1}(R_{T-1}\xi_T). \end{aligned}$$

Proof Denote the benefit-to-go at stage t by

$$f_t(V_t) = \max_{\pi_{t-1},\cdots,\pi_{T-1}} E[\sum_{s=t}^{T}(\lambda V_s - \omega V_s^2)\xi_s \mid \mathcal{F}_{t-1}],$$

for $t = 1, 2, \cdots, T$. Note that $f_t(V_t)$ can be further expressed as

$$\begin{aligned} f_t(V_t) =& \max_{\pi_{t-1},\cdots,\pi_{T-1}} E\{(\lambda V_t - \omega V_t^2)\xi_t + E[\sum_{s=t+1}^{T}(\lambda V_s - \omega V_s^2)\xi_s \mid \mathcal{F}_t] \mid \mathcal{F}_{t-1}\} \\ =& \max_{\pi_{t-1}} E[(\lambda V_t - \omega V_t^2)\xi_t + f_{t+1}(V_{t+1}) \mid \mathcal{F}_{t-1}], \end{aligned}$$

for $t = 1, 2, \cdots, T-1$, and the boundary condition is

$$f_T(V_T) = \max_{\pi_{T-1}} E[(\lambda V_T - \omega V_T^2)\xi_T \mid \mathcal{F}_{T-1}].$$

The dynamic programming algorithm starts from stage T. For given $\mathcal{F}_{T-1}$, the optimization problem is as follows:

$$\begin{aligned} & f_T(V_T) \\ =& \max_{\pi_{T-1}} E_{T-1}[(\lambda V_T - \omega V_T^2)\xi_T] \\ =& \max_{\pi_{T-1}} E_{T-1}\{[\lambda(V_{T-1}r^0_{T-1} + R'_{T-1}\pi_{T-1})\xi_T] - \omega(V_{T-1}r^0_{T-1} + R'_{T-1}\pi_{T-1})^2\xi_T\} \\ =& \max_{\pi_{T-1}} E_{T-1}\{[\lambda V_{T-1}r^0_{T-1}\xi_T - \omega V_{T-1}^2(r^0_{T-1})^2\xi_T] \\ &+ [\lambda\xi_T R'_{T-1}\pi_{T-1} - 2\omega V_{T-1}r^0_{T-1}\xi_T R'_{T-1}\pi_{T-1} - \omega\xi_T\pi'_{T-1}R_{T-1}R'_{T-1}\pi_{T-1}]\}. \end{aligned}$$

Maximization of the above function with respect to π_{T-1} yields,

$$\pi^*_{T-1} = E_{T-1}(R_{T-1}R'_{T-1}\xi_T)^{-1}[\frac{\lambda}{2\omega}E_{T-1}(R_{T-1}\xi_T) - V_{T-1}E_{T-1}(r^0_{T-1}R_{T-1}\xi_T)].$$

Substituting π^*_{T-1} back to $f_T(V_T)$ yields to optimal benefit-to-go at given $\mathcal{F}_{T-1}$,

$$f^*_T(V_T) = \lambda A_{T-1}V_{T-1} - \omega B_{T-1}V^2_{T-1} + C_{T-1},$$

where

$$\begin{aligned}
A_{T-1} =& E_{T-1}(r^0_{T-1}\xi_T) \\
&- E_{T-1}(r^0_{T-1}R_{T-1}\xi_T)'E_{T-1}(R_{T-1}R'_{T-1}\xi_T)^{-1}E_{T-1}(R_{T-1}\xi_T), \\
B_{T-1} =& E_{T-1} \\
&- E_{T-1}(r^0_{T-1}R_{T-1}\xi_T)'E_{T-1}(R_{T-1}R'_{T-1}\xi_T)^{-1}E_{T-1}(r^0_{T-1}R_{T-1}\xi_T), \\
C_{T-1} =& \frac{\lambda^2}{4\omega}E_{T-1}(R_{T-1}\xi_T)'E_{T-1}(R_{T-1}R'_{T-1}\xi_T)^{-1}E_{T-1}(R_{T-1}\xi_T).
\end{aligned}$$

Therefore, the benefit-to-go at stage $T-1$ is

$$\begin{aligned}
&f_{T-1}(V_{T-1}) \\
&= \max_{\pi_{T-2}} E_{T-2}[(\lambda V_{T-1} - \omega V^2_{T-1})\xi_{T-1} + f^*_T(V_T)] \\
&= \max_{\pi_{T-2}} E_{T-2}\{\lambda(\xi_{T-1} + A_{T-1})V_{T-1} - \omega(\xi_{T-1} + B_{T-1})V^2_{T-1} + C_{T-1}\} \\
&= \max_{\pi_{T-2}} E_{T-2}\{[\lambda V_{T-2}r^0_{T-2}(\xi_{T-1} + A_{T-1}) - \omega V^2_{T-2}(r^0_{T-2})^2(\xi_{T-1} + B_{T-1})] \\
&\quad + C_{T-1} + [\lambda(\xi_{T-1} + A_{T-1})R'_{T-2}\pi_{T-2} - 2\omega V_{T-2}r^0_{T-2}(\xi_{T-1} + B_{T-1})R'_{T-2}\pi_{T-2} \\
&\quad - \omega(\xi_{T-1} + B_{T-1})\pi'_{T-2}R_{T-2}R'_{T-2}\pi_{T-2}]\},
\end{aligned}$$

which has a similar structure as the original utility function at stage T.

Assume that the derived utility function has a similar form at stage t, $1 < t < T-2$, to the original utility function at stage T. The benefit-to-go at stage t is

$$\begin{aligned}
f_t(V_t) &= \max_{\pi_{t-1}} E_{t-1}[(\lambda V_t - \omega V^2_t)\xi_t + f_{t+1}(V_{t+1})] \\
&= \max_{\pi_{t-1}} E_{t-1}\{\lambda(\xi_t + A_t)V_t - \omega(\xi_t + B_t)V^2_t + C_t\} \\
&= \max_{\pi_{t-1}} E_{t-1}\{[\lambda V_{t-1}r^0_{t-1}(\xi_t + A_t) - \omega V^2_{t-1}(r^0_{t-1})^2(\xi_t + B_t)] \\
&\quad + C_t + [\lambda(\xi_t + A_t)R'_{t-1}\pi_{t-1} - 2\omega V_{t-1}r^0_{t-1}(\xi_t + B_t)R'_{t-1}\pi_{t-1} \\
&\quad - \omega(\xi_t + B_t)\pi'_{t-1}R_{t-1}R'_{t-1}\pi_{t-1}]\}.
\end{aligned}$$

Maximizing the above function derives the optimal policy at given $\mathcal{F}_{t-1}$,

$$\pi_{t-1}^* = E_{t-1}(R_{t-1}R_{t-1}'(\xi_t + B_t))^{-1} \times [\frac{\lambda}{2\omega}E_{t-1}(R_{t-1}(\xi_t + A_t)) - V_{t-1}E_{t-1}(r_{t-1}^0 R_{t-1}(\xi_t + B_t))],$$

and the cost-to-go $f_{t-1}(V_{t-1})$ is

$$\begin{aligned} f_{t-1}(V_{t-1}) &= \max_{\pi_{t-1}} E_{t-1}[(\lambda V_{t-1} - \omega V_{t-1}^2)\xi_{t-1} + \lambda A_{t-1}V_{t-1} - \omega B_{t-1}V_{t-1}^2 + C_{t-1}] \\ &= \max_{\pi_{t-2}} E_{t-2}\{[\lambda V_{t-2}r_{t-2}^0(\xi_{t-1} + A_{t-1}) - \omega V_{t-2}^2(r_{t-2}^0)^2(\xi_{t-1} + B_{t-1})] \\ &\quad + C_{t-1} + [\lambda(\xi_{t-1} + A_{t-1})R_{t-1}'\pi_{t-2} - 2\omega V_{t-2}r_{t-2}^0(\xi_{t-1} + B_{t-1})R_{t-2}'\pi_{t-2} \\ &\quad - \omega(\xi_{t-1} + B_{t-1})\pi_{t-2}'R_{t-2}R_{t-2}'\pi_{t-2}]\}, \end{aligned}$$

where

$$\begin{aligned} A_{t-1} &= E_{t-1}[r_{t-1}^0(\xi_t + A_t)] \\ &\quad - E_{t-1}[R_{t-1}(\xi_t + A_t)]'E_{t-1}[R_{t-1}R_{t-1}'(\xi_t + B_t)]^{-1}E_{t-1}[r_{t-1}^0 R_{t-1}(\xi_t + B_t)], \\ B_{t-1} &= E_{t-1}[(r_{t-1}^0)^2(\xi_t + B_t)] \\ &\quad - E_{t-1}[r_{t-1}^0 R_{t-1}(\xi_t + B_t)]'E_{t-1}(R_{t-1}R_{t-1}'(\xi_t + B_t))^{-1}E_{t-1}[r_{t-1}^0 R_{t-1}(\xi_t + B_t)], \\ C_{t-1} &= C_t + \\ &\quad \frac{\lambda^2}{4\omega}E_{t-1}(R_{t-1}(\xi_t + A_t))'E_{t-1}(R_{t-1}R_{t-1}'(\xi_t + B_t))^{-1}E_{t-1}(R_{t-1}(\xi_t + A_t)). \end{aligned}$$

It is obvious that the second term in $\pi_t^*(V_t, \gamma)$ is linear with respect to the wealth V_t and is independent of γ, and the first term is a linear function of γ. Substituting (12.12) into the equation of wealth dynamics yields the dynamics of the wealth under policy $\pi_t^*(V_t, \gamma)$,

$$V_{t+1}(\gamma) = (r_t^0 - K_t'R_t)V_t(\gamma) + \frac{\gamma}{2}R_t'u_t. \tag{12.19}$$

V_t ($t = 1, 2, \cdots, T$) can be derived by solving the above dynamics as follows:

$$V_t = M_t v_0 + \frac{\gamma}{2}N_t, \quad t = 1, 2, \cdots, T, \tag{12.20}$$

where

$$M_t = \prod_{i=0}^{t-1}(r_i^0 - K_i'R_i), \tag{12.21}$$

$$N_t = \sum_{i=0}^{t-1}[\prod_{j=i+1}^{t-1}(r_j^0 - K_j'R_j)]u_i'R_i. \tag{12.22}$$

Squaring both sides of(12.19) yields

$$V_{t+1}^2(\gamma) = (r_t^0 - K_t'R_t)^2 V_t^2(\gamma) + \gamma(r_t^0 - K_t'R_t)V_t(\gamma)R_t'u_t + \frac{\gamma^2}{4}u_t'R_tR_t'u_t. \quad (12.23)$$

Similarly, we derive $V_t^2(\gamma)$ for $t = 1, 2, \cdots, T$ as follows by solving the above dynamics:

$$V_t^2(\gamma) = I_t v_0^2 + \gamma J_t v_0 + \frac{\lambda^2}{4}L_t, \quad (12.24)$$

where

$$I_t = \prod_{i=0}^{t-1}(r_i^0 - K_i'R_i)^2, \quad (12.25)$$

$$J_t = \sum_{i=0}^{t-1}\prod_{j=i+1}^{t-1}(r_j^0 - K_j'R_j)^2 u_i'R_i \prod_{s=0}^{i}(r_s^0 - K_s'R_s), \quad (12.26)$$

$$L_t = \sum_{i=0}^{t-1}\prod_{j=i+1}^{t-1}(r_j^0 - K_j'R_j)^2(u_i'R_i)^2 + 2\sum_{i=1}^{t-1}\prod_{j=i+1}^{t-1}(r_j^0 - K_j'R_j)^2 u_i'R_i[\sum_{s=0}^{i-1}\prod_{l=s+1}^{i} t(r_l^0 - K_l'R_l)u_s'R_s]. \quad (12.27)$$

Notice that $E(V_{T\wedge\tau}) = E(\sum_{t=1}^T V_t\xi_t)$ and $E(V_{T\wedge\tau}^2) = E(\sum_{t=1}^T V_t^2\xi_t)$, the expectation of terminal wealth $V_{T\wedge\tau}$ and $V_{T\wedge\tau}^2$ are

$$E(V_{T\wedge\tau}) = v_0 E(\sum_{t=1}^T M_t\xi_t) + \frac{\gamma}{2}E(\sum_{t=1}^T N_t\xi_t), \quad (12.28)$$

$$E(V_{T\wedge\tau}^2) = v_0^2 E(\sum_{t=1}^T I_t\xi_t) + \gamma v_0 E(\sum_{t=1}^T J_t\xi_t) + \frac{\gamma^2}{4}E(\sum_{t=1}^T L_t\xi_t). \quad (12.29)$$

The variance of the uncertain terminal wealth under portfolio policy $\pi_t^*(V_t, \gamma)$ can be expressed in the terms of γ by using (12.28) and (12.29),

$$Var(V_{T\wedge\tau}(\gamma)) = E(V_{T\wedge\tau}^2) - E^2(V_{T\wedge\tau}) = v_0^2 I + v_0\gamma J + \frac{\gamma^2}{4}L, \quad (12.30)$$

where

$$I = E(\sum_{t=1}^{T} I_t\xi_t) - (E(\sum_{t=1}^{T} M_t\xi_t))^2, \tag{12.31}$$

$$J = E(\sum_{t=1}^{T} J_t\xi_t) - E(\sum_{t=1}^{T} M_t\xi_t)E(\sum_{t=1}^{T} N_t\xi_t), \tag{12.32}$$

$$L = E(\sum_{t=1}^{T} L_t\xi_t) - (E(\sum_{t=1}^{T} N_t\xi_t))^2. \tag{12.33}$$

Rewrite $E(V_{T\wedge\tau})$ as follows:

$$E(V_{T\wedge\tau}) = v_0 M + \frac{\gamma}{2} N, \tag{12.34}$$

where

$$M = E(\sum_{t=1}^{T} M_t\xi_t), \tag{12.35}$$

$$N = E(\sum_{t=1}^{T} N_t\xi_t). \tag{12.36}$$

Note that the expected uncertain terminal wealth $E(V_{T\wedge\tau}(\gamma))$ is an increasing linear function of γ while the variance $Var(V_{T\wedge\tau}(\gamma))$ is a quadratic function of γ. We express $U(E(V_{T\wedge\tau}), Var(V_{T\wedge\tau}))$ as a function of γ,

$$\begin{aligned} & U(E(V_{T\wedge\tau}), Var(V_{T\wedge\tau})) \\ & = v_0 M + \frac{\gamma}{2} N - \omega(v_0^2 I + v_0\gamma J + \frac{\gamma^2}{4} L). \end{aligned} \tag{12.37}$$

It can be seen that U is a concave function of γ. Differentiating (12.37) with respect to γ yields

$$\frac{\partial U}{\partial \gamma} = N/2 - \omega(v_0 J + L\gamma/2). \tag{12.38}$$

The optimal γ must satisfy the optimality condition of $\frac{\partial U}{\partial \gamma} = 0$, that is,

$$\gamma^* = -\frac{2J}{L} v_0 + \frac{N}{L\omega}. \tag{12.39}$$

Notice that $\lambda^* = \omega\gamma^*$ satisfies the condition that $\lambda^* = 1 + 2\omega E(V_{T\wedge\tau}) \mid_{\phi^*}$. Actually, the necessary condition for optimal λ^* is $\frac{\partial U}{\partial \lambda}|_{\lambda^*} = 0$. Since $\gamma = \frac{\lambda}{\omega}$ and ω is given, the necessary condition is equivalent to $\frac{\partial U}{\partial \gamma}|_{\gamma^*} = 0$.

Substituting the optimal γ^* in (12.39) into Eq. (12.12) yields the optimal multi-period portfolio policy for $(P3(\omega))$.

Theorem 4 *The optimal multi-period portfolio policy for $(P3(\omega))$ is specified by the following analytical form:*

$$\begin{aligned}\pi_t^* &= -E_t(R_tR_t'(\xi_{t+1} + B_{t+1}))^{-1}E_t(r_t^0R_t(\xi_{t+1} + B_{t+1}))V_t \\ &+(-\frac{J}{L}v_0 + \frac{N}{2L\omega})E_t(R_tR_t'(\xi_{t+1} + B_{t+1}))^{-1}E_t(R_t(\xi_{t+1} + A_{t+1})),\end{aligned}$$

and the mean-variance efficient frontier can be specified by the following analytical form:

$$Var(V_{T\wedge\tau}) = L(\frac{E(V_{T\wedge\tau}) - v_0M}{N} + \frac{v_0J}{L})^2 + v_0^2I - \frac{v_0^2J^2}{L},$$

where A_t, B_t, L, J, M, N are defined in (12.16), (12.17), (12.33), (12.32), (12.35), and (12.36).

12.5 Special Cases of Stochastic Investment Horizon

12.5.1 State-Independent Uncertain Exit Time

When the uncertain exit time is state-independent, the stochastic process ξ_t defined in (12.3) satisfies

$$\xi_t := P(\tau = t|\mathcal{A}_t) = P(\tau = t|\mathcal{A}_\infty) = P(\tau = t),$$

which is $\mathcal{A}$-independent. Hence ξ_t and R_{t-1} are independent for any t.

In this case, the parameters defined in Eqs. (12.14)–(12.17) now take the following forms:

$$\begin{aligned}u_{t-1} &= \frac{\xi_t + A_t}{\xi_t + B_t}E(R_{t-1}R_{t-1}')^{-1}E(R_{t-1}),\\ K_{t-1} &= E(R_{t-1}R_{t-1}')^{-1}E(r_{t-1}^0R_{t-1}),\\ A_{t-1} &= (\xi_t + A_t)[E(r_{t-1}^0) - E(r_{t-1}^0R_{t-1})'E(R_{t-1}R_{t-1}')^{-1}E(R_{t-1})],\\ B_{t-1} &= (\xi_t + B_t)[E((r_{t-1}^0)^2) - E(r_{t-1}^0R_{t-1})'E(R_{t-1}R_{t-1}')^{-1}E(r_{t-1}^0R_{t-1})].\end{aligned}$$

The expectation and variance of the uncertain terminal wealth under portfolio policy π_t^* can be expressed as follows:

$$E(V_{T\wedge\tau}) = v_0 M + \frac{\gamma}{2} N,$$

$$Var(V_{T\wedge\tau}(\gamma)) = v_0^2 I + v_0 \gamma J + \frac{\gamma^2}{4} L,$$

and the expressions for M, N, I, J, and L in (12.35)–(12.36) and (12.31)–(12.33) can be simplified to the following forms:

$$M = \sum_{t=1}^{T} \xi_t \prod_{i=0}^{t-1} Y_t^1,$$

$$N = \sum_{t=1}^{T} \xi_t \sum_{i=0}^{t-1} \prod_{j=i+1}^{t-1} Y_j^1 Y_i^0 \frac{\xi_{i+1} + A_{i+1}}{\xi_{i+1} + B_{i+1}},$$

$$I = \sum_{t=1}^{T} \xi_t \prod_{i=0}^{t-1} Y_i^2 - (\sum_{t=1}^{T} \xi_t \prod_{i=0}^{t-1} Y_i^1)^2,$$

$$J = -\sum_{t=1}^{T} \xi_t \prod_{i=0}^{t-1} Y_i^1 \times \sum_{t=1}^{T} \xi_t \sum_{i=0}^{t-1} \prod_{j=i+1}^{t-1} Y_j^1 Y_i^0 \frac{\xi_{i+1} + A_{i+1}}{\xi_{i+1} + B_{i+1}},$$

$$L = \sum_{t=1}^{T} \xi_t \sum_{i=0}^{t-1} \prod_{j=i+1}^{t-1} Y_j^2 Y_i^0 (\frac{\xi_{i+1} + A_{i+1}}{\xi_{i+1} + B_{i+1}})^2 - [\sum_{t=1}^{T} \xi_t \sum_{i=0}^{t-1} \prod_{j=i+1}^{t-1} Y_j^1 Y_i^0 \frac{\xi_{i+1} + A_{i+1}}{\xi_{i+1} + B_{i+1}}]^2,$$

with

$$Y_t^0 = E(R_t)' E(R_t R_t')^{-1} E(R_t),$$
$$Y_t^1 = E(r_t^0) - E(r_t^0 R_t)' E(R_t R_t')^{-1} E(R_t),$$
$$Y_t^2 = E[(r_t^0)^2] - E(r_t^0 R_t)' E(R_t R_t')^{-1} E(r_t^0 R_t).$$

The optimal multi-period mean-variance portfolio policy when uncertain exit time is state-independent can be formulated in the following forms:

$$\pi_t^* = -E(R_t R_t')^{-1} E(r_t^0 R_t) V_t + (-\frac{J}{L} v_0 + \frac{N}{2L\omega}) \frac{\xi_{t+1} + A_{t+1}}{\xi_{t+1} + B_{t+1}} E(R_t R_t')^{-1} E(R_t).$$

This result is consistent with the result of Guo and Hu (2005).

12.5.2 Deterministic Exit Time

If we define the stochastic process ξ_t as

$$\xi_t := \begin{cases} 0 & 0 \le t \le T-1, \\ 1 & t = T, \end{cases}$$

then the multi-period portfolio selection model with stochastic investment horizon reduces to the case with a deterministic investment horizon. The result we derived in the last section can be simplified with this specific ξ_t.

When the exit time is certain, the parameters defined in Eqs. (12.14)–(12.17) now take the following forms:

$$\begin{aligned} u_{t-1} &= \frac{A_t}{B_t} E(R_{t-1}R'_{t-1})^{-1} E(R_{t-1}), \\ K_{t-1} &= E(R_{t-1}R'_{t-1})^{-1} E(r^0_{t-1}R_{t-1}), \\ A_{t-1} &= A_t[E(r^0_{t-1}) - E(R_{t-1})' E(R_{t-1}R'_{t-1})^{-1} E(r^0_{t-1}R_{t-1})], \\ B_{t-1} &= B_t[E((r^0_{t-1})^2) - E(r^0_{t-1}R_{t-1})' E(R_{t-1}R'_{t-1})^{-1} E(r^0_{t-1}R_{t-1})]. \end{aligned}$$

The expectation and variance of the uncertain terminal wealth under portfolio policy π^*_t can be expressed as follows:

$$E(V_T) = v_0 M + \frac{\gamma}{2} N,$$

$$Var(V_T) = v_0^2 I + v_0 \gamma J + \frac{\gamma^2}{4} L,$$

and the expressions for M, N, I, J, and L in (12.35)–(12.36) and (12.31)–(12.33) can be simplified to the following forms:

$$\begin{aligned} M =& E(\prod_{i=0}^{T-1}(r^0_i - K'_i R_i)) = \prod_{i=0}^{T-1} Y^1_t, \\ N =& \sum_{i=0}^{T-1}(\prod_{j=i+1}^{T-1} Y^1_j) Y^0_i \frac{\prod_{j=i+1}^{T-1} Y^1_j}{\prod_{j=i+1}^{T-1} Y^2_j}, \\ I =& \prod_{i=0}^{T-1} Y^2_i - (\prod_{i=0}^{T-1} Y^1_i)^2, \\ J =& -\prod_{i=0}^{T-1} Y^1_t \times \sum_{i=0}^{T-1}(\prod_{j=i+1}^{T-1} Y^1_j) Y^0_i \frac{\prod_{j=i+1}^{T-1} Y^1_j}{\prod_{j=i+1}^{T-1} Y^2_j}, \\ L =& \sum_{i=0}^{T-1}(\prod_{j=i+1}^{T-1} Y^2_j) Y^0_i (\frac{\prod_{j=i+1}^{T-1} Y^1_j}{\prod_{j=i+1}^{T-1} Y^2_j})^2 - [\sum_{i=0}^{T-1}(\prod_{j=i+1}^{T-1} Y^1_j) Y^0_i \frac{\prod_{j=i+1}^{T-1} Y^1_j}{\prod_{j=i+1}^{T-1} Y^2_j}]^2, \end{aligned}$$

with

$$\begin{aligned} Y_t^0 &= E(R_t)'E(R_tR_t')^{-1}E(R_t), \\ Y_t^1 &= E(r_t^0) - E(r_t^0R_t)'E(R_tR_t')^{-1}E(R_t), \\ Y_t^2 &= E[(r_t^0)^2] - E(r_t^0R_t)'E(R_tR_t')^{-1}E(r_t^0R_t). \end{aligned}$$

The optimal multi-period mean-variance portfolio policy is given in the following forms:

$$\begin{aligned} \pi_t^* &= -E(R_tR_t')^{-1}E(r_t^0R_t)V_t \\ &\quad +(-\frac{2J}{L}v_0 + \frac{N}{L\omega})\frac{\prod_{i=t+1}^{T-1} Y_i^1}{\prod_{i=t+1}^{T-1} Y_i^2}E(R_tR_t')^{-1}E(R_t). \end{aligned}$$

The above result is just consistent with the result of Li and Ng (2000).

12.5.3 Illustrative Examples

The following examples illustrate the effect of time risk on the mean-variance efficient frontier.

Example 2.1 Consider a financial market with investment horizon $T = 3$. There are one risk free asset and two risky assets in the market. Suppose that the economy has a discrete sample space Ω (totally 54 samples). The riskless return is 1.08 and risk returns are listed in Table 12.1. An investor enters the financial market and invests her money among these three assets. At the same time, she is waiting for a gold mining opportunity. If this opportunity is available and is more profitable than market portfolio (a portfolio with half on each asset), she will exit the market. The failure rate of gold mining opportunity $\widehat{\lambda}_t$ is also listed in Table 12.1.

The failure rate λ_t is pre-determined for given $\widehat{\lambda}_t$ (the failure rate increases as the market portfolio return decreases). The corresponding failure rate λ_t are showed in Table 12.2. So the cumulative probability P_t can be calculated by using formulation (12.2). Therefore, the probability of stochastic investment horizon can be determined by using (12.3), and they are also listed in Table 12.2.

Example 2.2 Consider the same economy as in Example 2.1 except that $\widehat{\lambda}_1 \equiv 0.18, \widehat{\lambda}_2 \equiv 0.24$. So that the sample space becomes $\Omega = \{\omega_1, \cdots, \omega_{27}\}$. An investor enters the financial market with one unit of wealth. She is trying to find the best allocation of her wealth among these three assets. At the same time, she is waiting for a gold mining opportunity. If this opportunity is available and is more profitable than market portfolio (a portfolio with half on each asset), she will exit the market. Hence the probability of exit time is listed in Table 12.2. The investor would like to maximize $E(x_{3\wedge\tau}) - 2Var(x_{3\wedge\tau})$, where $x_{3\wedge\tau}$ is the wealth at the exit time.

Table 12.1 Risky returns r_1, r_2, r_3 and failure rate $\widehat{\lambda}_1, \widehat{\lambda}_2$ in Example 2.1

ω	$r_1(1)$	$r_2(1)$	$r_3(1)$	$r_1(2)$	$r_2(2)$	$r_3(2)$	$\widehat{\lambda}_1$	$\widehat{\lambda}_2$
ω_1	1.1697	1.2258	0.8997	1.2776	0.9934	1.0292	0.18	0.24
ω_2	1.1697	1.2258	1.1548	1.2776	0.9934	1.5623	0.18	0.24
ω_3	1.1697	1.2258	1.0399	1.2776	0.9934	1.2076	0.18	0.24
ω_4	1.1697	0.8409	1.2362	1.2776	1.6678	1.3599	0.18	0.24
ω_5	1.1697	0.8409	1.2234	1.2776	1.6678	1.2717	0.18	0.24
ω_6	1.1697	0.8409	1.3665	1.2776	1.6678	1.0603	0.18	0.24
ω_7	1.1697	1.1455	1.2334	1.2776	1.4281	1.0825	0.18	0.24
ω_8	1.1697	1.1455	1.0842	1.2776	1.4281	1.3757	0.18	0.24
ω_9	1.1697	1.1455	1.2080	1.2776	1.4281	0.9684	0.18	0.24
ω_{10}	1.2771	1.3005	1.0401	1.2406	1.4795	1.4743	0.18	0.24
ω_{11}	1.2771	1.3005	1.1596	1.2406	1.4795	1.4123	0.18	0.24
ω_{12}	1.2771	1.3005	1.1562	1.2406	1.4795	1.0059	0.18	0.24
ω_{13}	1.2771	1.1058	1.1620	1.2406	1.4110	1.1684	0.18	0.24
ω_{14}	1.2771	1.1058	1.1236	1.2406	1.4110	0.8989	0.18	0.24
ω_{15}	1.2771	1.1058	1.2943	1.2406	1.4110	0.6024	0.18	0.24
ω_{16}	1.2771	1.3692	0.9356	1.2406	1.1561	1.5342	0.18	0.24
ω_{17}	1.2771	1.3692	1.2137	1.2406	1.1561	1.0944	0.18	0.24
ω_{18}	1.2771	1.3692	1.2702	1.2406	1.1561	1.3417	0.18	0.24
ω_{19}	0.9607	0.8504	1.2503	1.3380	1.6880	1.3144	0.18	0.24
ω_{20}	0.9607	0.8504	1.2318	1.3380	1.6880	1.2523	0.18	0.24
ω_{21}	0.9607	0.8504	1.1669	1.3380	1.6880	0.9526	0.18	0.24
ω_{22}	0.9607	1.2514	1.2438	1.3380	1.3195	0.9692	0.18	0.24
ω_{23}	0.9607	1.2514	1.2307	1.3380	1.3195	1.1366	0.18	0.24
ω_{24}	0.9607	1.2514	1.1311	1.3380	1.3195	0.8994	0.18	0.24
ω_{25}	0.9607	1.2176	1.1164	1.3380	1.2615	0.9374	0.18	0.24
ω_{26}	0.9607	1.2176	1.1262	1.3380	1.2615	1.6763	0.18	0.24
ω_{27}	0.9607	1.2176	0.9838	1.3380	1.2615	1.2623	0.18	0.24
ω_{28}	1.1697	1.2258	0.8997	1.2776	0.9934	1.0292	0.21	0.26
ω_{29}	1.1697	1.2258	1.1548	1.2776	0.9934	1.5623	0.21	0.26
ω_{30}	1.1697	1.2258	1.0399	1.2776	0.9934	1.2076	0.21	0.26
ω_{31}	1.1697	0.8409	1.2362	1.2776	1.6678	1.3599	0.21	0.26
ω_{32}	1.1697	0.8409	1.2234	1.2776	1.6678	1.2717	0.21	0.26
ω_{33}	1.1697	0.8409	1.3665	1.2776	1.6678	1.0603	0.21	0.26
ω_{34}	1.1697	1.1455	1.2334	1.2776	1.4281	1.0825	0.21	0.26
ω_{35}	1.1697	1.1455	1.0842	1.2776	1.4281	1.3757	0.21	0.26
ω_{36}	1.1697	1.1455	1.2080	1.2776	1.4281	0.9684	0.21	0.26
ω_{37}	1.2771	1.3005	1.0401	1.2406	1.4795	1.4743	0.21	0.26
ω_{38}	1.2771	1.3005	1.1596	1.2406	1.4795	1.4123	0.21	0.26
ω_{39}	1.2771	1.3005	1.1562	1.2406	1.4795	1.0059	0.21	0.26
ω_{40}	1.2771	1.1058	1.1620	1.2406	1.4110	1.1684	0.21	0.26
ω_{41}	1.2771	1.1058	1.1236	1.2406	1.4110	0.8989	0.21	0.26

(continued)

Table 12.1 (continued)

ω	$r_1(1)$	$r_2(1)$	$r_3(1)$	$r_1(2)$	$r_2(2)$	$r_3(2)$	$\widehat{\lambda}_1$	$\widehat{\lambda}_2$
ω_{42}	1.2771	1.1058	1.2943	1.2406	1.4110	0.6024	0.21	0.26
ω_{43}	1.2771	1.3692	0.9356	1.2406	1.1561	1.5342	0.21	0.26
ω_{44}	1.2771	1.3692	1.2137	1.2406	1.1561	1.0944	0.21	0.26
ω_{45}	1.2771	1.3692	1.2702	1.2406	1.1561	1.3417	0.21	0.26
ω_{46}	0.9607	0.8504	1.2503	1.3380	1.6880	1.3144	0.21	0.26
ω_{47}	0.9607	0.8504	1.2318	1.3380	1.6880	1.2523	0.21	0.26
ω_{48}	0.9607	0.8504	1.1669	1.3380	1.6880	0.9526	0.21	0.26
ω_{49}	0.9607	1.2514	1.2438	1.3380	1.3195	0.9692	0.21	0.26
ω_{50}	0.9607	1.2514	1.2307	1.3380	1.3195	1.1366	0.21	0.26
ω_{51}	0.9607	1.2514	1.1311	1.3380	1.3195	0.8994	0.21	0.26
ω_{52}	0.9607	1.2176	1.1164	1.3380	1.2615	0.9374	0.21	0.26
ω_{53}	0.9607	1.2176	1.1262	1.3380	1.2615	1.6763	0.21	0.26
ω_{54}	0.9607	1.2176	0.9838	1.3380	1.2615	1.2623	0.21	0.26

Using the result derived in Sect. 12.4.2, we can get the efficient frontiers of wealth at $t = 1, 2, 3$, which are showed in Fig. 12.1. It is obvious that the longer the investment horizon is, the higher the efficient frontier is. We also compare the efficient frontier of terminal wealth in the certain-exit-time case to that of the uncertain investment horizon case in Fig. 12.2. We can see that the efficient frontier under the certain-exit-time case is above that of the uncertain case. Uncertain investment horizon actually adds more risk in the investment.

Example 2.3 Consider an investor enters the financial market with one unit of wealth at the very beginning. She plan to stay in the financial market at most $T = 4$ period. However, she will be forced to exit the market for some market-independent exogenous reason. Suppose the uncertain exit time τ has pre-given exit probability $P(\tau = i) = 0.1i$ for $i = 1, 2, 3, 4$. We use the same market data as Example 2 in Li and Ng (2000). The investor is trying to find the best allocation of her wealth among three risky securities, A, B, C and a risk free security D. The expected returns for risky securities, A, B, and C are $E(r_t^A) = 1, 162$, $E(r_t^B) = 1.246$, and $E(r_t^C) = 1.228$, $t = 1, 2, 3, 4$, and the return for risk free asset D is 1.04. The covariance of $r = [r_t^A, r_t^B, r_t^C]'$ is

$$Cov(r) = \begin{bmatrix} 0.0146 \; 0.0187 \; 0.0145 \\ 0.0187 \; 0.0854 \; 0.0104 \\ 0.0145 \; 0.0104 \; 0.0289 \end{bmatrix}, \quad t = 1, 2, 3, 4.$$

The investor would like to maximize $E(x_{4\wedge\tau}) - 2Var(x_{4\wedge\tau})$. We can calculate

$$M = 0, 1088, \quad N = 0, 8917,$$

$$I = 0.1229, \quad J = -0.097, \quad L = 0.0879.$$

Table 12.2 Failure rate λ, cumulative probability P, and exit probability ξ in Example 2.2

ω	λ_1	λ_2	P_1	P_2	ξ_1	ξ_2	ξ_3
ω_1	0.20	0.20	0.1813	0.3297	0.1813	0.1484	0.6703
ω_2	0.20	0.20	0.1813	0.3297	0.1813	0.1484	0.6703
ω_3	0.20	0.20	0.1813	0.3297	0.1813	0.1484	0.6703
ω_4	0.20	0.15	0.1813	0.2953	0.1813	0.1140	0.7047
ω_5	0.20	0.15	0.1813	0.2953	0.1813	0.1140	0.7047
ω_6	0.20	0.15	0.1813	0.2953	0.1813	0.1140	0.7047
ω_7	0.20	0.10	0.1813	0.2592	0.1813	0.0779	0.7408
ω_8	0.20	0.10	0.1813	0.2592	0.1813	0.0779	0.7408
ω_9	0.20	0.10	0.1813	0.2592	0.1813	0.0779	0.7408
ω_{10}	0.15	0.05	0.1393	0.1813	0.1393	0.0420	0.8187
ω_{11}	0.15	0.05	0.1393	0.1813	0.1393	0.0420	0.8187
ω_{12}	0.15	0.05	0.1393	0.1813	0.1393	0.0420	0.8187
ω_{13}	0.15	0.15	0.1393	0.2592	0.1393	0.1199	0.7408
ω_{14}	0.15	0.15	0.1393	0.2592	0.1393	0.1199	0.7408
ω_{15}	0.15	0.15	0.1393	0.2592	0.1393	0.1199	0.7408
ω_{16}	0.15	0.12	0.1393	0.2366	0.1393	0.0973	0.7634
ω_{17}	0.15	0.12	0.1393	0.2366	0.1393	0.0973	0.7634
ω_{18}	0.15	0.12	0.1393	0.2366	0.1393	0.0973	0.7634
ω_{19}	0.30	0.10	0.2592	0.3297	0.2592	0.0705	0.6703
ω_{20}	0.30	0.10	0.2592	0.3297	0.2592	0.0705	0.6703
ω_{21}	0.30	0.10	0.2592	0.3297	0.2592	0.0705	0.6703
ω_{22}	0.30	0.20	0.2592	0.3935	0.2592	0.1343	0.6065
ω_{23}	0.30	0.20	0.2592	0.3935	0.2592	0.1343	0.6065
ω_{24}	0.30	0.20	0.2592	0.3935	0.2592	0.1343	0.6065
ω_{25}	0.30	0.16	0.2592	0.3687	0.2592	0.1095	0.6313
ω_{26}	0.30	0.16	0.2592	0.3687	0.2592	0.1095	0.6313
ω_{27}	0.30	0.16	0.2592	0.3687	0.2592	0.1095	0.6313
ω_{28}	0.25	0.25	0.2212	0.3935	0.2212	0.1723	0.6065
ω_{29}	0.25	0.25	0.2212	0.3935	0.2212	0.1723	0.6065
ω_{30}	0.25	0.25	0.2212	0.3935	0.2212	0.1723	0.6065
ω_{31}	0.25	0.18	0.2212	0.3495	0.2212	0.1283	0.6505
ω_{32}	0.25	0.18	0.2212	0.3495	0.2212	0.1283	0.6505
ω_{33}	0.25	0.18	0.2212	0.3495	0.2212	0.1283	0.6505
ω_{34}	0.25	0.14	0.2212	0.3229	0.2212	0.1017	0.6771
ω_{35}	0.25	0.14	0.2212	0.3229	0.2212	0.1017	0.6771
ω_{36}	0.25	0.14	0.2212	0.3229	0.2212	0.1017	0.6771
ω_{37}	0.18	0.1	0.1647	0.2442	0.1647	0.0795	0.7558
ω_{38}	0.18	0.1	0.1647	0.2442	0.1647	0.0795	0.7558
ω_{39}	0.18	0.1	0.1647	0.2442	0.1647	0.0795	0.7558
ω_{40}	0.18	0.18	0.1647	0.3023	0.1647	0.1376	0.6977
ω_{41}	0.18	0.18	0.1647	0.3023	0.1647	0.1376	0.6977

(continued)

Table 12.2 (continued)

ω	λ_1	λ_2	P_1	P_2	ξ_1	ξ_2	ξ_3
ω_{42}	0.18	0.18	0.1647	0.3023	0.1647	0.1376	0.6977
ω_{43}	0.18	0.16	0.1647	0.2882	0.1647	0.1235	0.7118
ω_{44}	0.18	0.16	0.1647	0.2882	0.1647	0.1235	0.7118
ω_{45}	0.18	0.16	0.1647	0.2882	0.1647	0.1235	0.7118
ω_{46}	0.32	0.14	0.2739	0.3687	0.2739	0.0948	0.6313
ω_{47}	0.32	0.14	0.2739	0.3687	0.2739	0.0948	0.6313
ω_{48}	0.32	0.14	0.2739	0.3687	0.2739	0.0948	0.6313
ω_{49}	0.32	0.22	0.2739	0.4123	0.2739	0.1384	0.5877
ω_{50}	0.32	0.22	0.2739	0.4123	0.2739	0.1384	0.5877
ω_{51}	0.32	0.22	0.2739	0.4123	0.2739	0.1384	0.5877
ω_{52}	0.32	0.2	0.2739	0.4055	0.2739	0.1316	0.5945
ω_{53}	0.32	0.2	0.2739	0.4055	0.2739	0.1316	0.5945
ω_{54}	0.32	0.2	0.2739	0.4055	0.2739	0.1316	0.5945

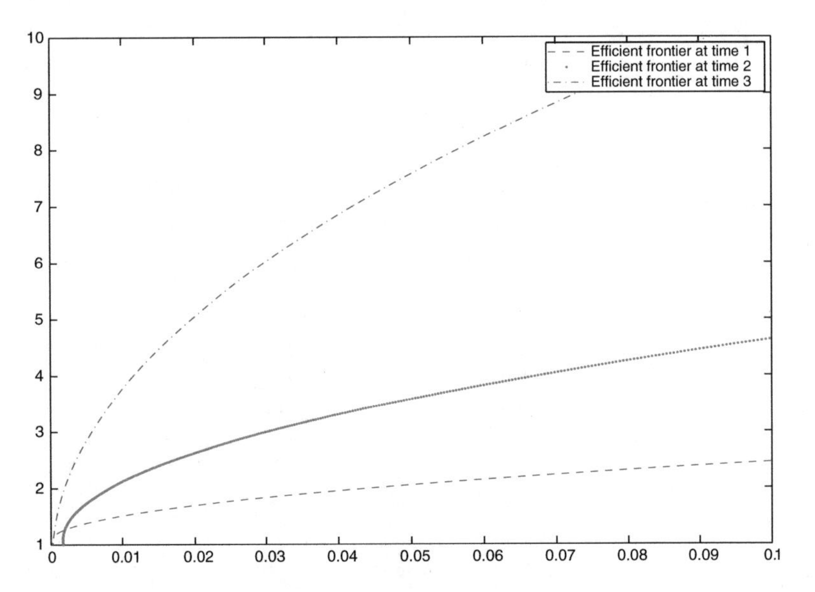

Fig. 12.1 Efficient frontiers when exit time is uncertain in Example 2.2

The mean-variance efficient frontier in this case is given as follows:

$$Var(x_{4\wedge\tau}) = 0.1105E^2(x_{4\wedge\tau}) - 0.239E(x_{4\wedge\tau}) + 0.1479.$$

The associated optimal portfolio policy is given as follows:

$$\pi_t^* = x_t - K_t V_t,$$

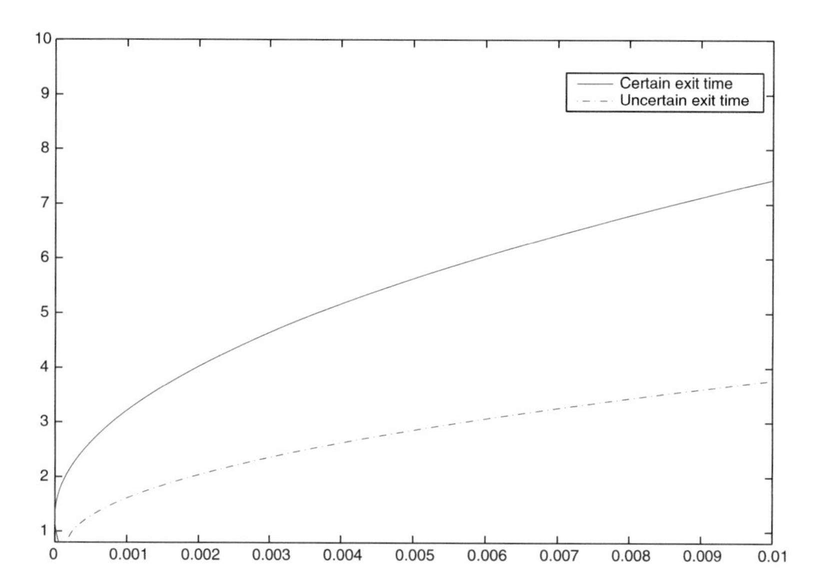

Fig. 12.2 Efficient frontiers with and without uncertain investment horizon in Example 2.2

where

$$K_t = \begin{bmatrix} 0.4004 \\ 0.6496 \\ 2.3133 \end{bmatrix}, \quad t = 1, 2, 3, 4.$$

$$x_1 = \begin{bmatrix} 1.3410 \\ 2.1755 \\ 7.7477 \end{bmatrix}, x_2 = \begin{bmatrix} 1.3588 \\ 2.2044 \\ 7.8505 \end{bmatrix}, x_3 = \begin{bmatrix} 1.3713 \\ 2.2247 \\ 7.9227 \end{bmatrix}, x_4 = \begin{bmatrix} 1.4013 \\ 2.2733 \\ 8.0960 \end{bmatrix}.$$

We compare the result with that of the certain-exit-time case in Example 2 of Li and Ng (2000),

$$\pi_t^* = x_t - K_t V_t,$$

where

$$K_t = \begin{bmatrix} 0.4004 \\ 0.6496 \\ 2.3133 \end{bmatrix}, \quad t = 1, 2, 3, 4.$$

$$x_1 = \begin{bmatrix} 3.5440 \\ 5.7494 \\ 20.4751 \end{bmatrix}, x_2 = \begin{bmatrix} 3.6858 \\ 5.9794 \\ 21.2941 \end{bmatrix}, x_3 = \begin{bmatrix} 3.8332 \\ 6.2185 \\ 22.1459 \end{bmatrix}, x_4 = \begin{bmatrix} 3.9865 \\ 6.4673 \\ 23.0317 \end{bmatrix}.$$

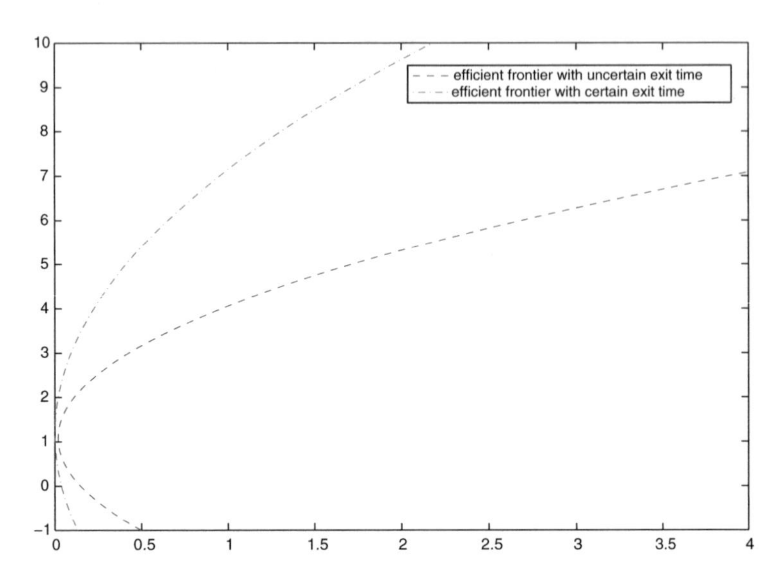

Fig. 12.3 Efficient frontiers with and without uncertain exit time in Example 2.3

It is obvious that the second part K_t of the optimal policy are the same for the two different cases. When the exit time is uncertain, the investor invests few wealth on risky assets than that of the certain-exit-time case. The mean-variance efficient frontiers of the two different cases are showed in Fig. 12.3.

12.6 Conclusion

In this work, multi-period mean-variance portfolio selection problem with a state-dependent uncertain exit time is introduced. This formulation is practically meaningful since most investors do not know exactly when they will exit the financial market at the beginning of their investment. Introducing state-dependent uncertain investment horizon is actually presenting a market-dependent exit strategy. However, this market-dependent exit strategy does not relate to investment policy.

The optimal policy of the original inseparable problem has been derived by solving a separable auxiliary problem, based on the embedding technique of Li and Ng (2000). The mean-variance efficient frontier under the optimal policy has also been derived. This work has also analyzed the special case where the exit time is state-independent, and has compared the result to the certain-exit-time case. It is found that introducing uncertain exit time adds extra risk to the investment.

References

C. Blanchet-Scaillet, N.E. Karoui, L. Martellini, Dynamic asset pricing theory with uncertain time-horizon. J. Econ. Dyn. Control. **29**(10), 1737–1764 (2005)

C. Blanchet-Scalliet, N.E. Karoui, M. Jeanblanc, L. Martellini, Optimal investment decisions when time-horizon is uncertain. J. Math. Econ. **44**, 1100–1113 (2008)

S. Browne, Risk-constrained dynamic active portfolio management. Manag. Sci. **46**(9), 1188–1199 (2000)

U. Çelikyurt, S. özekici, Multiperiod portfolio optimization models in stochastic markets using the mean-variance approach. Eur. J. Oper. Res. **179**, 186–202 (2007)

M.C. Chiu, D. Li, Asset and liability management under a continuous-time mean-variance optimization framework. Insur. Math. Econ. **39**, 330–355 (2006)

O.L.V. Costa, M.V. Araujo, A generalized multi-period mean-variance portfolio optimization with Markov switching parameters. Automatica **44**, 2487–2497 (2008)

X.Y. Cui, D. Li, S.Y. Wang, S.S. Zhu, Better than dynamic mean-variance: time inconsistency and free cash flow stream. Math. Financ. **22**(2), 346–378 (2012)

X.Y. Cui, J.J. Gao, X. Li, D. Li, Optimal multi-period mean-variance policy under no-shorting constraint. Eur. J. Oper. Res. **234**(2), 459–468 (2012)

X.Y. Cui, X. Li, D. Li, Unified framework of mean-field formulations for optimal multi-period mean-variance portfolio selection. IEEE Trans. Autom. Control **59**(7), 1833–1844 (2013)

W.J. Guo, Q.Y. Hu, Multi-period portfolio optimization when exit time is uncertain. J. Manag. Sci. China **8**(2), 14–19 (2005)

N.H. Hakansson, Optimal investment and consumption strategies under risk, an uncertain lifetime, and insurance. Int. Econ. Rev. **10**(3), 443–446 (1969)

R. Josa-fombellida, J.P. Rincón-zapatero, Mean-variance portfolio and contribution selection in stochastic pension funding. Eur. J. Oper. Res. **187**, 120–137 (2008)

I. Karatzas, H. Wang, Utility maximization with discretionary stopping. SIAM J. Control Optim. **39**(1), 306–329 (2000)

D. Li, W.L. Ng, Optimal dynamic portfolio selection: multiperiod mean-variance formulation. Math. Financ. **4**(2), 387–406 (2000)

Z.F. Li, J. Yao, D. Li, Behavior patterns of investment strategies under Roy's safety-first principle. Q. Rev. Econ. Finance **50**, 167–179 (2010)

J.F. Liang, S.Z. Zhang, D. Li, Optioned portfolio selection: models and analysis. Math. Financ. **18**, 569–593 (2008)

H. Markowitz, *Portfolio Selection: Efficient Diversification of Investment* (Wiley, New York, 1959)

L. Martellini, B. Urosevic, Static mean-variance analysis with uncertain time horizon. Manag. Sci. **52**(6), 955–964 (2006)

R.C. Merton, Optimal consumption and portfolio rules in a continuous-time model. J. Econ. Theory **3**(4), 373–413 (1971)

R.C. Merton, An analytic derivation of the efficient portfolio. J. Financ. Quant. Anal. **7**(4), 1852–1872 (1972)

M.E. Yaari, Uncertain lifetime, life insurance, and the theory of the consumer. Rev. Econ. Stud. **32**(2), 137–150 (1965)

L. Yi, Z.F. Li, D. Li, Multi-period portfolio selection for asset-liability management with uncertain investment horizon. J. Ind. Manag. Optim. **4**(3), 535–552 (2008)

L. Yi, X.P. Wu, X. Li, X.Y. Cui, A mean-field formulation for optimal multi-period mean-variance portfolio selection with an uncertain exit time. Oper. Res. Lett. **42**(8), 489–494 (2014)

X.Y. Zhou, D. Li, Continuous-time mean-variance portfolio selection: a stochastic LQ framework. Appl. Math. Optim. **42**, 19–33 (2000)

S.S. Zhu, X.T. Cui, X.L. Sun, D. Li, Factor-risk-constrained mean-variance portfolio selection: formulation and global optimization solution approach. J. Risk **14**(2), 51–89 (2011)

Part IV
Operations Analysis

Chapter 13
A New Model and Method for Order Selection Problems in Flow-Shop Production

Jun Wang, Xiaoxia Zhuang, and Baiyi Wu

Abstract As the economic growth of China gradually slows down in recent years, the flow-shop production enterprises pay more and more attention to the production capacity planning problem. The order selection problem plays a central role in the production capacity planning of flow-shop production enterprises. Traditional order selection models separate the processes of production scheduling and order selection. The performance of the order selection depends entirely on production scheduling. In this paper we study the relationship between the processes of order selection and production scheduling, and propose a new nonlinear 0–1 programming model aiming at profit maximization. Our new model considers simultaneously order selection and production scheduling and we will demonstrate that our new model generates a production schedule that is much better than that from traditional models. We solved the new model using Lingo 11.0 and numerical results show that the optimal solution can be obtained within an hour on a personal computer when the order size is less than 16.

Keywords Flow-shop production • Order selection • Production scheduling • Nonlinear 0–1 programming

13.1 Introduction

As the economic growth of China gradually slows down in recent years, the economy of China has entered a "new normal" era, where the economy has shifted gear from the previous high speed to a medium-to-high speed growth and the economic structure is constantly improved and upgraded. Along with the new

J. Wang (✉) • X. Zhuang
Department of Management Science and Engineering, Business School, Qingdao University, Shandong, People's Republic of China
e-mail: jwang@qdu.edu.cn; zhuang826@qq.com

B. Wu
School of Finance, Guangdong University of Foreign Studies, Guangzhou, People's Republic of China

T.-M. Choi et al. (eds.), *Optimization and Control for Systems in the Big-Data Era*, International Series in Operations Research & Management Science 252,
DOI 10.1007/978-3-319-53518-0_13

economic condition, production enterprises, especially the flow-shop production enterprises, pay more and more attention to the process of production capacity planning, to which the order selection problem is most crucial.

Many flow-shop production enterprises, such as those from the chemical industry, the metallurgical industry, and the feed industry, have a very high transition cost from one order to another due to their special production specification. This means that once an order is put on production, it will be finished before another order is put on production. Also, under the capacity planning framework, a limited capacity is assumed. Delaying an order fulfillment entails additional costs or penalties, while finishing an order too early can lead to an excess inventory cost. Because of these complications, the processes of order selection and production scheduling are of great importance.

The order selection problem has been actively studied for the last 20 years. In the usually conventional approach, before the order selection, all the orders are sorted according to a certain ranking scheme, such as first-come-first-serve (FCFS), earliest-due-date (EDD), shortest-processing-time (SPT), or just sorted by profits from different orders. Then order selection is conducted on the sorted order list. In the literature, Wester et al. (1992) proposed an order acceptance strategy under a single-machine environment with setup times. Slotnick and Morton (2007) considered the situation with a limited capacity and delay cost and studied how the orders such that can be selected the overall profit is maximized. Song and Ma (2007) studied the inter relationship of the flow line balancing, production scheduling, and the makespan of flow line. They proposed a co-optimizing genetic algorithm to optimize the makespan of the mixed-model assembly flow line. Liao et al. (2011) studied the order selection strategy by the analytic hierarchy process (AHP) and fuzzy comprehensive evaluation approaches. They chose indicators from the aspects of delivery time, order size, the importance of customer, etc., and then used AHP to determine the weights. With a comment rating scale, they defined the fuzzy evaluation matrix. Finally the scores of each grading level are summed up to form the basis for order priority. Li and Wang (2014) incorporated the factors of out-sourcing and reputation cost into their order selection model and studied the order acceptance problem with an aim of maximizing overall profit of the production enterprise.

Traditional order selection models separate the processes of production scheduling and order selection. The performance of the order selection depends entirely on production scheduling. However, in flow-shop production, production scheduling has a direct impact on the decision of order selection. On the other hand, the outcome of order selection can also change the final production schedule. Xu et al. (2014) studied the order selection problem in a flow-shop environment and proposed a mixed-integer programming model to simultaneously optimize the production scheduling and order selection. Their model would allocate those orders that are not selected into a delayed group, and the delay cost and the machine idle cost are incorporated into the objective of the model.

In this paper we study the relationship between the processes of order selection and production scheduling and propose a new nonlinear 0–1 programming model aiming at profit maximization. Our new model is a generalization and improvement of Slotnick-and-Morton's model (Slotnick and Morton 2007) as we consider simultaneously the optimal order selection and production scheduling.

13.2 Slotnick's Order Selection Model

The model proposed by Slotnick and Morton (2007) has the following assumptions:

- The production capacity is limited.
- Only a part of the orders to fulfill is selected in order to maximize the overall profit.
- The set of orders to be selected from is given at time zero, with the complete specification such as processing time, delivery due date, delay cost, and profit for each order.
- Linear delay cost.

For the i-th order, let q_i be its profit if it is finished on time; p_i be the processing time; d_i be the delivery due date; w_i be the delay cost for the i-th order. Then Slotnick-and-Morton's order selection model is formulated as follows:

$$\max \sum_{i=1}^{n} x_i \left[q_i - w_i(c_i - d_i)^+\right]$$

$$\text{s.t.} \quad c_i = \sum_{j=1}^{i} x_j p_j, \; x_i \in \{0, 1\}, \; i = 1, \ldots, n,$$

where the decision variable c_i is the actual delivery date of the i-th order and the decision variable x_i is binary: $x_i = 1$ or 0 means i-th order is accepted or not. This model aims at maximizing the overall profit and if a selected order is finished earlier than the delivery due date, there would be no additional profit to the production enterprise, i.e., it is of no use finishing an order too early.

One key issue in Slotnick-and-Morton's model is how to rank the n orders before the order selection. Given different ranking schemes, the optimal solutions for the order selection problem would be different. This fact is demonstrated in the following example.

Example 1 In this example, there are four orders to be selected from. Their specifications are in Table 13.1. If the orders are ranked using the FCFS scheme, the ranking would be $(1, 2, 3, 4)$ and the optimal solution of Slotnick-and-Morton's model is to select orders $1, 2, 3$ and forego order 4, resulting a profit of 60. If the orders are ranked using the SPT scheme, the ranking would be $(2, 3, 1, 4)$ and the optimal solution of Slotnick-and-Morton's model is to select orders $2, 3, 1$ and forego order 4, resulting a profit of 64.

Table 13.1 The data of Example 1

Order i	Processing time p_i	Due date d_i	Delay cost coefficient w_i	Profit q_i
1	8	22	5	19
2	5	20	3	28
3	7	19	4	17
4	9	15	6	32

Slotnick-and-Morton's order selection model separates the processes of production scheduling and order selection. Under different order ranking schemes, the optimal solutions could be different. This has been shown in Example 1. Thus the true global optimal solution cannot be guaranteed by solving this model. This motivates us to find a global optimization model that integrates the processes of production scheduling and order selection together.

13.3 New Order Selection Model

In flow-shop production, production scheduling has a direct impact on the decision of order selection. On the other hand, the outcome of order selection can also change the final production schedule. In this section, to generalize and improve Slotnick-and-Morton's model, we study the relationship between the processes of order selection and production scheduling and propose a new order selection model that integrates these two processes.

13.3.1 Basic Assumptions

Our new model takes the following additional assumptions except for those of Slotnick-and-Morton's model:

- The order selection is conducted for a single planning period.
- The n orders to be selected have been ranked from 1 to n so that the i-th order must be fulfilled before the $(i + 1)$-th order.
- No inventory cost.

13.3.2 Model Description

Index the orders from $j = 1, \ldots, n$, where n is the total number of orders to be selected from. For order j, let q_j be its profit if it is finished on time; p_j be the processing time; d_j be the delivery due date; w_j be the delay cost coefficient for the

i-th order. Index the position in production schedule from $i = 1 \cdots, n$. Then our new order selection model is formulated as follows:

$$\max \quad \sum_{j=1}^{n}\sum_{i=1}^{n} x_{ij}\left[q_j - w_j(c_j - d_j)^+\right]$$

$$\text{s.t.} \quad \sum_{i=1}^{n} x_{ij} \le 1,\ j = 1, \ldots, n, \tag{3.1}$$

$$\sum_{j=1}^{n} x_{ij} \le 1,\ i = 1, \ldots, n, \tag{3.2}$$

$$s_0 = 0,\ s_i = s_{i-1} + \sum_{j=1}^{n} p_j x_{ij},\ i = 1, \ldots, n, \tag{3.3}$$

$$c_j = \sum_{i=1}^{n} s_i x_{ij},\ j = 1, \ldots, n, \tag{3.4}$$

$$x_{ij} \in \{0, 1\},\ i, j = 1, \ldots, n, \tag{3.5}$$

where the decision variables are explained as follows:

- When x_{ij} is set to 1, then order j is selected and it is the i-th order to be fulfilled. When x_{ij} is set to 0, then order j is given up.
- s_i is the finished time for the i-th order to be fulfilled.
- c_j is the actual delivery date of order j.

Constraint (3.1) ensures that each order is fulfilled at most once. When it is not binding, some order is not selected and thus given up. Since the production enterprise has a limited capacity, in order to maximize the profit, some of the orders will be foregone. Constraint (3.2) ensures that the i-th order to be fulfilled is unique, because we do not allow for fulfilling two orders at the same time. Constraint (3.3) ensures that s_i is the finished time for the i-th order to be fulfilled. Constraint (3.4) ensures that c_j is the actual delivery date of order j.

Our new model still aims at maximizing the overall profit. But we have integrated the processes of order selection and production scheduling, and thus no prior ranking scheme is needed. In the following example, we continue to use the problem in Example 1 to demonstrate the effectiveness of our new model.

Example 2 To tackle the problem in Example 1, we use Lingo 11.0 to solve our new model. The resulting optimal solution is displayed in Table 13.2, This table shows that the optimal solution for our new model is to fulfill orders 4, 2, 1 sequentially. And the overall profit for this optimal solution is 79. We can show that this is indeed the globally optimal solution by enumerating all the combinations.

Table 13.2 The optimal solution of Example 2

	x_{ij}	j			
		1	2	3	4
i	1	0	0	0	1
	2	0	1	0	0
	3	1	0	0	0
	4	0	0	0	0

Table 13.3 The computational results of the new model

Number of orders	Worse case	Best case	Average computational time
4	14	1	8.6
8	385	9	178.2
12	2154	47	900.6
16	3593	232	1842.8

13.3.3 Model Complexity

Because our new model is a nonlinear mixed-integer programming problem, its computational complexity grows exponentially as the number of orders increases. To test the computational effectiveness of our model, we randomly generate four group instances with $n = 4, 8, 12, 16$. Each group contains 10 instances. Each instance is solved by Lingo 11.0 on a 2.20 Ghz thread with 2G memory. The computational time (in seconds) is summarized in Table 13.3. The above table shows that when the number of orders is less than 16, the order selection problem can be solved to optimality within an hour by our new model.

13.4 Conclusion

Traditional order selection models separate the processes of production scheduling and order selection. The performance of the order selection depends entirely on production scheduling. In this paper we have proposed a new nonlinear 0–1 programming model that integrates the processes of order selection and production scheduling. Our new model can find better solutions than traditional models in terms of profit maximization. We have solved the new model using Lingo 11.0 and numerical results show that the optimal solution can be obtained within an hour on a personal computer when the order size is less than 16. Because the problem is NP-hard in general, when the number of orders increases, the computational complexity grows exponentially. In future research, we will apply the exact methods (Wang et al. 2007; Duan Li et al. 2007) or develop heuristics for our new model targeting larger problem sizes, which would be critical to effectively solve the problem in the big data era when the amounts of orders are massive. On the other hand, the big data technology could provide a possible way to deal with the estimation errors of parameter in the new model.

References

D. Li, J. Wang, X.L. Sun, Computing exact solution to nonlinear integer programming: convergent Lagrangian and objective level cut method, journal of global optimization. J. Glob. Optim. **39**(1), 127–154 (2007)

J. Li, J. Wang, Research of order acceptance with process production based on outsourcing. Logist. Manag. **37**(2), 68–70 (2014)

Z. Liao, S. Hu, B. Zhang, J. Ou, Strategy for priority of order based on AHP and fuzzy comprehensive evaluation. Sci. Technol. Managem. Rese. **31**(24), 191–194 (2011)

S.A. Slotnick, T.E. Morton, Order acceptance with weighted tardiness. Comput. Oper. Res. **34**(10), 3029–3042 (2007)

H. Song, S. Ma, Co-optimization for mixed-model assembly flow line to minimize makespan. Syst. Eng. Theory Pract. **27**(2), 153–160 (2007)

J. Wang, D. Li, X.L. Sun, A revised Taha's algorithm for polynomial 0–1 programming. Optimization **56**(5), 699–713 (2007)

F. Wester, J. Wijngaard, W.R.M. Zijm, Order acceptance strategies in a production-to-order environment with setup times and due-dates. Int. J. Prod. Res. **30**(6), 1313–1326 (1992)

S. Xu, T. Li, B. Wang, L. Bai, Repair-based constraint satisfaction algorithm for an order release problem in flow shop environment. Manuf. Autom. **36**(3), 1–6 (2014)

Chapter 14
Quick Response Fashion Supply Chains in the Big Data Era

Tsan-Ming Choi

Abstract The quick response strategy has been widely adopted in the fashion industry. With a shortened lead time, quick response allows fashion supply chain members to conduct forecast information updating which helps to reduce demand uncertainty. In the big data era, forecast information updating is even more effective as more data points can be collected easily to improve forecasting. In this paper, after reviewing the related literature, we explore how the quick response strategy with n observations can improve the whole fashion supply chain's performance. We study how the number of observations affects the expected values of quick response for the fashion supply chain, the fashion retailer, and the fashion manufacturer. Then, we analytically how the robust win–win coordination can be achieved in the quick response fashion supply chain using the commonly seen wholesale pricing markdown contract. Insights are generated.

Keywords Bayesian information updating • Quick response • Supply chain coordination • Supply chain optimization • Use of information

14.1 Introduction and Related Literature

Quick response is a well-established strategy in fashion supply chain management. The first proposal on the implementation of quick response started in the USA in the 1980s by the fashion manufacturers (Fisher and Raman 1996; Iyer and Bergen 1997; Choi et al. 2003; Choi and Chow 2008). After decades of evolution, quick response is now a critical measure to achieve business models such as fast fashion (Cachon and Swinney 2011).

One basic element of all quick response programs is the reduction of lead time (Choi et al. 2004, 2006). As a matter of fact, by reducing lead time, more market signals (Shaltayev and Sox 2010) can be observed and incorporated into

T.-M. Choi (✉)
Institute of Textiles and Clothing, The Hong Kong Polytechnic University, Hung Hom, Kowloon, Hong Kong
e-mail: jason.choi@polyu.edu.hk

T.-M. Choi et al. (eds.), *Optimization and Control for Systems in the Big-Data Era*, International Series in Operations Research & Management Science 252,
DOI 10.1007/978-3-319-53518-0_14

the demand forecasting process. This leads to a more accurate forecast[1]. Based on a more accurate forecast, inventory planning becomes more precise and the respective supply chain system is more efficient. This is an especially important measure for industries which face highly volatile demand, such as fashion apparel.

In the big data era, a massive amount of data is available (Chan et al. 2016). Data collection is also made easier (Choi et al. 2016a). As a result, in quick response, fashion companies can easily make use of the large amount of related data to improve forecast. This would make quick response an even more significant measure to improve the supply chain system's performance. However, two challenges exist: (1) Even though quick response can significantly enhance the supply chain's profitability, the supply chain system itself will not be optimal by itself owing to the double marginalization problem (Donohue 2000; Chiu et al. 2011; Choi 2016c); (2) it is a well-known fact that quick response implementation need not be win–win to the supplier and the buyer (Iyer and Bergen 1997).

In light of these two challenges and the convenience of having a lot of market observations (i.e., n observations) under quick response[2], this paper is developed. The focal points of this paper include: (1) Examining how the number of observations affects the expected values of quick response for the fashion supply chain, the fashion retailer, and the fashion manufacturer; (2) uncovering analytically how the win–win coordination outcome can be achieved after implementing quick response by the wholesale pricing markdown contract. Both academic and managerial insights are developed.

For the related literature, as quick response in supply chain management is a big topic, we examine some recently published related papers as follows and refer readers to the review by Choi and Sethi (2010) for the other older studies. First, in the operations-marketing interface, Cachon and Swinney (2011) explore how the quick response strategy supports the fast fashion business model in the presence of forward looking consumers. Yang et al. (2011) explore the supply chain with a single retailer and two suppliers with different lead times. Forecast updating is feasible before the retailer orders from the short lead time (i.e., quick response) supplier. The authors reveal how the supply chain with forecast updating can be coordinated. Then, in a competitive environment, Lin and Parlakturk (2012) study via a game-theoretic model how competition affects the performance of quick response. Choi (2013) studies the impacts of imposing carbon emission tax on fashion quick response systems and argues that carbon emission tax is an effective means to entice retailers to source locally. Liu and Nagurney (2013) explore a global sourcing problem. The authors consider both demand and cost uncertainties and discuss how quick response performs. Other recent studies related to quick response include: a study on the quick response system in the presence of loss-averse strategic

[1]This concept is in line with the "advance booking scheme" (see, e.g., McCardle et al. 2004).

[2]Notice that different from Choi (2007), this paper considers the situation when n observations can be collected all within the same period of time whereas Choi (2007) considers multiple observations at different time durations.

consumers (Lee et al. 2015), an exploration on the coordination challenge in supply chains with multiple shipments during the season (Chen et al. 2016b), an analysis of the risk averse behaviors on quick response systems (Choi 2016a), an investigation on how social media information affects the performance of quick response system in the presence of boundedly rational retailers (Choi 2016b), and a case study on how quick response manufacturing complements lean supply chain operations (Fernando J. Gómez and Filho 2016).

Notice that similar to all the above papers, this paper focuses on quick response. Similar to most reviewed papers, such as Iyer and Bergen (1997), Choi et al. (2003, 2004), Kim (2003), Choi (2007, 2016a, b, c), this paper also employs the Bayesian normal conjugate pair model in the analysis. However, different from all of them, this paper considers the case when n observations can be collected for the forecast updating and investigates how the large number of observations affects the performance of quick response. In this regard, this paper is closest to the paper by Chan et al. (2015) which also considers the problem of having multiple observations. However, in Chan et al. (2015), observations are expensive and hence the authors discuss the optimal number of sampling whereas in this paper, the number of observations "n" is taken as a parameter and we examine the performance of quick response when n increases and when it goes to infinity (and hence follows the trend in the "big data" era).

This paper is organized as follows. First, we present the basic inventory model and the Bayesian information updating model in Sect. 14.2. Then, we explore the performance of the centralized supply chain system under quick response in Sect. 14.3. After that, we investigate the impacts brought by quick response in the decentralized supply chain in Sect. 14.4. We report how the win–win coordination scenario can be achieved by using the wholesale pricing markdown contract in Sect. 14.5. Finally, we conclude the paper with a discussion of future research in Sect. 14.6.

14.2 Basic Model

14.2.1 Inventory Model

We employ the newsvendor problem (Chen et al. 2016a) to model the inventory model for the fashion product. To be specific, before the selling season starts, we consider the case where a fashion retailer needs to order a certain quantity q of a seasonal fashion product (e.g., a colorful tee) from its supplier, which is a fashion manufacturer, with a unit product ordering cost c. The fashion supplier operates as a follower and it will start production only after receiving the order from the fashion retailer. The fashion retailer sells the product in the market with a unit retail selling price r, and the supplier produces the product at a unit cost p. For the unsold product, for the sake of simplicity, we assume that the holding cost and the salvage value together would lead to a net salvage value v. To avoid trivial cases, we have $r > c > p > v$. The seasonal fashion product's demand is uncertain and follows a distribution which is described in the next sub-section.

14.2.2 Bayesian Information Updating

In this sub-section, we present the demand distribution. As we consider the quick response strategy in this paper, we would consider two time points. To be specific, suppose that the fashion retailer used to order from the supplier with a long lead time at Time 0. In this ordering time point, as the fashion retailer possesses relatively rough forecast regarding the real seasonal demand for the product, the respective demand uncertainty is high. Now, if the supplier allows the fashion retailer to order at a time point with a shorter lead time, called Time 1, the fashion retailer can make use of market observations to improve its forecast. In this paper, we consider the case when the fashion retailer can collect and use a sufficient amount of market information so that the demand uncertainty is much reduced.

Following the standard definition of quick response (see Choi and Chow 2008; Choi 2016a, b, c), we refer the ordering at Time 1 as the one under quick response (QR) whereas the ordering at Time 0 is under slow response (SR).

To model the above relationship, we employ the Bayesian theory (Iyer and Bergen 1997; Kim 2003) with the normal conjugate pair. First, we denote the predicted demand of the product at Time 0 by x_0. Following the basic demand uncertainty structure as shown in the literature (see Iyer and Bergen 1997; Choi and Chow 2008; Choi 2016a, b, c) we model the distribution of x_0 as a normal distribution with mean θ and variance δ in the following: $x_0 \; N(\theta, \delta)$. Notice that δ represents the inherent demand volatility of the seasonal fashion product and it is not reducible by market observation. For θ, the mean of demand at Time 0, we model it as a random variable which also follows a normal distribution, with mean μ_0 and variance d_0:

$$\theta \sim N(\mu_0, d_0).$$

At Time 0, with the above formulation, it is known that the unconditional distribution of x_0 is a normal distribution with mean μ_0 and variance $(d_0 + \delta)$,

$$x_0 \sim N(\mu_0, d_0 + \delta).$$

The above demand model captures the demand distribution of the seasonal fashion product if the fashion retailer orders at Time 0, i.e., under SR.

We now explore QR and denote the predicted demand at Time 1 by x_1. In the big data era, collecting data is easy and we assume that the fashion retailer can quickly and conveniently collect a massive amount of related market observations of products following the same normal process (P.S.: we represent the number of observations as n) between Time 0 and Time 1. Using the Bayesian conjugate pair theory, this leads to the following unconditional distribution for x_1 as follows (e.g., see Chan et al. 2015):

$$x_1 \sim N\left(\mu_1(n), d_1(n) + \delta\right),$$

where

$$\mu_1(n) = \left(\frac{\delta\mu_0 + nd_0o}{nd_0 + \delta}\right),$$

$$d_1(n) = \frac{d_0\delta}{nd_0 + \delta},$$

and o is the mean of the n observations.

Notice that the Bayesian conjugate pair demand model proposed above is not new and it is a standard result in the literature. This paper simply follows and uses it for further analysis. From the above model, we have Lemma 2.1.

Lemma 2.1. *(a) Comparing the levels of demand uncertainty (measured by demand variance) under QR and SR, with n-observation based information updating, the demand uncertainty under QR is smaller and the reduction is increasing in n.*

(b) When $n \to \infty$, demand uncertainty under QR (measured by demand variance) becomes δ.

Proof of Lemma 2.1. All proofs are placed in the appendix.

Lemma 2.1 shows that the market observation can improve forecasting via the Bayesian information updating process. In addition, when we take more observations and incorporate them into the forecast revision, the significance of QR is higher. At the extreme, when the number of observations goes to infinity, all the reducible demand uncertainty vanishes and the remaining demand uncertainty is simply equal to the inherent demand uncertainty δ which cannot be reduced further.

For a notational purpose: We employ $\varphi(\cdot)$, $\Phi(\cdot)$, and $\Psi(x) = \int_x^\infty (y-x)\,\varphi(y)dy$ to represent the standardized normal density function, standardized normal cumulative distribution function, and linear loss function with the standard normal distribution, respectively. The inverse function of $\Phi(\cdot)$ is represented by $\Phi^{-1}(\cdot)$. Table 14.1 shows a summary of the major notation employed in this paper.

14.3 Centralized Supply Chain

From Sect. 14.2, we have already reviewed and presented the model with the consideration of Bayesian information updating. In the following, we conduct analysis on the performance of QR focusing on the centralized supply chain system. The analysis result will be used as a benchmark for the further exploration in Sect. 14.4.

Table 14.1 Notation

Notation	Meaning
QR	Quick response
SR	Slow response
R	Fashion retailer
S	Fashion supplier
SC	Fashion supply chain
EP	Expected profit
q	Quantity
EVQR	Expected value of QR
r	Unit retail price
c	Unit wholesale price
p	Unit production cost
v	Unit net salvage value
n	Number of observations
WPM	Wholesale pricing markdown
Ω	WPM contract
$\widehat{c}$	Unit wholesale price in Ω
$\widehat{m}$	Unit markdown sponsor in Ω

First of all, adopting the same approach as in Iyer and Bergen (1997), we can easily derive the fashion supply chain (SC)'s optimal ordering quantity if ordering is placed at Time 0 (i.e., under SR) as follows:

$$q_{0,SC*} = \mu_0 + \sqrt{d_0 + \delta}\Phi^{-1}\left[(r-p)/(r-v)\right]. \quad (14.1)$$

Notice that (14.1) follows the standard "critical fractile" expression as in the standard newsvendor problem.

Denote $s_{SC} = (r-p)/(r-v)$. With (14.1), the corresponding optimal fashion supply chain's expected profit when the ordering is placed at Time 0 (i.e., under SR) can be found to be the following:

$$EP_{0,SC*} = (r-p)\mu_0 - \sqrt{d_0 + \delta}T_{SC}(s_{SC}), \quad (14.2)$$

where

$$T_{SC}(s_{SC}) = (p-v)\Phi^{-1}(s_{SC}) + (r-v)\Psi\left[\Phi^{-1}(s_{SC})\right]. \quad (14.3)$$

If the ordering is placed at Time 1 with the updated mean of demand $\mu_1(n)$, the optimal fashion supply chain quantity and the corresponding optimal fashion supply chain expected profit are shown below:

$$q_{1,SC*}\Big|\mu_1(n) = \mu_1(n) + \sqrt{d_1(n) + \delta}\Phi^{-1}\left[(r-p)/(r-v)\right]. \quad (14.4)$$

$$EP_{1,SC*}\Big|\mu_1(n) = (r-p)\mu_1(n) - \sqrt{d_1(n) + \delta}T_{SC}(s_{SC}), \quad (14.5)$$

Un-conditioning (14.4) and (14.5) with respect to $\mu_1(n)$ yields the following expected optimal fashion supply chain quantity and optimal fashion supply chain expected profit at Time 0:

$$q_{1,SC*} = \mu_0 + \sqrt{d_1(n) + \delta}\Phi^{-1}\left[(r-p)/(r-v)\right], \tag{14.6}$$

$$EP_{1,SC*} = (r-p)\,\mu_0 - \sqrt{d_1(n) + \delta}T_{SC}\left(s_{SC}\right). \tag{14.7}$$

Define the expected value of quick response for the fashion supply chain system as follows: $EVQR_{SC}(n) = EP_{1,SC_*} - EP_{0,SC_*}$. We have the following expression for $EVQR_{SC}(n)$ and Lemma 3.1:

$$EVQR_{SC}(n) = \left(\sqrt{d_0 + \delta} - \sqrt{d_1(n) + \delta}\right)T_{SC}\left(s_{SC}\right). \tag{14.8}$$

Lemma 3.1. *(a) From the centralized supply chain perspective, adopting QR is always beneficial because $EVQR_{SC}(n) > 0$.*

(b) $EVQR_{SC}(n)$ is increasing in n.

(c) When $n \to \infty$, $EVQR_{SC}(n \to \infty) = \left(\sqrt{d_0 + \delta} - \sqrt{\delta}\right)T_{SC}\left(s_{SC}\right)$.

Lemma 3.1 shows that in the centralized fashion supply chain, QR is always a beneficial measure and it gives a positive expected gain in profit to the supply chain. In addition, when the number of observations increases, the expected gain by using QR for the fashion supply chain is even higher. When the number of observations goes to infinity, the maximum amount of expected gain for the supply chain by using QR is shown in Lemma 3.1c.

14.4 Decentralized Supply Chain

In Sect. 14.3, we have examined the centralized case and revealed that QR is always beneficial to the supply chain system. In this section, we explore the expected profits for the fashion retailer and the fashion manufacturer under a decentralized setting.

First, it is straightforward to find the fashion retailer (R)'s optimal ordering quantity if the order is placed at Time 0 (i.e., under SR):

$$q_{0,R*} = \mu_0 + \sqrt{d_0 + \delta}\Phi^{-1}\left[(r-c)/(r-v)\right]. \tag{14.9}$$

Denote $s_R = (r-c)/(r-v)$. With (14.9), if the order is placed at Time 0, the corresponding optimal fashion retailer's expected profit can be derived to be the following:

$$EP_{0,R*} = (r-c)\,\mu_0 - \sqrt{d_0 + \delta}T_R\left(s_R\right), \tag{14.10}$$

where

$$T_R(s_R) = (c - v)\,\Phi^{-1}(s_R) + (r - v)\,\Psi\left[\Phi^{-1}(s_R)\right]. \tag{14.11}$$

The fashion supplier's expected profit is listed as follows if the fashion retailer orders at Time 0:

$$EP_{0,S*} = (c - p)\left(\mu_0 + \sqrt{d_0 + \delta}\Phi^{-1}(s_R)\right). \tag{14.12}$$

If the ordering is placed at Time 1 with the updated mean of demand $\mu_1(n)$, the optimal fashion retailer's ordering quantity, and the corresponding optimal fashion retailer's expected profit and the fashion supplier's expected profit are listed below:

$$q_{1,R*}\Big|\mu_1(n) = \mu_1(n) + \sqrt{d_1(n) + \delta}\Phi^{-1}(s_R), \tag{14.13}$$

$$EP_{1,R*}\Big|\mu_1(n) = (r - c)\,\mu_1(n) - \sqrt{d_1(n) + \delta}T_R(s_R), \tag{14.14}$$

$$EP_{1,S*}\Big|\mu_1(n) = (c - p)\left(\mu_1(n) + \sqrt{d_1(n) + \delta}\Phi^{-1}(s_R)\right). \tag{14.15}$$

Un-conditioning (14.13)–(14.15) with respect to $\mu_1(n)$ yields the following:

$$q_{1,R*} = \mu_0 + \sqrt{d_1(n) + \delta}\Phi^{-1}(s_R), \tag{14.16}$$

$$EP_{1,R*} = (r - c)\,\mu_0 - \sqrt{d_1(n) + \delta}T_R(s_R), \tag{14.17}$$

$$EP_{1,S*} = (c - p)\left(\mu_0 + \sqrt{d_1(n) + \delta}\Phi^{-1}(s_R)\right). \tag{14.18}$$

Define the expected values of quick response for the fashion retailer and the fashion supplier as follows:

$$EVQR_R(n) = EP_{1,R*} - EP_{0,R*},$$

$$EVQR_S(n) = EP_{1,S*} - EP_{0,S*}.$$

After simplification, we have:

$$EVQR_R(n) = \left(\sqrt{d_0 + \delta} - \sqrt{d_1(n) + \delta}\right)T_R(s_R), \tag{14.19}$$

$$EVQR_S(n) = -(c - p)\left(\sqrt{d_0 + \delta} - \sqrt{d_1(n) + \delta}\right)\Phi^{-1}(s_R). \tag{14.20}$$

The properties of $EVQR_R(n)$ and $EVQR_S(n)$ are summarized in Lemmas 4.1 and 4.2.

Lemma 4.1. *(a) In the decentralized supply chain, adopting QR is always beneficial for the fashion retailer because* $EVQR_R(n) > 0$.

(b) $EVQR_R(n)$ *is increasing in* n.

(c) When $n \to \infty$, $EVQR_R(n \to \infty) = \left(\sqrt{d_0 + \delta} - \sqrt{\delta}\right) T_R\left(s_R\right)$.

Lemma 4.2. *(a) In the decentralized supply chain, adopting QR is NOT always beneficial for the fashion supplier because* $EVQR_S(n) > 0$ *if and only if* $s_R < 0.5$, *and* $EVQR_S(n) \leq 0$ *if and only if* $s_R \geq 0.5$.

(b) $EVQR_S(n)$ *is increasing in* n *if and only if* $s_R < 0.5$, *and* $EVQR_S(n)$ *is decreasing in* n *if and only if* $s_R \geq 0.5$.

(c) When $n \to \infty$, $EVQR_S(n \to \infty) = -\left(c - m\right)\left(\sqrt{d_0 + \delta} - \sqrt{\delta}\right) \Phi^{-1}\left(s_R\right)$.

Lemmas 4.1 and 4.2 show that in the decentralized supply chain, QR is always a beneficial measure to the fashion retailer but it may not be good for the fashion supplier. For the fashion retailer, the case is similar to the centralized supply chain case: when the number of observations increases, the expected value of QR for the fashion retailer becomes higher and the maximum value is equal to $\left(\sqrt{d_0 + \delta} - \sqrt{\delta}\right) T_R\left(s_R\right)$. For the fashion supplier, whether QR is beneficial or not depends on the inventory service level s_R (which is the same as what the literature shows (see, e.g., Iyer and Bergen (1997); Choi (2016c)). When the inventory service level is very low (i.e., $s_R < 0.5$), QR is beneficial to the fashion supplier; otherwise, QR hurts the fashion supplier's expected profit. Depending on the value of inventory service level, the effect brought by the number of observations n varies. To be specific, as shown by Lemma 4.2b, QR is a more beneficial measure when n increases if the inventory service level is very low (i.e., $s_R < 0.5$); otherwise, QR is less beneficial when n increases. Moreover, when the number of observations goes to infinity, Lemma 4.2c shows that the expected value of QR for the fashion supplier is finite and it becomes $-\left(c - m\right)\left(\sqrt{d_0 + \delta} - \sqrt{\delta}\right) \Phi^{-1}\left(s_R\right)$.

Furthermore, by checking the quantity decisions in (14.1), (14.6), (14.9), and (14.16), we have Lemma 4.3.

Lemma 4.3. *When* $c > p$, *we have:* $q_{0,R*} < q_{0,SC*}$ *and* $q_{1,R*}|\mu_1(n) < q_{1,SC*}|\mu_1(n)$.

Lemma 4.3 indicates that for every time point, the fashion retailer's optimal ordering quantity (under the decentralized setting) is different from the fashion supply chain's optimal ordering quantity (under the centralized setting). This is an intuitive result because the fashion retailer faces a different and lower profit margin compared to the fashion supply chain system. As a result, it is natural for the existence of different optimal quantities in which the fashion retailer will order a smaller amount compared to the supply chain system counterpart. This follows the classic double marginalization theory (Spengler 1950) in the literature.

14.5 Win–Win Coordination

In Sects. 14.3 and 14.4, we have found that QR is a good strategy to improve the fashion supply chain's performance. However, in a decentralized setting, not only does the supply chain fail to be optimal, but the fashion supplier may also suffer a loss after adopting QR. In this section, we propose how a commonly seen wholesale pricing markdown (WPM) contract can be applied to achieve win–win coordination for the fashion supply chain upon its implementation of QR.

Under the WPM contract, we consider the case when the fashion supplier offers a unit wholesale price $\widehat{c}$ and a unit markdown sponsor $\widehat{m}$ to the fashion retailer at Time 1 under QR. We denote this WPM contract by: $\Omega\left(\widehat{c}, \widehat{m}\right)$. To be specific, in the presence of the WPM contract, when the fashion retailer has leftover, the fashion supplier is willing to provide a unit sponsor of $\widehat{m}$ for all the product leftover. This sponsor directly reduces the risk faced by overstocking and can entice the fashion retailer to order more. Notice that the WPM contract is commonly seen in the fashion industry (see Shen et al. 2016).

In the presence of $\Omega\left(\widehat{c}, \widehat{m}\right)$, the unconditional expected profit of the fashion retailer, the fashion retailer's optimal ordering quantity under QR, and the unconditional expected profit of the fashion supplier are given as follows:

$$EP^{\Omega}_{1,R*} = \left(r - \widehat{c}\right)\mu_0 - \sqrt{d_1(n) + \delta} T^{\Omega}_R\left(s^{\Omega}_R\right), \tag{14.21}$$

where

$s^{\Omega}_R = \left(r - \widehat{c}\right) / \left(r - \widehat{m} - v\right), and,$ and $T^{\Omega}_R\left(s^{\Omega}_R\right) = \left(\widehat{c} - \widehat{m} - v\right)\Phi^{-1}\left(s^{\Omega}_R\right) + \left(r - \widehat{m} - v\right)\Psi\left[\Phi^{-1}\left(s^{\Omega}_R\right)\right]$,

$$q^{\Omega}_{1,R*}\Big|\mu_1(n) = \mu_0 + \sqrt{d_1(n) + \delta}\Phi^{-1}\left(s^{\Omega}_R\right),$$

$$\begin{aligned} EP^{\Omega}_{1,S*} = {} & \left(\widehat{c} - p\right)\left[\mu_0 + \sqrt{d_1(n) + \delta}\Phi^{-1}\left(s^{\Omega}_R\right)\right] \\ & - \widehat{m}\sqrt{d_1(n) + \delta}\left[\Phi^{-1}\left(s^{\Omega}_R\right) + \Psi\left\{\Phi^{-1}\left(s^{\Omega}_R\right)\right\}\right]. \end{aligned}$$

Define the following:

$$EVQR^{\Omega}_R(n) = EP^{\Omega}_{1,R*} - EP_{0,R*},$$

$$EVQR^{\Omega}_S(n) = EP^{\Omega}_{1,S*} - EP_{0,S*}.$$

After simplification, we have:

$$EVQR^{\Omega}_R(n) = \left(c - \widehat{c}\right)\mu_0 + \sqrt{d_0 + \delta} T_R\left(s_R\right) - \sqrt{d_1(n) + \delta} T^{\Omega}_R\left(s^{\Omega}_R\right), \tag{14.22}$$

$$EVQR_S^{\Omega}(n) = (\widehat{c}-c)\,\mu_0 + (\widehat{c}-p)\,\sqrt{d_1(n)+\delta}\,\Phi^{-1}\left(s_R^{\Omega}\right) - (c-p)\,\sqrt{d_0+\delta}\,\Phi^{-1}\left(s_R\right) - \widehat{m}\sqrt{d_1(n)+\delta}\left[\Phi^{-1}\left(s_R^{\Omega}\right) + \Psi\left\{\Phi^{-1}\left(s_R^{\Omega}\right)\right\}\right]. \tag{14.23}$$

To achieve win–win coordination in the supply chain after adopting QR, we have to ensure the following three conditions are met:

$$EVQR_R^{\Omega}(n) > 0, \tag{14.24}$$

$$EVQR_S^{\Omega}(n) > 0, \tag{14.25}$$

$$q_{1,R*}^{\Omega}\,|\mu_1(n) = q_{1,SC*}|\,\mu_1(n). \tag{14.26}$$

Observe that the conditions in (14.24) and (14.25) guarantee that the win–win outcome appears because both the fashion retailer and the fashion supplier are benefited under QR. The condition of (14.26) ensures the fashion retailer will order the quantity which is the best for the whole supply chain system.

In order to derive the win–win coordinating WPM contract, we define the following and present Lemma 5.1:

$$m* = (\widehat{c} - p)\,(r - v)\,/\,(r - p)\,, \tag{14.27}$$

$$c_{upper} = \arg_{\widehat{c}}\left\{EVQR_R^{\Omega}\left(n|\widehat{m} = m*\right) = 0\right\}, \tag{14.28}$$

$$c_{lower} = \arg_{\widehat{c}}\left\{EVQR_S^{\Omega}\left(n|\widehat{m} = m*\right) = 0\right\}. \tag{14.29}$$

Lemma 5.1. *(a)* $\widehat{m} = m*$ *if and only if* $q_{1,R*}^{\Omega}\Big|\mu_1(n) = q_{1,SC_*}|\mu_1(n)$.

(b) $EVQR_R^{\Omega}\left(n|\widehat{m} = m*\right)$ *is a decreasing function of* $\widehat{c}$, *and* $EVQR_S^{\Omega}\left(n|\widehat{m} = m*\right)$ *is an increasing function of* $\widehat{c}$.

Lemma 5.1 gives the important result and structural properties for deriving the coordinating WPM contract. First, Lemma 5.1a shows that by setting $\widehat{m} = m*$ under QR, supply chain coordination is achieved in which the fashion retailer will order a quantity the same as the optimal quantity for the whole supply chain system. Thus, by substituting $\widehat{m} = m*$ into $EVQR_R^{\Omega}(n)$ and $EVQR_S^{\Omega}(n)$ helps to reduce the dimension of setting the coordinating WPM contract from two dimensions $(\widehat{c}, \widehat{m})$to one dimension $(\widehat{c})$. Second, Lemma 5.1b shows the monotonic structural properties of $EVQR_R^{\Omega}\left(n|\widehat{m} = m*\right)$ and $EVQR_S^{\Omega}\left(n|\widehat{m} = m*\right)$. From them, we know that $c_{upper}$and c_{lower} give the upper and lower bound for setting the wholesale price $(\widehat{c})$of the win–win coordinating WPM contract (and the corresponding $\widehat{m}$ can be found by using (14.28)).

Based on Lemma 5.1, we present Lemma 5.2 on the setting of contract parameters to achieve win–win coordination for QR implementation by using the WPM contract.

Lemma 5.2. *After the QR implementation, win–win coordination can be achieved by setting* $\widehat{m} = m*$ *and* $c_{lower} < \widehat{c} < c_{upper}$.

From Lemma 5.2, we can see that: (1) there exist an infinite number of the WPM contracts which can achieve win–win coordination (because any $\widehat{c}$ in the range of $c_{lower} < \widehat{c} < c_{upper}$, together with the corresponding $\widehat{m} = m*$ will do). The specific setting depends on the bargaining power of the fashion retailer and the fashion supplier. When $\widehat{c}$ is set closer to c_{upper}, the fashion retailer's expected gain from QR drops whereas the fashion supplier's expected gain from QR increases. When $\widehat{c}$ is set closer to c_{lower}, the opposite happens. Thus, we see that the proposed WPM contract is rather robust which can divide the expected gain from QR in the supply chain system flexibly between the fashion retailer and the fashion supplier.

14.6 Conclusion

QR is a well-established and important industrial practice in the fashion industry. Motivated by the importance of QR and the availability of a huge amount of data, we have examined in this paper the value of QR (with forecast information updating) in fashion supply chains in the big data era. We have proven how the quick response strategy with n observations can help improve the whole fashion supply chain's performance. We have shown how the number of observations affects the expected values of QR for the fashion supply chain, the fashion retailer, and the fashion manufacturer. To be specific, we have demonstrated analytically that under QR, if the number of observations increases, the expected values of QR for the fashion supply chain (under the centralized model) and the fashion retailer (under the decentralized model) will both increase and reach the finite maximum when the number of observations goes to infinity. For the fashion supplier, the situation depends on the inventory service level and a larger number of observations can lead to an increase or a reduction of the expected value of QR (see Table 14.2).

After that, we have discussed how the win–win coordination after implementing QR can be achieved using the WPM contract. The proper setting of the contract

Table 14.2 Impacts of n on EVQRs

	Centralized	Decentralized	
	$EVQR_{SC}(n)$	$EVQR_R(n)$	$EVQR_S(n)$
$n\uparrow$	$\uparrow$	$\uparrow$	$\uparrow$ *iff* $s_R < 0.5$ $\downarrow$ *iff* $s_R \geq 0.5$
$n \to \infty$	$\left(\sqrt{d_0 + \delta} - \sqrt{\delta}\right) T_{SC}(s_{SC})$	$\left(\sqrt{d_0 + \delta} - \sqrt{\delta}\right) T_R(s_R)$	$-(c-m)\left(\sqrt{d_0 + \delta} - \sqrt{\delta}\right) \Phi^{-1}(s_R)$

parameters as well as the analytical bounds has been derived. We have revealed that the win–win coordinating WPM contract is quite robust as it not only can guarantee the achievability of win–win coordination, but it also can divide the expected gain from QR of the supply chain system flexibly between the seller (i.e., the fashion supplier) and the buyer (i.e., the fashion retailer).

For future research, one can extend the model to cover the case when there are multiple products and explore the corresponding coordination challenges. In addition, the consideration of social media data for QR is also interesting and Choi (2016b) provides a good reference for further investigation. Finally, one may conduct a multi-methodological research (Choi et al. 2016b) on QR supply chain systems with real data analyses. This provides further insights into the real world applicability of QR and some probable extensions of the analytical model.

Acknowledgments The author thanks the reviewers for their comments on the earlier draft of this paper. This paper is partially supported by The Hong Kong Polytechnic University's funding (grant number: G-YBGR).

A.1 Appendix: All Proofs

Proof of Lemma 2.1. (a) By directly comparing between the cases of QR and SR, the demand uncertainty under QR (with n-observation based information updating) is always smaller than the demand uncertainty under SR. By differentiation, we can find that $d_1(n)$ is decreasing in n, which implies the demand uncertainty reduction $((d_0+\delta)-(d_1(n)+\delta))$ is increasing in n. (b) When $n\to\infty$, $\lim_{n\to\infty} d_1(n) = \frac{d_0\delta}{nd_0+\delta} = 0$ and hence the demand uncertainty under QR $(d_1(n)+\delta)$ becomes δ. (Q.E.D.)

Proof of Lemma 3.1. (a) First of all, from (14.8), we have $EVQR_{SC}(n)= \left(\sqrt{d_0+\delta}-\sqrt{d_1(n)+\delta}\right) T_{SC}(s_{SC})$. From (14.3), we have:

$$T_{SC}(s_{SC}) = (p-v)\,\Phi^{-1}(s_{SC}) + (r-v)\,\Psi\left[\Phi^{-1}(s_{SC})\right]$$

$$= (p-v)\,\Phi^{-1}(s_{SC}) + (p-v)\,\Psi\Big[\Phi^{-1}(s_{SC})\Big] + (r-p)\,\Psi\left[\Phi^{-1}(s_{SC})\right]. \quad (14.30)$$

At Time 0, since the expected product leftover by the end of the season can be expressed as $\sqrt{d_0+\delta}\left\{\Phi^{-1}(s_{SC}) + \Psi\left[\Phi^{-1}(s_{SC})\right]\right\}$, which must be non-zero in the model we considered in this paper, we thus have:

$$\left[\Phi^{-1}(s_{SC}) + \Psi\left[\Phi^{-1}(s_{SC})\right]\right] \geq 0. \quad (14.31)$$

Put (14.31) into (14.30) implies that $T_{SC}(s_{SC})>0$. Since $\left(\sqrt{d_0+\delta}-\sqrt{d_1(n)+\delta}\right) > 0$, from (14.31), we have: $EVQR_{SC}(n)=\left(\sqrt{d_0+\delta}-\sqrt{d_1(n)+\delta}\right) T_{SC}(s_{SC}) > 0$.

(b) Differentiate $EVQR_{SC}(n) = \left(\sqrt{d_0 + \delta} - \sqrt{d_1(n) + \delta}\right) T_{SC}(s_{SC})$ with respect to n reveals that $dEVQR_{SC}(n)/dn > 0$.

(c) When $n \to \infty$, from Lemma 2.1, we have: $d_1(n) = 0$, and hence $EVQR_{SC}(n \to \infty) = \left(\sqrt{d_0 + \delta} - \sqrt{\delta}\right) T_{SC}(s_{SC})$. (Q.E.D.)

Proof of Lemma 4.1. Similar to the Proof of Lemma 3.1. (Q.E.D.)

Proof of Lemma 4.2. Notice that $\Phi^{-1}(s_R) < 0$ if and only if $s_R < 0.5$ and $\Phi^{-1}(s_R) \geq 0$ if and only if $s_R \geq 0.5$. Then, Lemma 4.2 can be proven by following the same approach as in the proofs of Lemmas 3.1 and 4.1. (Q.E.D.)

Proof of Lemma 4.3. When $c > p$, by direct observations from the analytical expressions, we have: $q_{0,R*} < q_{0,SC*}$ and $q_{1,R*}|\mu_1(n) < q_{1,SC*}|\mu_1(n)$. (Q.E.D.)

Proof of Lemma 5.1. (a) By equating $q^{\Omega}_{1,R*}\Big|\mu_1(n)$ and $q_{1,SC*}|\mu_1(n)$, we know that $q^{\Omega}_{1,R*}\Big|\mu_1(n) = q_{1,SC*}|\mu_1(n)$ if and only if $\widehat{m} = m*$. (b) Checking the first order derivative reveals that $EVQR^{\Omega}_{R}(n|\widehat{m} = m*)$ is a decreasing function of $\widehat{c}$, and $EVQR^{\Omega}_{S}(n|\widehat{m} = m*)$ is an increasing function of $\widehat{c}$. (Q.E.D.)

Proof of Lemma 5.2. Directly implied by using Lemma 5.1. (Q.E.D.)

References

G. Cachon, R. Swinney, The value of fast fashion: quick response, enhanced design, and strategic consumer behavior. Manag. Sci. **57**, 778–795 (2011)

H.L. Chan, T.M. Choi, C.L. Hui, S.F. Ng, Quick response healthcare apparel supply chains: value of RFID and coordination. IEEE Trans. Syst. Man Cybernet. Syst. **45**, 887–890 (2015)

H.K. Chan, T.M. Choi, X. Yue, Big data analytics: risk and operations management for industrial applications. IEEE Trans. Ind. Inf. **12**, 1214–1218 (2016)

R.R. Chen, T.C.E. Cheng, T.M. Choi, Y. Wang, Novel advances in applications of the newsvendor model. Decis. Sci. **47**(1), 8–10 (2016a)

S. Chen, H. Lee, K. Moinzadeh, Supply chain coordination with multiple shipments: the optimal inventory subsidizing contracts. Oper. Res. **64**, 1320–1337 (2016b)

C.H. Chiu, T.M. Choi, C.S. Tang, Price, rebate, and returns supply contracts for coordinating supply chains with price dependent demands. Prod. Oper. Manag. **20**, 81–91 (2011)

T.M. Choi, Pre-season stocking and pricing decisions for fashion retailers with multiple information updating. Int. J. Prod. Econ. **106**, 146–170 (2007)

T.M. Choi, Local sourcing and fashion quick response system: the impacts of carbon footprint tax. Transp. Res. Part E **55**, 43–54 (2013)

T.M. Choi Impacts of retailer's risk averse behaviors on quick response fashion supply chain systems. Ann. Oper. Res. (2016a), in press. doi:10.1007/s10479-016-2257-6

T.M. Choi, Incorporating social media observations and bounded rationality into fashion quick response supply chains in the big data era. Transp. Res. Part E (2016b), in press. doi:10.1016/j.tre.2016.11.006

T.M. Choi, Inventory service target in quick response fashion retail supply chains. Serv. Sci. **8**(4), 406–419 (2016c)

T.M. Choi, P.S. Chow, Mean-variance analysis of quick response programme. Int. J. Prod. Econ. **114**, 456–475 (2008)

T.M. Choi, S. Sethi, Innovative quick response programs: a review. Int. J. Prod. Econ. **127**, 1–12 (2010)

T.M. Choi, D. Li, H. Yan, Optimal Two-stage ordering policy with Bayesian information updating. J. Oper. Res. Soc. **54**(8), 846–859 (2003)

T.M. Choi, D. Li, H. Yan, Optimal single ordering policy with multiple delivery modes and Bayesian information updates. Comput. Oper. Res. **31**, 1965–1984 (2004)

T.M. Choi, D. Li, H. Yan, Quick response policy with Bayesian information updates. Eur. J. Oper. Res. **170**, 788–808 (2006)

T.M. Choi, H.K. Chan, X. Yue, Recent development in big data analytics for business operations and risk management. IEEE Trans. Cybernet. **47**, 81–92 (2016a). doi:10.1109/TCYB.2015.2507599

T.M. Choi, T.C.E. Cheng, X. Zhao, Multi-methodological research in operations management. Prod. Oper. Manag. **25**(3), 379–389 (2016b)

K.L. Donohue, Efficient supply contract for fashion goods with forecast updating and two production modes. Manag. Sci. **46**, 1397–1411 (2000)

P. Fernando J. Gómez, M.G. Filho, Complementing lean with quick response manufacturing: case studies. Int. J. Adv. Manuf. Technol. 2016, in press. doi:10.1007/s00170-016-9513-4

M. Fisher, A. Raman, Reducing the cost of demand uncertainty through accurate response to early sales. Oper. Res. **44**, 87–99 (1996)

A.V. Iyer, M.E. Bergen, Quick response in manufacturer-retailer channels. Manag. Sci. **43**, 559–570 (1997)

H.S. Kim, A Bayesian analysis on the effect of multiple supply options in a quick response environment. Nav. Res. Logist. **50**, 1–16 (2003)

C.H. Lee, T.M. Choi, T.C.E. Cheng, Selling to strategic and loss-averse consumers: stocking, procurement, and product design policies. Nav. Res. Logist. **62**, 435–453 (2015)

Y.T. Lin, A. Parlakturk, Quick response under competition. Prod. Oper. Manag. **21**(3), 518–533 (2012)

Z. Liu, A. Nagurney, Supply chain networks with global outsourcing and quick-response production under demand and cost uncertainty. Ann. Oper. Res. **208**(1), 251–289 (2013)

K. McCardle, K. Rajaram, C.S. Tang, Advance booking discount programs under retail competition. Manag. Sci. **50**, 701–708 (2004)

D.S. Shaltayev, C.R. Sox, The impact of market state information on inventory performance. Int. J. Inventory Res. **1**, 93–124 (2010)

B. Shen, T.M. Choi, K.Y. Lo, Markdown money policy in the textile and clothing industry: China vs U.S.A. Sustainability **8**(1), 31 (2016)

J.J. Spengler, Vertical restraints and antitrust policy. J. Polit. Econ. **58**, 347–352 (1950)

D. Yang, T.M. Choi, T. Xiao, T.C.E. Cheng, Coordinating a two-supplier and one-retailer supply chain with forecast updating. Automatica **47**, 1317–1329 (2011)

Part V
Concluding Remarks

Chapter 15
Optimization and Control for Systems in the Big Data Era: Concluding Remarks

Tsan-Ming Choi, Jianjun Gao, James H. Lambert, Chi-Kong Ng, and Jun Wang

Abstract In the big data era, new research opportunities and challenges exist for systems optimization and control problems. In this concluding chapter, we share several probable related areas which may lead to fruitful research in the future. We also summarize the future research directions proposed by papers featured in this book.

Keywords Concluding remarks • Big data • Optimization • Control • Future research • Opportunities • Challenges

---

T.-M. Choi (✉)
Institute of Textiles and Clothing, The Hong Kong Polytechnic University, Hung Hom, Kowloon, Hong Kong
e-mail: jason.choi@polyu.edu.hk

J. Gao
School of Information Management and Engineering, Shanghai University of Finance and Economics, Shanghai, People's Republic of China
e-mail: gao.jianjun@shufe.edu.cn

J.H. Lambert
Department of Systems and Information Engineering, University of Virginia, Charlottesville, VA, USA
e-mail: lambert@virginia.edu

C.-K. Ng
Department of Systems Engineering and Engineering Management, The Chinese University of Hong Kong, Shatin, N.T., Hong Kong
e-mail: ckng@se.cuhk.edu.hk

J. Wang
Department of Management Science and Engineering, Business School, Qingdao University, Shandong, People's Republic of China
e-mail: jwang@qdu.edu.cn

T.-M. Choi et al. (eds.), *Optimization and Control for Systems in the Big-Data Era*, International Series in Operations Research & Management Science 252,
DOI 10.1007/978-3-319-53518-0_15

15.1 Optimization and Control: Challenges and Opportunities in the Big Data Era

Big data optimization and control is an important topic. It is influential to not only business operations but also the society and science (Stefanowski and Japowicz 2016). One underlying principle of big data analysis is that: Having the big data, more useful information can be found than when we separate the big data into smaller datasets. This means we simply cannot separate the datasets and solve the big data problems properly in many cases (Sparks et al. 2016). In addition, in the presence of big data, we can explore more complex systems with a goal of improving their performance or even achieving optimization (Chan et al. 2016; Choi et al. 2016a, b). However, this goal is easier said than done because it involves a lot of technical challenges. In light of the challenging issues around big data optimization and control problems, based on the various Vs associated with big data, we propose a few areas for future research as shown below.

Volume The most basic element of big data is the huge volume of data. This is also related to high dimensionality of the datasets. This means the traditional optimization and control methods which work well for small datasets may not function well anymore. To cope with this challenge, the first proposal is to see if the big data problem can be solved by decomposing it. Even though not all big data problems can be solved by decomposition, some problems probably can. This approach is usually called the divide-and-conquer (Wang and He 2016). Moreover, new methods based on the parallelization approach (Daneshmand et al. 2015; Facchinei and Scutari 2015) can also be developed so that the smaller sized problems from the big problem can be processed in parallel. These provide rich opportunities for future research.

Velocity In optimization and control problems associated with big data, the emergence of data is quick and we also need to solve the problem in a timely manner. This creates challenges such as how we can achieve real time (or almost real time) optimization. New research on novel heuristics is probably needed to address this issue.

Variety The data available can be complex and not all of them are given in numbers. For example, in studying financial markets (Yao and Li 2013), on one hand, we have data such as historical records of indices and stock prices (Gao and Li 2013; Shi et al. 2015). On the other hand, we also have some expert advice, news, reports which are not directly expressed in numbers. How to combine them to formulate the optimal portfolio is a challenge and deserves deeper future research.

Veracity The big dataset may include missing information which means the data quality is not good. This creates uncertainty in the dataset and makes the respective optimization and control problem incomplete. New techniques (including dual control (Li et al. 2002), evolutionary optimization (Bhattacharya et al. 2016)) and even research methodologies (e.g., the multi-methodological approach

(Choi et al. 2016b), new framework (Boone et al. 2016)) need to be developed so that the optimization and control method can learn from "uncertainty" in the big data. Wang and He (2016) provide a recent discussion on the related research opportunities.

Value Big data potentially can yield a high or low value to the decision makers. It is hence important to measure the performance by conducting a proper performance analysis. Li et al. (2016) provide a review on the topic with a discussion on various performance tools for big data analytics. New research can be conducted to examine the proper performance tools and to help quantify the value of big data and the corresponding optimization methods.

15.2 Summary of Future Research Directions Highlighted in this Book

From the papers featured in this paper, various future research directions have been proposed. We summarize them in Table 15.1.

15.3 Concluding Remarks

In this concluding chapter, we have first discussed various challenges and research opportunities on optimization and control for systems in the big data era. We have already reviewed and reported the research opportunities proposed by the technical papers of this book. We hope the discussions can spark and inspire new research in optimization and control in the presence of big data.

Last but not least, we dedicate this book to our mentor Professor Duan Li, a true and distinguished scholar, a kind gentleman, an excellent professor, and an outstanding teacher who has made huge contribution to the advance of optimization and control theories. His theoretical works are also very influential and have been widely applied in financial engineering, industrial engineering, systems engineering, as well as many operations research related areas such as biomedical studies, production scheduling, and supply chain management, to name a few. We are very proud of being his students.

Table 15.1 Future research opportunities proposed in this book

Chapter	Paper	Future research opportunities	The related "Vs"
2	Fu	Dual controls with probing features are helpful to regulate complex systems in the big data era and deserve further explorations	Volume
3	Shi and Cui	In the big data era, more and more data-driven dynamic optimization problems are present, and many of them are time inconsistent (Cui et al. 2012) which create great computation challenges. This creates new research opportunities	Variety, velocity, veracity
4	Wu and Jiang	The quadratic convex reformulation can be further generalized based on the objective functions, constraints. The quadratic convex reformulation in the big data era is even more important than before because finding the best reformulation in the quadratic convex reformulation approach reduces to a stochastic dynamic programming problem, which is convex	Volume, velocity
5	Pei and Zhu	In exploring financial contagion, the methodologies have to consider timely and frequent collection, updating and integration of information emerged in the market. Future related research should address the above problems and many existing big data optimization techniques may be applicable to improve and conquer these challenges	Volume, value
6	Chiu	Applying the proposed method in asset liability management in the presence of big data would be a challenging task for future research	Value
7	Lu	Future research can be conducted on the quantum cryptosystems. Equipped with the rapid development of quantum theory and quantum computer, new insights on how objects behave at the microscopic level, can be learned from the big data available	Velocity, variety
8	Li, Li, Zhao	1. The channel coordination contract design under supply uncertainty 2. The effect of supply uncertainty on firm profitability should be evaluated in the context of the horizontal market competition 3. Other promising features to be explored in future research with the consideration of big data include information asymmetry, behaviors of decision makers, channel power and cooperation, and supply risk assessment	Veracity, value
9	Li, Li, Wu, Yao	Future research can be conducted on using the parameterized method to solve the portfolio selection problem when the returns are correlated in every period. New factors such as the presence of probability constraints, uncertain exit time, and Markov jumps will also be considered	Volume, veracity

10	Gao and Wu	Studying the stability issue of the out-of-sample test for the multiperiod mean-CVaR portfolio optimization problem is an interesting and challenging topic for future research	Volume, velocity
11	Liang	Future research can be carried out on the model statements and analysis. In particular, in order to improve the accuracy of the proposed models, new estimation methods for a large amount of parameters can be applied in the big data era	Volume, velocity
12	Yi	As introducing uncertain exit time adds extra risk to the investment, new measures have to be derived to deal with this challenge. It is an important future research direction	Volume, value
13	Wang, Zhuang, and Wu	In the big data era, it is important to examine the performance of the exact methods, and new heuristics may need to be developed to address larger sized problems. The big data technology could also provide a possible way to deal with the estimation errors of parameter in the new model, which also calls for more in-depth explorations	Volume, veracity
14	Choi	1. Examine the case with multiple products and the corresponding coordination challenges 2. Consider how social media data can be incorporated into quick response supply chain management 3. Adopt a multi-methodological approach with real data analyses	Volume, value
15	Choi et al.	1. Parallel optimization 2. Real time optimization 3. New methods to combine structured and unstructured data for optimization and control 4. New optimization and control techniques and methodologies to deal with data quality problems 5. New performance tools to quantify the value of big data	Volume, velocity, variety, veracity, value

References

M. Bhattacharya, R. Islam, J. Abawajy, Evolutionary optimization: a big data perspective. J. Netw. Comput. Appl. **59**, 416–426 (2016)

C.A. Boone, B.T. Hazen, J.B. Skipper, R.E. Overstreet. A framework for investigating optimization of service parts performance with big data. Ann. Oper. Res. (2016), in press

H.K. Chan, T.M. Choi, X. Yue, Big data analytics: risk and operations management for industrial applications. IEEE Trans. Ind. Inf. **12**, 1214–1218 (2016)

T.M. Choi, H.K. Chan, X. Yue, Recent development in big data analytics for business operations and risk management. IEEE Trans. Cybernet. 2016a, in press. doi:10.1109/TCYB.2015.2507599

T.M. Choi, T.C.E. Cheng, X. Zhao, Multi-methodological research in operations management. Prod. Oper. Manag. **25**(3), 379–389 (2016b)

X.Y. Cui, D. Li, S.Y. Wang, S.S. Zhu, Better than dynamic mean-variance: time inconsistency and free cash flow stream. Math. Financ. **22**(2), 346–378 (2012)

A. Daneshmand, F. Facchinei, V. Kungurtsev, G. Scutari, Hybrid random/deterministic parallel algorithms for convex and nonconvex big data optimization. IEEE Trans. Signal Process. **63**(15), 3914–3929 (2015)

F. Facchinei, G. Scutari, Parallel selective algorithms for nonconvex big data optimization. IEEE Trans. Signal Process. **63**(7), 1874–1889 (2015)

J.J. Gao, D. Li, Optimal cardinality constrained portfolio selection. Oper. Res. **61**(3), 745–761 (2013)

D. Li, F.C. Qian, P.L. Fu, Variance minimization approach for a class of dual control problems. IEEE Trans. Autom. Control **47**(12), 2010–2020 (2002)

Y. Li, Q. Guo, G. Chen, in *Performance Tools for Big Data Optimization*. ed. By A. Emrouznejad. Big Data Optimization: Recent Developments and Challenges, Series on Studies in Big Data (Springer, Switzerland, 2016), pp. 71–96

Y. Shi, X.Y. Cui, J. Yao, D. Li, Dynamic trading with reference point adaptation and loss aversion. Oper. Res. **63**(4), 789–806 (2015)

R. Sparks, A. Ickowicz, H.J. Lenz, in *An Insight on Big Data Analytics*. ed. By N. Japkowicz, J. Stefanowski. Big Data Analysis: New Algorithms for a New Society (Springer, Switzerland, 2016), 33–48

J. Stefanowski, N. Japowicz, Final remarks on big data analysis and its impact on society and science. in *Big Data Analysis: New Algorithms for a New Society* ed. By Japkowicz N and Stefanowski J (Springer, Switzerland, 2016), pp. 305–329

X. Wang, Y. He, Learning from uncertainty for big data. IEEE Syst. Man. Cybernet. Mag. **2**(2), 26–32 (2016)

J. Yao, D. Li, Prospect theory and trading patterns. J. Bank. Financ. **37**, 2793–2805 (2013)

Index

T.-M. Choi et al. (eds.), *Optimization and Control for Systems in the Big-Data Era*, International Series in Operations Research & Management Science 252,
DOI 10.1007/978-3-319-53518-0

Printed in the United States
By Bookmasters